Case studies on the effects of transferable fishing rights on fleet capacity and concentration of quota ownership

FAO
FISHERIES
TECHNICAL
PAPER

412

Edited by
Ross Shotton
Fishery Resources Officer
Marine Resources Service
Fishery Resources Division
FAO Fisheries Department

Food
and
Agriculture
Organization
of
the
United
Nations

Rome, 2001

ISBN 92-5-104659-X

FOREWORD

This FAO Fisheries Technical Paper has been compiled to complement the discussions of its companion volume: Case Studies on the Allocation of Transferable Quota Rights in Fisheries Management[1]. It thus provides a third volume of case studies on fisheries management practices published by FAO, which started with the collection of papers describing the management of elasmobranch fisheries[2]. Further, it continues the series of publications on the use of Rights-based fisheries management undertaken by FAO's Fisheries Department, which, together with the publication of this volume brings these publications to six in total[3]. These reflect the growing importance of this topic to contemporary fisheries management.

The topic selected for this study, as with that for its companion volume, the Allocation of Transferable Quota Rights in Fisheries Management, arose from my oft-encountered experience when discussing this issue with fishermen and other 'stakeholders' in the fishing industry. Uppermost in the minds of fishermen facing this possible form of management is "How much quota will I get?" Immediately this question is resolved (of course providing an answer is never a trivial exercise!), the next most frequent question is, "What is there to stop someone buying up all the quota and forming a monopoly?" Less frequent has been the question from administrators, "Will introducing transferable fishing quota into the fishery solve the fleet overcapacity problems?" Never heard from fishermen is a question, which I have always thought was emminently reasonable: "Will there be any constraints on my rights to sell my quota holdings, for example, to whoever I wish?"

One common response I have encountered from fishermen (which would perplex a strictly utility-maximizing economist) is that they do not think access to the fisheries they exploit should even be limited, though no such views on open-access existed, for example, in relation to their farms or woodlots whose potential harvests were obviously governed by a different cognitive rationality.

Another reason for compiling these case studies was to provide a factual basis for evaluating the claims, commonly made, that introducing individual transferable quotas (ITQs) leads to monopolies and the exclusion of small operators from fisheries. Often, these claims are made with little or no substantiation and in journals whose editors and referees should know better. In any event, the papers in this volume should partly remove the excuse for such non-substantiated claims. I say "partly remove" because what many of the papers show is the great difficulty in accurately identifying what happens in terms of fleet-capacity when a transferable rights-based management approach is adopted.

Much of what happens to the fleet depends on factors not directly associated with the fishery undergoing the management change. A further complication is that determining if, all of the changes that occur in a fishery after the introduction of a Rights-based Management system are the result of the new management regime, is practically impossible. Fisheries, despite the suggestions of those who promote experimental adaptive management and because of their social and biological complexity, are too important to administer as an academic experiment. Demand changes, supply changes, resource productivity changes (with consequential changes in operator's revenues), changes in factor costs (which usually only rise), superimposed on the business cycle and consumer product-substitution, make unequivocal conclusions about the effects of a particular management regime change a rarity. While some changes may be unequivocally attributed to a new rights-based management regime, many others, particularly those relating to efficiency, will not. However, what is clear is that without monitoring, or at least good documentation, of both the pre- and post-management situations, unequivocal assertions (positive or negative) about the effects of transferable fishing-rights on fleet capacity can be dangerously misleading.

---

[1] Shotton, R. (Ed.) 2001. Case studies of the allocation of transferable quota rights in fisheries. FAO Fish. Tech. Pap. No. 411. 373 pp.

[2] Shotton, R. (Ed.) 1999. Case studies of the management of elasmobranch fisheries. FAO Fish. Tech. Pap. No. 378, Vols 1 and 2.

[3] Earlier FAO publications are:
   Christy, F. 1982. Territorial use rights in marine fisheries: Definitions and Conditions. FAO Fish. Tech. Pap. No 227.
   Morgan, G.R. 1997. Individual quota management in fisheries. Methodologies for determining catch quotas and initial allocations. FAO Fish. Tech. Pap. No. 371. 41pp.
   Shotton, R. 2000. Use of property rights in fisheries management. Proceedings of the FishRights99 Conference, Fremantle, Western Australia. 11-19 November 1999. Vol.1: Mini-course lectures and Core Conference presentations. FAO Fish. Tech. Pap. 404/1 (342pp). Vol 2: Workshop presentations. FAO Fish. Tech. Pap. 404/2 (468pp).

While the interest of FAO's Fisheries Department in the costs and benefits of introducing rights-based approaches to fisheries management was a major stimulus for undertaking the compilation and publication of these papers, there was another compelling reason to address this topic. The twenty-second Session of the FAO Committee on Fisheries (COFI), held in 1997, had urged that the issues of excessive fishing-capacity and fishing-effort leading to overfishing should be given consideration by FAO. As a consequence the Fisheries Department organized a technical working group to review technical guidelines and consider an international plan of action for the management of fishing-capacity. As part of the process of addressing this issue much work was focused on how to determine the "capacity" of fishing fleets and detailed actions were identified as to how individual countries could address this issue.

Little or no detailed consideration has yet been given to the role that rights-based approaches to fisheries management might provide in reducing fleet-capacity, though some countries reported at the twenty-third Session of COFI that individual transferable quotas were being used to reduce or prevent increases in fishing-capacity. By undertaking the documentation of experiences at the national and fishery level I hope this collection of case studies will provide a factual basis on which to consider rights-based management approaches as a possible solution for solving problems of excess fishing capacity. Indeed, in my view, for many of the cases examined, a rights-based approach will provide the most practical remedy to the problems that exist, and at the same time will contribute to other desirable management objectives.

As those involved in managing fisheries are aware, day-to-day exigencies rarely allow the luxury of collecting detailed data that will permit the proper evaluation of new programmes (much less their soon-to-be-replaced predecessors) and, as the papers in this volume show, this situation has been the norm. Yet further, characterizing the "capacity" of a fishing fleet is a complex and difficult[4], if not fruitless, task. And, rarely have the authors had access to the detailed fleet-registry records that provide sufficient details on year of vessel construction, dates of vessel conversions or upgrades, design changes, engine upgrades, *etc.*, all of which change the fishing capacity of individual vessels and thus the total for the fleet. As has been well documented[5], statistics based on aggregate fleet statistics can be meaningless in trying to predict the changes that may occur when individual vessels enter or leave a fishing fleet. These vessels usually are the statistical outliers – those from the lower tail of the distribution of fishing success for vessels of similar dimensions. Not surprisingly, the challenge of obtaining these detailed data has proven difficult for many of the studies and authors have often been forced to rely on other approaches.

Of great interest will be the accounts that relate to fisheries where the measure of effort has not been most appropriately a number of fishing vessels or a corresponding proxy. Transferable quotas have also been introduced into other fisheries such as those for abalone where the effort is indicated by the number of divers. Readers should be interested to read of the experiences in these fisheries, which are most notably found in Australia.

The challenge of discerning overall or sectoral trends from the information provided in these papers is left to the reader. But, so diverse are the fisheries that now use transferable quotas as a management tool, that no longer can writers be excused for making simplistic assertions based on poorly-conceived and subjective opinions derived from one or two examples of such management. Such analysis ought be fisheries-specific and depend on sector-based examination. Enabling this is one of the objectives of this Technical Paper.

The contributing authors were asked to follow a common format so as to facilitate comparative analysis of different practices. But, at the same time they were asked not to let this request limit their treatment of the topic. I emphasized that I would rather receive an appropriate treatment of the topic justified in terms of the problems of the unique fishery they were describing, than an account that was limited by attempting to follow my suggested structure for the paper. Readers must understand the various conceptual elements that were involved and interpret for themselves the individual accounts in this light.

Observant readers of the companion volume, Case Studies on the Allocation of Transferable Quota Rights in Fisheries (FAO Fish. Tech. Pap. No. 411), will note that some fisheries covered in that volume have, for various reasons, not been addressed in this volume. Among these lacunae is that for the South East Fishery of Australia. However, I can direct interested readers to a recent publication, "Indicators of the effectiveness of quota markets: the South East Trawl Fishery of Australia", in Volume 52(4) of **Marine and Freshwater**

---

[4] For example, see Gréboval, D. 1999. Managing fishing capacity. Selected papers on underlying concepts and issues. FAO Fish. Tech. Paper No. 386. 206pp.

[5] For example, see Shotton, R. 1989. An Analysis of Factors Affecting Catch Rates of Sub-65' Groundfish fishing vessels in 4X/Sub-Area 5 of Southwest Nova Scotia. Can. Tech. Rep. Fish. Aquat. Sci. 1707. 129pp.

Research. This paper by Robin Connor and Dave Alden should provide much of the information about that fishery, which this volume attempts to address in other fisheries.

Once again, I must thank my secretary, **Marie-Thérèse Magnan**, for her enormous effort in preparing this paper for publication – her fourth in this series; my colleague, **Mike Mann**, in ensuring that the editorial quality of the papers is again of the highest standard, and **Françoise Schatto**, Publication Assistant, Fishery Information, Data and Statistics Unit for the difficult and unenviable responsibility of transforming the manuscript into the final document. I also thank those who have generously made photographs available, usually to illustrate a paper that is not their own – I believe that these illustrations have done much to bring these reports "to life". Credit for the design and preparation of the cover goes entirely to **Emanuela D'Antoni** of our Service.

Ross Shotton
Marine Resources Service
FAO, Rome.

Shotton, R. (ed.)
Case studies on the effects of transferable fishing rights on fleet capacity and concentration of quota ownership.
FAO Fisheries Technical Paper. No. 412. Rome, FAO. 2001. 238p.

ABSTRACT

This report, consisting of 16 national, or national fishery, studies, describes how the introduction of transferable fishing (effort) or fish (catch) quotas has affected the capacity of the fleet prosecuting the target fishery for which the harvesting rights apply.

The case studies include two from the European Union (the U.K. and the Netherlands) and for Iceland. Two studies are presented for fisheries along the eastern seaboard of the United States. Seven accounts are included from Australia, two of which describe fisheries managed by the Commonwealth Government through the Australian Offshore Constitutional Settlement (the Northern Prawn Fishery and the fishery for southern bluefin tuna). The other five accounts of Australian experiences describe the (unique?) Pilbara Trap Fishery in the northern region of Western Australia, Western Australia's rock lobster fishery and the fishery for the same species and that for abalone and pilchards in South Australia. In Tasmania an account is given for the rock lobster fishery while for New South Wales, a description is given for another invertebrate fishery, that for abalone. An omnibus account is given for the situation in New Zealand. In the Western Pacific, accounts are given for the Pacific Halibut and Sablefish fisheries in Alaska, the marine trawl fisheries of British Columbia and for Patagonian toothfish in Chile.

Keywords: Fisheries Management, Property Rights, ITQs, Individual Transferable Quotas, Fisheries Policy, Fleet Capacity, Fleet Capacity Reduction

TABLE OF CONTENTS

FISHING RIGHTS AND STRUCTURAL CHANGES IN THE UK FISHING INDUSTRY

A. Hatcher* and A. Read**
* Centre for the Economics and Managementof Aquatic Resources (CEMARE)
University of Portsmouth, Portsmouth PO4 8JF, United Kingdom
<Aaron.Hatcher@port.ac.uk>
** Danbrit Ship Management Ltd
8 Abbey Walk, Grimsby DN31 1NB, United Kingdom

1. INTRODUCTION
1.1 The UK fishing industry

The United Kingdom has a long history of fishing, reflecting its position as an island with a relatively long coastline and proximity to the productive fishing grounds of the European continental shelf (notably the North Sea, the English Channel and the West of Scotland).

The UK's fisheries are heterogeneous and this is reflected in its complex fleet structure. The shape of the modern UK fleet is the product of technological and market changes together with political developments, in particular the loss of access to traditional distant-water grounds (particularly Iceland and Greenland) in the 1970s and the development of the Common Fisheries Policy (CFP) by the European Community (EC) (which the UK joined in 1972). Under the CFP (see below) there have been national quotas for most stocks since the early 1980s, coupled with a succession of fleet-reduction programmes (the so-called MAGPs) or multi-annual guidance programmes.

There are currently just over 8000 fishing vessels in the UK, although nearly three-quarters of these are inshore boats of less than 10m in length. Table 1 shows the trend in vessel numbers for the period 1994 to 1999 by vessel type (the sector shown corresponds to the classification used by the EC for measuring fleet size).

Table 1
Number and type of UK fishing-vessels 1994-99

Segment	1994	1995	1996	1997	1998	1999
Pelagic	68	67	58	49	50	46
Beam trawl	212	220	215	153	123	114
Demersal trawl	1 644	1 549	1 451	1 428	1 318	1 235
Lines/nets	300	267	224	214	187	172
Shellfish mobile	206	194	265	227	241	243
Shellfish fixed	305	283	339	352	311	301
Distant-water	13	12	15	13	14	12
Others	355	263	0	0	2	2
Inshore (≤ 10m)	7 195	6 320	5 606	5 474	6 027	5 920
Total	10 298	9 175	8 173	7 910	8 273	8 045

Source: MAFF UK Sea Fisheries Statistics.

The contraction of most of the over-10m sector of the fleet in terms of number, is not matched by an equivalent decrease in capacity as vessels have become bigger and more powerful, particularly in the pelagic, beam trawl and demersal trawl segments. Hatcher and Read (2001) consider the changes in fleet capacity in the context of the UK's attempts to comply with MAGP targets.

Employment in fishing has declined somewhat during the 1990s from around 21 000 to 18 000 jobs, but most of the decline has been in part-time employment. Full-time employment in fishing has been rather stable in recent years at around 15 000.

Tables 2 and 3 show the total landings by weight and (nominal) value made by UK vessels during 1993-1999.

Table 2
Volume of landings by UK vessels ('000 tonnes)

	1993	1994	1995	1996	1997	1998
Demersal	359.2	371.6	386.0	407.7	426.1	456.7
Pelagic	393.8	388.9	396.3	343.9	323.2	334.4
Shellfish	104.6	114.4	129.5	140.6	142.0	132.7
Total	857.6	874.9	911.8	892.3	891.3	923.8

Cod, haddock and whiting are the main whitefish species caught by UK vessels and together they make up about 45% of the total demersal landings by weight and 36% by value. Other species landed in large quantities include ling, anglerfish, plaice, sand eels and blue whiting. Although classed as demersal the latter two species are fished by pelagic vessels and are the only species caught for reduction to meal. High-value demersal species caught in smaller quantities include sole, hake and megrim. Herring and mackerel account for roughly 80% of pelagic landings by weight and value. The most important shellfish species are scampi or Dublin Bay prawn (*Nephrops*) (which now rivals cod as the most valuable catch), crabs and scallops, which together account for nearly three-quarters of the value of shellfish landings.

Table 3
Value of landings by UK vessels (£ millions nominal)

	1993	1994	1995	1996	1997	1998
Demersal	356.4	364.8	369.4	383.5	368.5	372.2
Pelagic	56.4	58.4	64.3	90.0	88.4	113.8
Shellfish	113.6	138.2	156.4	163.0	165.0	175.4
Total	526.0	561.4	590.1	636.5	621.9	661.5

Source: MAFF UK Sea Fisheries Statistics.

An increasing proportion of landings by UK vessels are made at ports in other EC or European Economic Area countries. In 1997 around half of all catches of herring and mackerel were landed abroad (mainly into Norway and Denmark), as were UK catches of hake (into Spain) and plaice (into the Netherlands)[1]. Overall, some 38% of UK landings were made into non-UK ports.

Scotland accounts for 70% by weight and 60% by value of UK landings. The majority of the Scottish fleet operates from the East coats ports of Fraserburgh, Peterhead and, to a lesser extent, Aberdeen; the Shetland Islands are an important base for the pelagic fleet. The major fishing activity of the Scottish fleet is demersal trawling for whitefish and *Nephrops*, and pelagic trawling for species such as herring and mackerel. Scallop dredging is locally important in south-west Scotland and the Isle of Man. Scottish vessels primarily operate in the Northern North Sea (ICES area IVa) and the West of Scotland grounds (ICES area VI). There has been considerable development of the Scottish fleet during the 1990s with significant investment in new vessels and technology. Major advances that have had an impact on the efficiency of the fleet are the development of twin-rig and pair trawling.

The rest of the UK fleet is spread throughout England, Wales and Northern Ireland. The Northern Irish contribution is fairly small, and although locally important, there are only four ports with sizeable landings. The fishing fleet in Northern Ireland has contracted significantly since the early 1990s

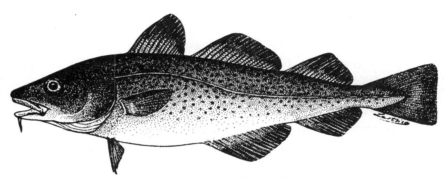

Atlantic Cod (*Gadus morhua*)
An endangered resource in the North Sea

with a 27% decrease in vessel numbers. A similar contraction has been seen elsewhere in the UK. The traditionally important ports of the Humber region on the north-east coast of England have been in steady decline and, although still very important in terms of transhipment and processing, the catching sector is barely represented in the local ports. The south-west of England, particularly Cornwall, has also seen reduced numbers of vessels during recent years, although the value of landings remains high in the region.

The UK fleet is diverse and there are fishing methods that are particular to certain regions. The predominant fishing methods in the South-West are beam trawling in the English Channel for sole and plaice, gill netting and trawling for high-value demersal species such as hake and megrim, and scallop dredging. The main markets are those of Newlyn, Brixham and Plymouth. Small-scale demersal trawling, gill-netting and crab/lobster potting is predominant in the English Channel. The North Sea fisheries of the East Coast include beam trawling for plaice and sole, but the predominant method is demersal trawling for whitefish such as cod and haddock. Also of great importance is the North Sea *Nephrops* fishery, fished mainly by small-scale demersal trawlers of under-24m. The key markets are in Lowestoft (for beam trawlers), Grimsby, and to a lesser extent, Scarborough and Whitby.

---

[1] In the case of hake and plaice the foreign landings reflect the foreign ownership of a significant part of the UK demersal fleet which is facilitated by the terms of EC membership.

1.2   International context and management responsibilities

The United Kingdom is a member of the European Community and its marine fisheries must therefore be managed within the framework of the EC's Common Fisheries Policy (CFP)[2]. Since 1983 the CFP has provided for the setting of annual total allowable catches (TACs) for most commercially important stocks within the overall zone of extended fisheries jurisdiction created by the 200-mile limits of those Member States bordering the North-East Atlantic and adjacent seas. The TACs are divided into national catch quotas according to an established allocation mechanism which gives each Member State a fixed percentage share each year (although a number of intergovernmental quota swaps are regularly and routinely undertaken).

Fishing opportunities in the waters of third countries and in international waters (such as the NAFO area in the North West Atlantic) are negotiated by the European Commission on behalf of the Community and are allocated to Member States in a similar way.

While national fleets fishing in Community waters are subject to certain common controls under the CFP (in particular technical conservation measures such as minimum mesh sizes) each Member State is able to determine the means for allocating its quotas to the national fleet and for regulating quota uptake[3]. EC rules nevertheless require all vessels of 10m or more in length to keep a logbook of their activities, which must include details of the quantities of TAC species caught and retained on board, and the time and location of capture[4]. Inshore vessels less than 10m long are not obliged to carry logbooks but Member States are still required to monitor their landings to ensure that national quota-limits are respected.

2.   USE RIGHTS AND THEIR TRANSFERABILITY

The historical development and detailed operation of the UK's licensing and quota management systems are described in Hatcher and Read (2001). UK-registered commercial fishing vessels require a licence appropriate to the type of vessel and to the stocks targeted. As a general rule, *quantitative* restrictions on landings are imposed as licence conditions, but the majority of the offshore fleet belong to producers' organisations (POs), which receive group-quota allocations from the Government (see below). For these vessels the licence conditions simply refer to the quota-management arrangements implemented by the PO to which they belong. The relatively small number of offshore vessels which operate independently have monthly landings-limits specified in their licences. Licences are issued annually by the Government Fisheries Departments[5] but licence-entitlements can be transferred between vessels, subject to certain conditions, and between ownerships.

There are four main categories of fishing-vessel licences: category "A" licences for offshore vessels over 10m authorise fishing for all the commercially important stocks subject to quotas under the European Community's common fisheries policy; category "B" licences authorise fishing by vessels over 10m for a smaller number of quota stocks[6]; category "C" licences authorise fishing by vessels over 10m for non-quota

---

[2] The CFP has four principal components: a common structural policy, a common market organization, a resource conservation and management system and an external policy (concerned with fisheries agreements with third countries). The common structural policy and the common organization of the market both date back to 1971 (two years before the UK joined the EEC). In addition to provisions for common structural actions (which include aids for fleet renewal as well as capacity reduction programmes) the structural regime lays down certain fundamental conditions for fishing, notably the principle of *equal access* of Member State's fishing fleets to each other's waters (beyond the six-mile territorial limit). The common market organization provides for a system of marketing standards, minimum prices and intervention arrangements (with compensation for products withdrawn from the market at minimum prices). In 1977 all the EC Member States in concert extended their fishery limits out to 200 miles (except in the Mediterranean). Negotiations then began on a system to regulate catches within Community waters. Because of the difficulty of reaching agreement on national TAC shares the "conservation and management" system was not finally adopted until 1983.

[3] Article 9 of Council Regulation (EEC) No 3760/92 establishing a Community system for fisheries and aquaculture (*Official Journal of the European Communities*, No L 389, 31.12.92, p.1).

[4] Article 6 of Council Regulation (EEC) No 2847/93 establishing a control system applicable to the common fisheries policy (*Official Journal of the European Communities*, No L 261, 20.10.93, p.1).

[5] In the UK responsibility for the day-to-day management of fisheries is shared between the Ministry of Agriculture, Fisheries and Food (for fisheries in England), the Agriculture and Fisheries Departments of the Scottish Executive, Welsh Assembly and Northern Ireland. MAFF, however, remains the "lead" Department and takes overall responsibility for national policy and international responsibilities.

[6] The stocks available to holders of category B licences include those which were considered to be subject to lighter exploitation when the equivalent of category A licences were introduced in 1984.

species only (principally shellfish). All vessels of 10m or under in length are issued with a special class of Category A licence[7].

The licence system also controls vessel-capacity, measured in terms of GRT (in the process of conversion to the GT measure) and rated engine-power (in kW). Each licence carries an entitlement to employ a number of "vessel capacity units" (VCUs) which must match the number of units calculated for the vessel according to a standard formula[8]. If a licence is transferred to another vessel a VCU "penalty" is incurred (unless the recipient vessel is identical or has the same number of VCUs with a lower tonnage). The system also allows licences from a number of vessels to be aggregated onto a larger or more powerful vessel, again with an overall VCU penalty.

The capacity penalties and other licence-transfer rules have been altered a number of times since they were introduced ten years ago. The main rules are currently as follows:

i.   in general, licences cannot be "upgraded" by transfer or aggregation, *i.e.* vessels under-10m to vessels over-10m, C to B, B to A, *etc*; also category A demersal species licences cannot be transferred onto pelagic vessels, and ordinary pelagic licences cannot be transferred to pelagic purse-seiners or freezer-trawlers[9]

ii.   a 10% penalty is imposed for most licence-transfers (except in the case of vessels 10m-and-under, and pelagic purser/freezer licences)

iii. a 20% penalty is imposed for aggregating two licences and 30% for aggregating three or more licences (except in the case of pelagic purser and freezer licences, for which the penalty is 10%)

iv.   no licence transfers or aggregations may result in any increase in either total tonnage or engine-power[10] and

v.   there is an exemption from the penalties, subject to certain conditions, for distant-water vessels.

A system for regulating the uptake of national quotas is linked to the licensing scheme. This involves the allocation of percentage quota-shares each year mainly to groups of vessels, although some individual vessel allocations are made by Government in the case of pelagic and distant-water stocks. Until 1998 these allocations were based on the track-record vessels' landings during the previous three years (the reference period for the majority of stocks), but in 1999 this was replaced by a system of fixed-quota allocations (FQAs). The allocations for 1999 and 1998 were based on track records over the period 1994-1996 which was the reference period for the 1997 allocations (in order to avoid any inflation of track record over a qualifying period)[11]. The 1994-1996 track records were converted into allocations of "quota units" with an initial value of 100kg to produce the 1999 allocations. Although a "one-off" reallocation of units between licences was allowed for the 1999 allocations to let the POs resolve outstanding quota deals (see below), the allocations for 2000 and thereafter, in theory at least, should be the same as the 1999 allocations on a percentage basis, *i.e.* adjusted only according to changes in the UK quota allocations[12].

In the case of the offshore (over-10m) fleet, each vessel, in effect, is given a set of *notional* individual quota allocations each year. By aggregating these individual notional quotas, group allocations are made by Government to Producers' Organisations (POs)[13], which are then allowed to manage those allocations as they wish, for example, by means of monthly landings-limits from a common quota pool, or by allocating annual individual vessel- or company-quotas[14]. There are now 20 recognised POs in the UK representing roughly two-thirds of the fleet over 10m. The POs together account for some three-quarters of total landings by UK vessels and over 95% (by weight) of UK quota allocations in Community waters.

---

[7] Within these main licence categories there are a number of specific licence types which apply to particular fisheries or which authorise fishing using certain types of vessel or gear. Category A or B beam trawl licences are required, for example, for the use of beam trawls by over 10m vessels in the North Sea and in Area VII (the English Channel and Western Waters) and specific licences are also required for the use of scallop dredges. There are special category A licences for pelagic purse-seiners and freezer-trawlers. There are also category "D" licences which authorise distant-water fishing.

[8] (overall length in metres x maximum breadth in metres) + (engine-power in kW x 0.45).

[9] However, there is a partial suspension of this rule for new pelagic freezer trawlers until 30 June 2001.

[10] In addition no aggregations of beam trawler licences may result in an engine-power in excess of 1,500kW and no aggregations of 10m and under licences may result in a VCU total of over 100.

[11] Sources indicate that the practice of artificially inflating track records, for example by over-recording landings or attributing catches to the wrong sea area (and hence stock), was widespread by 1996.

[12] The UK Government, however, reserves the right to issue or withdraw quota units. Because of the growing trading in quota between and within a number of the POs there is increasing pressure on the Government to allow annual reallocations.

[13] Producers' Organisations are vessel owners' associations recognised under EC law. Their primary role is the orderly marketing of fish and the implementation of common marketing rules and standards but they are encouraged to take on resource management responsibility.

[14] The management of quota allocations by the various UK producers' organisations was surveyed in Hatcher, A.C. 1997, *Producers' organizations and devolved fisheries management in the United Kingdom: collective and individual quota systems, Marine Policy* 21(6): 519-534.

The quota-shares allocated to vessels that are not members of a PO (the so-called "non-sector") are managed directly by the Government Fisheries Departments by monthly landings-limits. Uptake of the quota shares reserved for the inshore (10m and under) sector is not normally regulated unless the level of estimated landings dictates an early fishery closure, although increasing pressure on some inshore fisheries means that the use of monthly catch-limits is likely to become routine for this sector too.

Quota is allocated in 100kg units and attached to the vessel's licence, but individual-quota allocations remain entirely notional except for members of POs operating an IQ system, where the PO usually allocates each vessel an individual-quota based on its FQA.

As licences are transferred and aggregated, so are the FQAs attached to them. Clearly, however, there is no incentive for vessels to acquire notional-quota unless they can realise an individual quota-entitlement. Acquiring quota through the licence market is therefore only valuable for vessels which belong to a PO that operates an internal IQ system, although a "quota pooling" PO may not accept a new member unless he carries an adequate number of units. It is also not possible for (notional) quota to be divisible in movements between individual vessels through the licence transfer/aggregation system: FQA is transferred in its entirety.

The feature of the management system that greatly facilitates quota-trading is the relative freedom the POs have to exchange quota between themselves. As the quota management-system developed, POs were allowed to swap quota but initially all swaps had to balance in terms of "cod equivalents" (by value). This was later relaxed to allow non-balancing swaps, and then to allow "gifts", *i.e.* one-way movements of quota. Although all quota movements are subject to Government approval, in practice POs can now trade quota between themselves in any quantity, and the Government takes no interest in any financial transactions that may accompany quota-exchanges or gifts.

Since quota can be traded between POs, it can effectively be traded between a member of one PO and a member of another PO, either "permanently" (a straight quota sale) or on an annual lease basis. Under the FQA system as it is presently configured, however, a sale of one tonne of quota from a vessel in one PO to a vessel in another PO necessitates the transfer between the POs of one tonne of quota each year *in perpetuity*. This assumes that no annual adjustments of vessels' FQAs are permitted to take account of quota movements either between, or indeed within, POs (where quota-trading is obviously much more straightforward), which is the Government's current stated policy. At the time of writing the Fisheries Departments have indicated that annual adjustments will not be allowed, but industry pressure may result in a change of policy.

A further significant feature of the UK quota management-system is that it is essentially an informal arrangement between Government and industry. Legislation provides for the issuing of fishing licences and for the attachment of certain conditions to those licences, including quantitative restrictions on landings. There is no legal basis, however, for the notional individual quota-allocations which are used to calculate group-allocations, they are merely an administrative tool used by the Government Fisheries Departments in the exercise of the Government's discretionary right to issue licences in order to regulate sea fishing. Fishermen have no legal right to receive a licence nor any legal title to a share of national quotas[15].

3. MEASUREMENT AND ADJUSTMENT OF FISHING CAPACITY
3.1 Objectives

UK policy on the measurement and adjustment of fishing-capacity is largely dictated by the fisheries structural policy of the EU[16]. This requires member states to adjust their fishing capacity in line with the EU's so-called *Multi-Annual Guidance Programmes* (MAGPs) which impose target levels of fleet capacity measured in terms of gross tonnage (GT) and engine-power (in kW). Although the MAGPs date back to 1984, it was the third MAGP for 1993-1996 that first imposed significant cuts in (nominal) fishing capacity on the EU member states. This section concentrates on the UK's attempts to meet the objectives of MAGP III through a combination of measures, including a series of annual decommissioning schemes and the licence "market". Reference is also made to MAGP IV, which covers the period 1997-2001. It should be appreciated that the nominal capacity-reductions required under the MAGPs are dictated simply (and more or less *pro rata*) by biological estimates of excess fishing mortality for the various exploited stocks.

The MAGPs divide fishing fleets up by sector, with each segment by and large being defined according to fishing method, although in some cases sectors are distinguished by vessel size or target species. Targets must be achieved for each sector as well as overall. For the UK the third MAGP, approved by the European Commission in

---

[15] Legal opinion, however, suggests that in the event of a challenge a fisherman may well have claim to a "legitimate expectation" of quota.

[16] See Hatcher, A.C. 2000, Subsidies for European fishing fleets: the European Community's structural policy for fisheries 1971-1999, *Marine Policy* 24(2): 129-140.

December 1992, required the UK Government to achieve an overall cut in the size of the registered fishing fleet from 214 733GRT and 1 228 922kW (the reported situation at 1 January 1992) to 173 455GRT and 995 627kW by 31 December 1996[17]. These figures represented overall reductions of 19.2% in GRT and 19.0% in engine-power (measured in kW)[18]. The UK's targets were however amended in June 1995: the reductions required for the "*Nephrops* trawl" and "shellfish mobile" fleet-segments were relaxed in the light of information supplied by the Government to the Commission on the rates of by-catch of demersal species by these segments[19]. As a result, the overall reductions required for the UK fleet as a whole were reduced slightly to 17.6% in GRT and 17.4% in engine-power.

Table 4 summarises the original and revised objectives for each of the UK fleet-segments under MAGP III. It can be seen that the biggest cuts were required for the "beam trawl" and the "demersal trawl and seiners" segments. For other segments the cuts required were intended only to take account of the effects of technical progress. Under the terms of MAGP III, up to 45% of the overall reductions required could be achieved by reductions in activity (measured in terms of days at sea)[20].

Table 4
UK objectives by fleet-segment under MAGP III

Fleet segment		Situation at 1.1.92	Original objective for 31.12.96	Initial % reduction required	Amended objective for 31.12.96	Final % reduction required
Pelagic trawl	GRT	25 178	22 633	10.1%	22 633	10.1%
	kW	80 858	72 060	10.9%	72 060	10.9%
Beam trawl	GRT	23 062	17 621	23.6%	17 621	23.6%
	kW	107 542	81 465	24.2%	81 465	24.2%
Demersal trawl and seiners	GRT	71 956	51 746	28.1%	51 746	28.1%
	kW	368 194	262 505	28.7%	262 505	28.7%
Nephrops trawl	GRT	18 140	13 860	23.6%	16 306	10.1%
	kW	100 142	75 859	24.2%	89 246	10.9%
Netters liners and other static gears	GRT	12 121	10 896	10.1%	10 896	10.1%
	kW	58 503	52 137	10.9%	52 137	10.9%
Shellfish mobile	GRT	6 007	4 320	28.1%	5 400	10.1%
	kW	34 725	24 757	28.7%	30 947	10.9%
Shellfish fixed	GRT	2 636	2 370	10.1%	2 370	10.1%
	kW	18 397	16 395	10.9%	16 395	10.9%
Distant-water	GRT	10 987	9 876	10.1%	9 876	10.1%
	kW	23 829	21 235	10.9%	21 236	10.9%
Mixed (non-trawlers) <10m	GRT	24 438	21 968	10.1%	21 968	10.1%
	kW	304 630	271 484	10.9%	271 484	10.9%
Others >10m	GRT	20 208	18 165	10.1%	18 165	10.1%
	kW	132 102	117 728	10.9%	117 728	10.9%
Total	GRT	214 733	173 455	19.2%	176 981	17.6%
	kW	1 228 922	995 627	19.0%	1 015 204	17.4%

Note: The figures for the fleet situation at 1.1.92 do not include vessels registered in the Isle of Man or the Channel Islands (estimated to represent an additional 2 500 GRT and 33 000kW at the time); the objectives however do include this part of the UK fleet.

3.2 Implementation
3.2.1 Technical measures

In February 1992 the UK Minister of Agriculture, Fisheries and Food announced a package of measures designed to "reduce fishing effort over the period up to 1996, to meet a target which we shall need to quantify and

---

[17] Commission Decision 92/593/EEC (OJ No L 401, 31.12.92, p33).

[18] The MAGPs adopted at the end of 1992 took into account the objectives for the transitional guidance programmes for 1992: see Commission Decision 92/363/EEC (OJ No L 193, 13.7.92, p25) in the case of the UK.

[19] Commission Decision 95/243/EC (OJ No L 166, 15.7.95, p21).

[20] In the 1995 amended MAGPs "fishing effort" was defined by the Commission in GT x days and kW x days, GT to be estimated by the Commission according to the provisions of Commission Decision 95/84/EC (OJ No L 67, 25.3.95, p33).

agree with the Commission in line with our Community obligation under the 1992-1996 Multi-Annual Guidance Programme"[21]. These measures were to include the following:

i.   the introduction of (tradeable) days-at-sea entitlements
ii.   a series of annual decommissioning schemes
iii.   increased capacity penalties for fishing-vessel licence-transfers and aggregations and
iv.   the extension of restrictive licensing to vessels of 10m-or-under.

At the time the Government estimated that around a 5-6% reduction in capacity would be achieved through the decommissioning scheme, and that other measures would achieve the balance of the reductions required to meet the 1996 MAGP targets.

3.2.2 Days-at-sea limitation

At the end of 1992 new primary legislation was enacted to enable days-at-sea entitlements to be attached to fishing-vessel licences, and secondary legislation to establish the general rules for days allocations was put in place in May 1993[22]. The Government had originally planned to freeze effort at 1991-levels in 1993, and then to reduce it as necessary in the years 1994-1996, depending on the contribution of other measures to overall capacity-reduction[23]. Under the terms of MAGP III, a reduction of 8.6% in overall fleet activity could have been made to contribute a 45% achievement of the final objectives.

However, opposition from the fishing industry led first to a postponement of the scheme's implementation until 1994, and then to an indefinite suspension following a legal challenge by the National Federation of Fishermen's Organisations, which was referred by the British High Court to the European Court of Justice. In October 1995 the Court found in favour of the UK Government, but the Government decided against trying to reintroduce the scheme at this time (ostensibly because of the introduction in 1996 - in principle at least - of effort-limits for all Community vessels over-18m fishing in Western waters).

3.2.3 Decommissioning schemes

It was originally intended that a decommissioning scheme would run for two years with a total (gross) budget of £25m. In June 1994 the scheme was extended for a third year, and in January 1995 it was extended until the 1997-98 financial year with the total budget being increased to £53m[24].

The key operational features of the decommissioning schemes were as follows: (a) vessel owners were invited to submit tenders for compensation, which were ranked in terms of £ per vessel-capacity unit withdrawn[25]; (b) successful applicants had to surrender their licences and the vessels had to be scrapped after de-registration (despite the other options under EC rules of disposal outside the Community, or use for purposes other than fishing); and (c), eligible vessels had to be UK-registered (not registered in the Isle of Man or Channel Islands), seaworthy (with appropriate safety certificates if necessary) and over 10m in length. In order to be consistent with EC rules, grants were also limited to vessels that were at least ten years old. It was the responsibility of the vessel owner to supply proof of scrapping by the required date (and so, by implication, to bear all of the costs involved in scrapping the vessel).

The first scheme was announced in May 1993. In addition to the general rules outlined above, the following criteria were applied:

i.   the vessel must have been fishing for at least 100 days in 1991 and in 1992 and
ii.   the vessel's licence had to be no less restrictive in terms of the stocks authorised than at 27.2.92 (the date of the first announcement of the decommissioning schemes).

The rules for the 1994-scheme were modified in order to target vessels fishing for the most sensitive stocks (and those in the fleet-segments requiring the largest cuts) and to exclude distant-water vessels. The detailed criteria were now as follows:

i.   the vessel must have been fishing in Community waters for at least 100 days in 1992 and in 1993 as a UK-registered vessel and
ii.   the vessel's licence had to be a "full pressure stock licence" (the old equivalent of a category A licence).

---

[21] MAFF News Release No 73/92, 27.2.92.
[22] Sea Fish (Conservation) Act 1992; Sea Fish Licensing (Time at Sea) (Principles) Order 1993.
[23] MAFF News Release No 73/92, 27.2.92.
[24] MAFF News Release No 227/94, 15.6.94, and No 23/95, 18.1.95.
[25] The Government rejected the option of simply paying flat-rate compensation payments up to the maxima allowed under EU rules as being unlikely to produce value for money.

For the 1995-scheme the prawn (*Nephrops*) trawlers were excluded as were other shellfish boats, because these segments were considered to have already met their MAGP targets. The detailed criteria were now:

i.   the vessel must have been fishing in Community waters (or Norwegian waters south of 62°N) for at least 100 days in 1993 and in 1994

ii.   the vessel's licence had to be a category A or category B licence (or category C with certain individual species entitlements) and

iii. the vessel should not have been predominantly involved in the *Nephrops* fishery in 1994.

The eligibility rules announced for the 1996 scheme were relaxed to include all vessels over 10m in length, including distant-water vessels and prawn trawlers. The criteria were now simply:

i.   the vessel must have been fishing for at least 75 days (the EC minimum) in 1994 and in 1995 and

ii.   the vessel had to have a valid licence.

During consultations on the terms of the 1996-scheme, the Government proposed that the owners of decommissioned vessels should be allowed to retain their landings track-records, to be transferred to another vessel or sold, thus encouraging lower bids. The industry, however, rejected this option at the time.

Although MAGP III finished at the end of 1996, the 1997-scheme was designed to make up any shortfall under MAGP III. The main changes in the eligibility criteria were that applications were once more restricted to vessels with category A licences, and that *Nephrops* trawlers were again excluded (the Government left open the option of excluding vessels from other segments which were found to have met their MAGP targets). Most significantly, the industry had by now agreed that landings track-records could be retained by owners who decommissioned their vessels.

3.2.4 Licensing

The UK Government clearly hoped that the fishing-vessel licensing-system would make a significant contribution to reducing fleet-capacity in line with the MAGP III targets, in particular through increased capacity penalties for licence transfers and aggregations, and through administrative changes designed to limit the expansion of certain fisheries.

The UK licensing-system was and remains complex, having evolved in largely a piece-meal manner since 1984 as new restrictions were introduced in order to control the growth of various fisheries (and, in some cases, to try and limit the size of the Spanish- and Dutch-owned sectors).

By 1992, restrictive licensing applied to all vessels over-10m but still did not apply to vessels of 10m-or under. Following a period in which there were rather complicated restrictions on transferability, most licences had since 1990 been transferable both between vessels and between owners, but capacity penalties (described in Section 2. above) were incurred whenever licences were transferred or aggregated. A number of licensing measures were introduced during this period as follows:

i.   In 1992 the aggregation of pressure stock licences onto beam trawlers in ICES Area IV was dis-allowed, and the VCU penalty for all transfers and aggregations was increased to 20%.

ii.   In 1993 restrictive-licensing was extended to the sector 10m-and-under. The VCU penalty was again reduced to 10% for aggregations where the increase in engine-power was no more than 15% and for all over-10m licence-transfers. There was to be no penalty applied to transfers of vessels 10m-and-under (no increase in VCUs was allowed) but no licence-transfers from vessels 10m-and-under to vessels over-10m, or aggregations combining both under-10m and over-10m licences were permitted. In addition, no more than two vessels in the 10-17m band could be involved in aggregations. From 1993 it was also no longer possible to retain indefinitely a licence entitlement that was not actually assigned to a vessel.

iii. In 1995 the overall structure of the licensing system was revised. All licences were now fully transferable independently

An early example of an early Scottish purse seiner that would have depended on a producer's organization for its quota entitlement

of vessels, but still as a general rule only similar licences could be aggregated. An exception was made for purse-seiners and freezer-trawlers which could receive demersal licences for engine modifications up to +15% with a 10% VCU penalty. Also in 1995 the penalty for aggregating three or more licences was increased to 30% (except in the case of purse-seiners and freezer-trawlers), aggregations of more than two licences between 10-17m were now permitted, and Area IV beam trawl licences could now be aggregated if the resultant engine-power did not exceed 1500kW. Another significant change at this time was that vessels' landings track-records (which were normally assessed over the previous three years) were now attached to the licence rather than the vessel.

iv.   Early in 1996 it was announced that no further licence transfers or aggregations would be allowed which would increase either tonnage or engine-power, and that no aggregations of vessels 10m-and-under would be allowed to result in a VCU total of more than 100.

3.3   Results

3.3.1 Progress of MAGPs

The data in Table 5 are taken from the European Commission's report on the progress of the MAGPs at the end of 1996[26]. They are based on the Community fleet register but take account of the UK Government's reallocation of vessels from the "others" segment[27]. From the Table it appears that at the end of 1996 for the UK fleet there was still an overall reduction-backlog of some 5.0% in GRT and 2.9% in engine-power, but for certain fleet-segments the deficit was more significant, in particular for the segments: beam trawl, demersal trawl, and shellfish (fixed). Because of discrepancies between the UK register and the Community register, however, which affected the figures for 1992 as well as 1996, this situation was a provisional result.

Table 5
UK progress towards MAGP III targets

Fleet segment		Situation at 1.1.92	Situation at 31.12.96	Revised objective for 31.12.96	% reduction still required
Pelagic trawl	GRT	25 178	27 132	23 541	13.2%
	kW	80 858	73 896	77 955	--
Beam trawl	GRT	23 062	20 966	18 393	12.3%
	kW	107 542	109 259	86 467	20.9%
Demersal trawl and seiners	GRT	71 956	64 111	57 559	10.2%
	kW	368 194	323 184	300 176	7.1%
Nephrops trawl	GRT	18 140	14 350	18 123	--
	kW	100 142	83 820	101 018	--
Netters liners and other static gears	GRT	12 121	14 588	12 712	12.9%
	kW	58 503	54 738	63 910	--
Shellfish mobile	GRT	6 007	8 110	8 125	--
	kW	34 725	45 558	48 606	--
Shellfish fixed	GRT	2 636	5 839	5 094	12.8%
	kW	18 397	36 539	34 054	6.8%
Distant-water	GRT	10 987	7 107	9 876	--
	kW	23 829	18 120	21 236	--
Mixed (non-trawlers) <10m	GRT	24 438	19 577	21 968	--
	kW	304 630	264 868	271 482	--
Others >10m	GRT	20 208	0	0	
	kW	132 102	0	0	
Unclassified	GRT	--	2 760	--	
	kW	--	24 518	--	
Total	GRT	214 733	184 539	175 391	5.0%
	kW	1 228 922	1 034 498	1 004 903	2.9%

Notes: 1992 figures from amended MAGP III for the UK (Commission Decision 95/243/EC); the 1996 figures are from the Community register and still do not include all vessels registered in the Isle of Man and the Channel Islands (see text).

---

[26] COM(97) 352 final, 11.7.97.

[27] The Community register at that time still did not contain a complete record of vessels registered in the Channel Islands and the Isle of Man.

More up-to-date figures on the situation at the end of MAGP III are given in the decision on MAGP IV, which was approved at the end of 1997[28] (although these figures are still subject to revision because of the transition from GRT to GT as the common measure of vessel tonnage). These figures are shown in Table 6. Direct comparison with Table 5 is difficult because of the change from GRT to GT, the grouping of certain fleet segments and further adjustments to the UK register. Nevertheless, it is clear that by this stage that the overall objectives for MAGP III had more or less been met, but significant deficits remained in certain fleet segments, most notably the beam trawl segment. Table 6 also shows the capacity figures for the beginning of 1998 and the objectives which the UK is supposed to meet by the end of 2001 under MAGP IV[29].

Table 6
UK progress towards MAGP III targets (MAGP IV figures)

Fleet segment		Situation at 1.1.97	Revised objective for 31.12.96	Situation at 1.1.98	Objective for 31.12.01 under MAGP IV
Pelagic trawl and purse seines	GT	37 453	34 876	41 220	34 876
	kW	71 876	82 168	69 757	82 168
Beam trawl	GT	28 240	26 062	26 323	26 062
	kW	117 616	103 054	106 143	103 054
Demersal trawls seines and *Nephrops* trawls	GT	116 581	120 630	115 468	120 630
	kW	400 127	422 876	390 150	422 876
Netters liners and other static gears	GT	16 431	15 854	16 282	14 538
	kW	51 977	67 364	51 550	61 744
Shellfish mobile	GT	11 766	11 615	10 197	11 552
	kW	55 648	51 232	46 872	50 958
Shellfish fixed	GT	6 413	6 267	7 305	6 242
	kW	44 463	35 895	49 512	35 768
Distant-water	GT	15 567	14 883	15 829	14 883
	kW	25 400	23 741	25 004	24 281
Small-scale coastal (<10m)	GT	20 120	21 901	19 991	21 901
	kW	286 367	286 154	287 554	286 154
Total	GT	252 571	252 088	252 615	250 684
	kW	1 054 474	1 072 484	1 026 542	1 066 463

Notes: the 1997 GT figures include some estimations and are therefore subject to revision; the 1996 GRT objectives were converted to GT according to the relationship between GT and GRT for the fleet at 1.1.97.

3.3.2 Decommissioning

The four decommissioning schemes operated during 1993-96 removed a total of 578 vessels over-10m from the UK fleet, representing 19% of the 1992 total of 3036 vessels over 10m. Table 7 shows the capacity withdrawn by segment, compared to the 1992 situation, and the 1996 objectives as specified in the 1995 amended MAGP for the UK. Because of the adjustments to the UK register during 1996-7, these are probably the most appropriate comparisons available.

It is apparent that while the decommissioning schemes removed around a half of the required tonnage, they removed only one third of the required engine-power. Certain fleet segments (*Nephrops*-trawlers, shellfish-mobile and shellfish-fixed) were clearly over-represented in the decommissioning process, while other segments were under-represented.

The total gross expenditure on the 1993-6 schemes was £36.24m. The 1997-scheme, which cost around £14.3m, removed a further 108 vessels and 4406 GRT[30]. At the time of writing details of the capacity removed under this last scheme are not currently available.

3.3.3 Fishing vessel licensing

The only data presently available on the effects of licence-transfers and aggregations over the period 1992-1996 come from a 1997 report on the decommissioning schemes undertaken for the UK Government[31]. No data are

[28] Commission Decision 98/124/EC (OJ No L 39, 12.2.98, p34).
[29] COM(99) 175 final, 27.04.99.
[30] MAFF News Release No 383/97, 3.12.97.
[31] *Economic Evaluation of the Fishing Vessels (Decommissioning) Schemes*. Report to the UK Fisheries Departments. Nautilus Consultants, Edinburgh, September 1997.

currently available on transactions concerning vessel-licences 10m-and-under in the inshore sector, or the results of licence-transfers and aggregations since 1996.

During the period a total of 397 transactions involving vessels over-10m (measured in terms of the number of recipient vessels) resulted in a reduction of 39 737kW (17%) from the donor licence total of 232 478kW. This represented some 19% of the overall reduction in engine-power required under MAGP III (based on the 1995 amended targets). The contribution of licence transfers and aggregations in terms of tonnage is not known because of changes from GRT to GT, and because the VCU system does not take account of vessel tonnage directly.

Table 7
Capacity withdrawn by decommissioning 1993-1996

Fleet segment		Situation at 1.1.92	Amended objective for 31.12.96	% reduction required	Nominal capacity with-drawn	% 1992 capacity with-drawn
Pelagic trawl	GRT	25 178	22 633	10.1%	437	1.7%
	kW	80 858	72 060	10.9%	850	1.1%
Beam trawl	GRT	23 062	17 621	23.6%	2 138	9.3%
	kW	107 542	81 465	24.2%	9 791	9.1%
Demersal trawl and seiners	GRT	71 956	51 746	28.1%	6 916	9.6%
	kW	368 194	262 505	28.7%	25 805	7.0%
Nephrops trawl	GRT	18 140	16 306	10.1%	5 174	28.5%
	kW	100 142	89 246	10.9%	17 134	17.1%
Netters liners and other static gears	GRT	12 121	10 896	10.1%	841	6.9%
	kW	58 503	52 137	10.9%	3 849	6.6%
Shellfish mobile	GRT	6 007	5 400	10.1%	713	11.9%
	kW	34 725	30 947	10.9%	4 539	13.1%
Shellfish fixed	GRT	2 636	2 370	10.1%	697	26.4%
	kW	18 397	16 395	10.9%	3 735	20.3%
Distant-water	GRT	10 987	9 876	10.1%	228	2.1%
	kW	23 829	21 236	10.9%	1 214	5.1%
Mixed (non-trawlers) <10m	GRT	24 438	21 968	10.1%	0	0.0%
	kW	304 630	271 484	10.9%	0	0.0%
Others >10m [non-active/unknown]	GRT	20 208	18 165	10.1%	500	2.5%
	kW	132 102	117 728	10.9%	3 209	2.4%
Total	GRT	214 733	176 981	17.6%	17 643	8.2%
	kW	1 228 922	1 015 204	17.4%	70 126	5.7%

Notes: 1992 figures and 1996 targets as in 1995 amended MAGP; data on decommissioned vessels from MAFF published in Nautilus Consultants (1997).

In terms of engine-power at least (again calculations for tonnage are difficult because of the re-measurement/ estimation in terms of GT and the lack of tonnage figures for the licence contribution) decommissioning and licence penalties together removed around 71% of the observed decrease in the size of the fleet over-10m from 1992 to 1996 (45% by decommissioning, 26% through licence transactions). It is difficult to say how the remaining 29% was achieved, although some of the apparent decrease is almost certainly due to administrative adjustments to the UK register.

4.   CONSEQUENCES OF TRANSFERABLE USE RIGHTS
4.1   Licence and quota trading

Obtaining comprehensive and reliable data on licence-trading in the UK is extremely difficult, since the Government takes no interest in the financial transactions that accompany licence-transfers, and all trading takes place privately either directly or via licence brokers. The Government also does not routinely produce statistics on licence-transfers and aggregations. Nevertheless, there is no doubt that the volume of licence and quota-trading has increased greatly since 1995.

The report on the UK decommissioning schemes [31] investigated trends in licence/track-record prices during the period of MAGP III. Data was obtained from licence-brokers showing that in early 1997 track-record prices (as a component of total licence values) ranged from £350/t for herring, to £1200/t for cod, and up to £10 000/t for sole. The figures are reproduced in Table 8 below. The consultants also obtained data on licence prices over the period

1993-1996. These figures are summarised in Table 9 (with the pre-1995 licence-types expressed in terms of their current equivalents).

It is difficult to interpret the apparently chaotic price movements during 1993-1995, but the big jump in the prices for category A licences (which authorise fishing for all quota stocks) from 1995 to 1996 reflects the marked strengthening of demand for licences which started to occur around this time.

Table 8
Average track-record prices in 1997

Species	£ per tonne
cod	1 200
plaice	1 200
saithe	2 500
hake	2 000
monkfish (anglers)	3 300
sole	6 000 to 10 000
herring	350
mackerel	700

Table 9
Average licence prices (per VCU) 1993-1997

Licence type category	£ per VCU			
	1993	1994	1995	1996
A	179	269	184	776
A purser	na	395	na	2 083
A beam trawl	700	391	276	1 222
B	62	189	88	182
C	47	na	na	184

The report (1997) considered only that there was "some evidence" of quota-leasing arrangements between PO members. We find that in recent years quota-leasing and selling within and between POs has grown significantly in importance. A report for the Scottish Whitefish Producers Association[32], which represents the majority of catching power in the Scottish demersal fleet, found that almost a third of the Scottish fleet had bought quota in some form or other during the conversion of rolling track records to FQAs. Leasing, although becoming more widespread, was less prevalent at the time, with less than 10% of vessels leasing quota. This number, however, is still much greater than seen in previous years, when the only leasing that had taken place was between Dutch-owned vessels in the North Sea Fishermen's Organisation and in the Fife Fish Producers Organisation, which had already had experience of this while operating under the Dutch ITQ system. The same report indicated that the ten or so quota-trading and leasing arrangements currently in operation would increase rapidly, as skippers coming up for retirement opt to keep their quota entitlements and lease them out.

The average quota-prices given in Table 10 were obtained from the records of a company representing a significant proportion of the quota-trading in the UK. This company remarked that the prices of a number of stocks were largely driven by purchases by foreign-owned vessels (so-called "quota-hoppers"), with Spanish interests buying quota for monkfish (angler) and hake, and Dutch-owned companies buying quota for North Sea sole and plaice.

4.2 Impact of trading on fleet structure

We estimated previously that from 1992-96 around 45% of the observed decrease in the capacity of the fleet over-10m (in terms of engine-power in kW) was attributable to decommissioning, while around 26% was achieved through the capacity-penalties applied to licence-transactions.

The figures for the UK fleet for the beginning of 1997 (from Table 6) and the latest figures available from the Ministry are shown in Table 11. There has been a 3% overall increase in registered tonnage and a 7% decrease in rated engine-power[33]. However, this result masks some significant changes in the capacity of certain fleet segments.

---

[32] *Problems Associated with trading in fish quota: solutions for the benefit of the fishery and dependent communities.* Report prepared for the SWFPA by Rodgers, P. The Centre for Fishery Economics Research, December 1999.

[33] At the time of writing efforts are underway to tighten up the measurement of engine-power which may result in an increase in the measured total.

For example, the tonnage of the pelagic sector increased by 27%, while the beam trawl sector decreased by 13% in GT and 22% in engine-power.

Table 10
Average quota-prices (per tonne) 1995-2000

Quota stock	£ per tonne						
	1995	1996	1997	1998	01/1999	08/1999	01/2000
NS cod	250	400	1 000	1 000	1 500	2 000	2 200
NS haddock				1 000	1 200	1 800	3 000
NS plaice		1 800	2 000	1 800	1 500	1 500	1 200
NS sole		10 000	10 000	8 000	8 000	8 000	10 000
NS anglers				2 500	2 400		2 200
NS *Nephrops*				1 500	2 500		3 500
WOS anglers				3 500	5 500		6 000
WOS *Nephrops*				700	1 800		2 000
VIIa sole					7 000		6 000
VIIe sole					7 000		9 000
VIIfg sole					8 000		9 000
VIIa plaice					1 000		1 000
VIIde plaice					1 200		1 200
VI/VII hake				2 500	3 500		4 500
VII anglers				2 500	3 500		5 000
VII *Nephrops*				1 000	1 500		1 700

NS = North Sea (ICES area IV); WOS = West of Scotland (ICES area VI); VII = ICES Area VII.

Despite the lack of data on licence-transfers and aggregations during 1997-99 it is apparent that the licence market has not resulted in any significant overall rationalisation of fleet capacity during the last two years, although there has been some reallocation of capacity-entitlements between sectors and some reduction in licence/vessel numbers (see Table 1). The ability to "strip" licences during the move to FQAs in 1998 may nevertheless have assisted the rationalisation of the UK fleet. Abuse of the VCU system, with many new vessels built since 1996 under-declaring their engine-power, is now accepted to be widespread. The Government, realising the widespread extent of the problem, recently announced steps to ensure compliance with the regulations. Vessel owners have until 30 June 2000 to admit to any power "irregularities", after which they will have 4 years to obtain the required VCUs. The existence of a number of "stripped" licences, bought from ageing vessels by quota-traders during the move to FQAs in 1998, has provided a ready source of VCUs for those vessels wishing to become legitimate, without the added difficulty of buying quota that they could not afford and did not require. The "disappearance" of the tonnage associated with these licences when they are amalgamated onto existing vessels, may well form a significant component of the tonnage-reduction required under MAGP IV.

The increasing trade in licences certainly had an effect on the operation of the 1992-97 decommissioning schemes, as did the decommissioning schemes on the licence market, by reducing the overall supply of licences and injecting liquid capital into the industry. It became clear during the schemes, that the fleet segments with the highest average licence-plus-vessel values were the least represented among the vessels successfully decommissioned. Given the current level of licence-prices, a tender-based decommissioning scheme, even based on EC maximum rates[34] would now be likely to remove few vessels from the most profitable fleet segments.

Whether the recent growth in quota-trading will start to have a significant effect on fleet capacity remains to be seen. Nevertheless, there are indications at least that the ability of many within the industry to separate quota and licence may result in a rationalisation of capacity.

For example, licences are currently allowed to lie unused as entitlements for 3 years. A number of vessel-owners have sold vessels, and for 2 years at least are renting their quotas, while keeping their licences and deciding whether to replace the vessels or leave the industry all together. In addition, the few deals that have taken place in the pelagic sector within the last 3 years have seen quota divided, in several cases, between purchasers. This is certainly having at least a short-term effect on capacity, as the licences are currently unused while the quota is being caught by others.

---

[34] See Annex III (point 1.1) to Council Regulation 3699/93 (OJ No L 346, 31.12.93, p1) as amended by Council Regulation 1624/95 (OJ No L 155, 6.7.95, p1).

Table 11
Changes in UK fleet capacity 1997-9

Fleet segment		Situation at 1.1.97	Situation at 31.12.99	% change
Pelagic trawl and purse seines	GT	37 453	47 661	27%
	Kw	71 876	77 209	7%
Beam trawl	GT	28 240	24 498	-13%
	kW	117 616	91 417	-22%
Demersal trawls, seines and *Nephrops* trawls	GT	116 581	116 752	0%
	kW	400 127	357 128	-11%
Netters, liners and other static gears	GT	16 431	15 046	-8%
	kW	51 977	43 968	-15%
Shellfish mobile	GT	11 766	12 317	5%
	kW	55 648	52 607	-5%
Shellfish fixed	GT	6 413	6 443	0%
	kW	44 463	43 353	-2%
Distant-water	GT	15 567	16 664	7%
	kW	25 400	25 015	-2%
Small-scale coastal (<10m)	GT	20 120	20 309	1%
	kW	286 367	288 239	1%
Total	GT	252 571	259 812	3%
	kW	1 054 474	979 473	-7%

5.   CONCLUDING REMARKS

The operation of a series of annual decommissioning schemes during the 1990s, and the changing nature of the possibilities for quota-trading under the UK's quota-management system, makes it difficult to attribute any changes in the size and capacity of the fleet to the emergence of transferable quota rights. From the figures that are available, the licence market did, however, contribute to the reduction in fleet size that was observed during the course of the decommissioning schemes.

Nevertheless, as might be expected, there are signs that the huge growth in quota-trading in the last couple of years may be starting to lead to some rationalisation of fleet capacity, if yet on a relatively small scale. Whether there will be a significant rationalisation in the future depends on the direction taken by the Government policy on quota-rights, and hence on the subsequent characteristics of the quota-market as well as the efficiency of quota-enforcement.

6.   LITERATURE CITED

Hatcher, A. and A. Read 2001. The Allocation of Fishing Rights in UK Fisheries: 1-14. *In:* Shotton, R.   (Ed.). Case Studies on the Allocation of Transferable Quota Rights in Fisheries. Fish. Tech. Pap. No. 411, FAO, Rome. 373 pp.

THE EFFECTS OF TRANSFERABLE PROPERTY RIGHTS ON THE FLEET CAPACITY AND OWNERSHIP OF HARVESTING RIGHTS IN THE DUTCH DEMERSAL NORTH SEA FISHERIES

W.P. Davidse
Agricultural Economics Research Institute, LEI
Burgemeester Patijnlaan 19, 2502 LS, The Hague, Netherlands
<w.p.davidse@lei.wag-ur.nl>

1. INTRODUCTION

This study considers the development of fleet capacity and harvesting rights in the Dutch demersal North Sea fishery since 1983. The Common Fisheries Policy (CFP) of the European Union was implemented in that year, which meant for this fishery a growing importance of harvesting rights. Individual vessel quota (IQs) for sole and plaice had already been introduced in 1976, within the framework of the North East Atlantic Fishery Convention (NEAFC)[1].

In the period 1976-84 these IQs were perceived by the vessel-owners as limitations rather than as rights. But enforcement of the quotas was rather weak in this period, so they were regarded as not much more than 'a piece of paper'. Transferability of the IQs was officially allowed from 1985. This, and also intensification of enforcement, gradually brought a transition of perceptions from individual limitations, towards the view that IQs for sole and plaice are valuable property rights.

The CFP of the European Union (EU) establishes annual Total Allowable Catches (TACs) for almost all commercial species landed by vessels of the member states. The Council of Ministers of the EU decides annually on these TACs, which are proposed by the European Commission. Each country operates its own management-system in order to fulfill its TAC obligations. The Dutch fishing sector is so far the only one within the EU that operates under an individual transferable quota (ITQ) system.

At the end of 1983 the Dutch demersal North Sea fishery consisted of 595 vessels owned by some 500 firms, and composed of four main segments:

i.   beam-trawlers, targetting sole and plaice; these were by far the most important segment. Most of these vessels have an engine-power that exceeds 800kW.
ii. roundfish-trawlers, concentrated in the 225-810kW range of engine-power
iii. vessels with an engine-power of 221kW, mostly operating in a variety of fisheries (beam-trawling for flatfish, demersal-trawling for cod and whiting, and shrimp fishing)
iv. vessels with engine-power under 221kW, which were generally specialised shrimp-trawlers. Some of these vessels operate in the Wadden Sea, in the north of the Netherlands.

Together, these four segments are known in the Netherlands as the "Cutter Fishery".

Table 1 shows some major characteristics of the demersal North Sea fishery. Recent figures have been added to demonstrate the important changes which have taken place. The next sections of this paper explain how transferable property rights have influenced the changes in fleet-capacity and ownership of rights.

An example of a beam trawler, common in the Dutch ITQ fishery for sole and plaice

---

[1] See Smit 2001.

Table 1
Characteristics of the Dutch demersal North Sea fishery in 1983 and 1998

	1983	1998
Annual quota (tonnes)		
Sole	15 400	14 600
Plaice	53 700	35 300
Cod	22 900	14 900
Financial results		
Proceeds (million NLG, deflated) [1]	840	607
Net Profit (million NLG, deflated) [1]	-44	39
Number of vessels	595	407
Value of harvesting rights per vessel (on average, NLG) [2]	150 000	5 000 000

[1] Deflated for 1983 on the basis of the NLG purchase power in 1998. 1 NLG= 0.45 EURO or 0.49 US$.
[2] Estimated on the basis of market prices.
Source: Ministry of Agriculture, Nature Management and Fisheries, Shipping Inspection, Dutch
Agricultural Economics Research Institute, LEI.

2. THE NATURE OF THE HARVESTING RIGHT

The transferability of the IQs for sole and plaice was officially allowed in 1985 by the Ministry of Agriculture and Fisheries after an informal trade in these 'rights' developed in the early 1980s. An extension of rights-based fishing came in 1994 when ITQs for cod were introduced, and in 1996 with the implementation of rights for herring and mackerel. As a result, today all quota species have been brought under an ITQ management-regime.

Co-management groups have pooled the quota (ITQs) of their members since 1993. This includes eight different management groups, whereby the board of each group is responsible for compliance with the group's quota. The ownership of the rights remains with the individual holders. These groups facilitate trade, hiring and renting of the ITQs between their members, which makes the system far more flexible. The rights can be used as a collateral for a loan. In fact, the ITQs have always served as a security for Banks when a loan was required, for example to finance a new vessel. Investments in ITQs used to be encouraged by a tax allowance for depreciation. This included a 12.5% annual depreciation on the purchase price of the right.

During the 1990s trade in quota led to very high prices for the ITQs. In fact they have become an important production factor for the firms, as indicated by the high value of the harvesting rights (Table 1). The sole and plaice ITQs are responsible for the major part of this value.

Apart from ITQs the Dutch rights-based fisheries management nowadays consists of a number of other individual rights:

i.   Licences, expressed in quantities of engine-power per vessel, were introduced in 1984. These transferable rights aim to limit the total engine-power of the sea-going fleet, and to give an entitlement to fish on quota species. This licence-scheme resulted from the first Multi-Annual Guidance Programme (MAGP), implemented in 1985 within the framework of the CFP. The target of the subsequent MAGPs has been the limitation of the capacity of fishing fleets in EU waters.

ii.   Transferable entitlements for shrimp fishing in the North Sea and in the Wadden Sea.

iii.   Entitlements to fish in the coastal zone, the so-called List I and II documents, which may also be transferred.

iv.   Limitation of gross tonnage (GT) of each vessel, implemented in 1998, which has led to rising values for transferable GT licences. This measure results from the Dutch obligations under MAGP lV (which runs from 1997-2001).

3. MEASUREMENT OF FLEET-CAPACITY
3.1 Characterizing fleet-capacity

As noted in the introduction, this study considers the development of fleet-capacity over the period 1983-98, *i.e.* since the EU Common Fisheries Policy started in 1983. Individual quota changed gradually from limitations on activities, towards valuable property rights in the early 1980s, and the transferability of these rights, officially allowed in 1985, marks this development.

Specialised beam-trawlers, fitted with engines exceeding 810kW (1100HP) comprised the most important part (65%) of the total fleet-capacity in terms of engine-power in 1983. These vessels target sole and plaice, but take turbot, cod and whiting as by-catch species. Their crew varies from 6-8 persons.

The medium-size trawlers, with engine-power ranging from 222-810kW, operate in various fisheries such as otter-trawling or pair-trawling on cod and whiting, herring pair-trawling and also beam-trawling. This fleet segment consisted of 173 vessels in 1983 accounting for 24% of the total engine-power of the fleet. The 221kW vessels mostly operate in the beam-trawl and shrimp-fishery, whereas most of the smallest vessels are specialised shrimp-vessels.

Engine-power is mostly used to express the fishing-capacity of the Dutch demersal fleet because this parameter is likely to have the main influence on the catches of the vessels. For the beam-trawlers in particular, this relationship is rather clear. Table 2 gives an overview of the fleet-capacity and its development.

Table 2

Dutch demersal North Sea fleet, number of vessels and total engine-power

	1983	1998
Total number of vessels	595	407
Number of vessels, by engine-power:		
0 - 190kW	141	82
191 - 221kW	78	142
222 - 1104kW	295	32
>1104kW	81	151
Total engine-power (kW)	367 000	319 000

Source: Ministry of Agriculture, Nature Management and Fisheries, Shipping Inspection, LEI.

The fishing effort expended by the fleet comprises fleet-capacity and fishing-time. For the Dutch demersal fleet it is usually expressed as engine-power times days-at-sea. Table 3 gives this effort for the different types of gear. Beam-trawling accounted for 77% of the total effort in 1983, followed by otter- and pair-trawling on roundfish (15%).

The major part of the fleet was rather young in 1983, with most vessels having an age of ten years or less. This was caused by a wave of investment in the period 1979-1983, that resulted in the addition of 126 new vessels to the fleet.

3.2 Changes in fleet-capacity in the period 1983-98

The number of vessels has decreased significantly over this fifteen-year period, and the fleet-composition has changed dramatically (Tables 2 and 5). The segment of medium-sized vessels almost disappeared and two other segments became far more important in 1998. These two segments, the 'Euro-cutters' and the larger beam-trawlers (more than 1104kW) nowadays count for about 90% of the total engine-power of the fleet. In terms of engine-power the capacity of the fleet fell by 13%, and the fishing-effort was at a level 7% lower than in 1998. Thus, the average number of days-at-sea per vessel has increased since 1983.

Table 3

Fishing effort of the Dutch demersal North Sea fleet (100 000 Horse-power days)

Fishing method	1983	1998
Beam trawl	656	703
Otter trawl and pair trawl, roundfish	126	32
Pair trawl, herring	35	6
Shrimp trawl	33	38
Other	5	12
Total	855	791

Source: LEI.

Table 4

Dutch demersal North Sea fleet, age profile of the vessels

Age	1983	1998
0 – 10 years	231	94
11 – 20 years	170	148
> 20 years	194	165
Total number of vessels	595	407

Source: Ministry of Agriculture. Nature Management and Fisheries; Shipping Inspection; LEI.

Another important change regards the age-composition of the fleet. The number of younger vessels (less than ten years old) decreased from 39% in 1983 to 23% in 1998. In contrast, the number of older vessels (more than twenty years old) rose from 33% in 1983, to 41% in 1998.

The change in vessel numbers is analyzed further in Table 6. It appears that different subsequent decommissioning schemes[2] have had an important impact on the fleet-capacity. The decommissioned vessels had to be scrapped or sold to third countries, *i.e.* countries outside the EU.

Some of the vessels under 'other withdrawals' in Table 6 have been re-flagged to other EU countries. This means that the fleet under Dutch ownership is in fact bigger than the previous tables suggest. These re-flagged vessels operate in Euro-waters and they are entitled to British, German and Belgian flatfish and cod quota. The re-flagged fleet accounted for about 20% of the demersal North Sea fishery under the Dutch flag in 1998, in terms

---

[2] The first decommissioning scheme started in 1988.

Table 5
Dutch demersal North Sea fishery, index of changes
in fleet capacity from 1983 to 1998 (1983=100)

	Index 1998
Total number of vessels	68
Number of smallest vessels (0-190kW)	58
Number of 'Euro-cutters' (191-221kW)	182
Number of mid-size vessels (222-1104kW)	11
Number of bigger vessels (>1104kW)	186
Total engine power (kW)	87
Total engine power (standard kWs) [1]	77
Fishing effort (in horse power/days):	
Beam trawl	107
Otter/pair trawl	25
Shrimp trawl	115
Total fishing effort	93

[1] Explanation see Section 3.3.
Source: Ministry of Agriculture, Nature Management and Fisheries,
Shipping Inspection, LEI.

of vessel number, engine-power and fishing-effort. Taking this into account the demersal North Sea fleet under Dutch ownership has more or less stabilized in the period 1983-98 from the viewpoint of total engine-power and fishing-effort.

The changes in fleet-capacity have been caused by a chain of several factors which are described below in the sequence they are presented. It has to be kept in mind that there are no simple cause-effect relationships, as causes may be effects from other points of view. Transferable harvesting rights have played a role among other factors.

Common Fisheries Policy

The implementation of the CFP in 1983 was the first and main factor influencing fleet-capacity, through the implementation of TACs within the framework of the conservation policy and the introduction of MAGPs resulting from the EU's structural policy. The CFP has led to several national measures which have caused major changes in the structure and scope of the Dutch demersal North Sea fleet.

TAC limitations

In the 1980s the national quota-levels for sole, plaice, cod and whiting caused a big imbalance between the capacity of many cutters and their fishing rights. A study by LEI (Salz *et al*.1988) pointed out that 70 000 – 100 000HP in the operating fleet would encounter liquidity problems in the next 2-4 years, due to this disproportion.

To comply with the TACs allocated to the Netherlands by the EU, a number of measures have been implemented, such as distribution of the national quota through ITQs, days-at-sea regulations, decommissioning, and stringent enforcement of the quota-system.

Horsepower licence system

In order to fulfill the obligations resulting from the first MAGP, the Dutch Ministry of Agriculture, Nature Management and Fisheries implemented a licence-scheme in 1984 which led to a total engine-power ceiling for the fleet. The total engine-power of the active fleet was allowed to increase until 1988 (due to orders for new vessels that were in the pipeline when the

Table 6
Dutch demersal North Sea fishery: additions to, and
withdrawals from, the fleet in the period 1988-1998

	Number of vessels
Fleet on 31 December 1987	611
Period 1988-1998:	
New constructions	+ 111
Second-hand, bought abroad	+ 22
Decommissioned	- 161
Other withdrawals [1]	- 176
Fleet on 31 December 1998	407

[1] Sold to other countries, re-flagged, changed to other activities,
scrapped, etc.
Source: Fisheries Directorate, Shipping Inspection, LEI.

licence-scheme came into force in 1984). The decrease of total engine-power over the period 1983-98 (Table 5) demonstrates the effectiveness of the engine-power scheme, since it prevented an expansion of fishing-effort after the profitable years of 1991-92.

Decommissioning schemes

The first decommissioning scheme started in 1988 and this was followed by subsequent programmes so that decommissioning grants could be obtained throughout nearly the whole period 1988-98.

Quota-limitations for cod and whiting have forced most of the owners of otter-trawlers and pair-trawlers to apply for decommissioning. This has been the main cause of the decline of the cutter fleet after 1988, in particular the dramatic decrease of the number of medium-size vessels. A total engine-power of 183 000HP (135 000kW) from 161 vessels, was withdrawn from the fleet in the period 1988-98. The majority of these decommissioned vessels (120) belonged to the medium-size group (222-1104kW).

Stringent enforcement

A major intensification of enforcement of the ITQs in 1988, through a systematic control of landings carried out by some hundred inspectors, made the over-capacity of the fleet visible, sensitive and in need of reallocation. This had a major impact in contributing to the effectiveness of the decommissioning schemes.

Limitations for the coastal zone

A limited number of vessels are entitled to fish within the 12-mile limit. This is an EU measure (Regulation No. 55/87) whereby the vessels concerned are registered under two separate files. The engine-power of the coastal vessels should not exceed 221kW. This special entitlement has been the main cause of the increase in the number of vessels just under the 221kW limit.

High prices of fishing rights

The stringent enforcement, mentioned earlier, led to a sharp rise in prices for flatfish ITQs in 1988. The good profitability of the cutters in 1991 and 1992 kept these prices at a high level, and even resulted in further price increases. The decommissioning process contributed significantly to the trade in ITQs over the period 1988-98. This enabled those who remained in the industry to adjust their fishing rights according to the available capacity of the vessel, by buying additional ITQs. The high Dutch prices for rights stimulated the purchase of different types of rights in other countries in the early 1990s, which has led to re-flagging of vessels.

Allowances on fiscal investment

A special law for stimulation of investment (for all industries) was introduced in 1978. This allowed the deduction of a certain percentage (12% at minimum) of the investment amount from taxable income. In fact, it diminished the income tax or corporate tax amount, and stimulated the construction of new fishing vessels during the period 1979-88, which contributed to an increase of total engine-power of the fleet up to 1988. This investment allowance was however abolished in 1988.

Economic performance

A good level of profitability in the years 1985-87 and 1991-92 stimulated the construction of new vessels, together with (over 1985-87 only) the investment allowances mentioned above. The existence of a second-hand market for vessels abroad enabled investors in new vessels to sell their 'old' ones at rather high prices and to transfer the engine-power licence from the vessel sold to the new one. In cases of upgrading the engine-power of existing vessels additional licence-units could be bought from those who withdrew their vessels from the Dutch fleet (other than by decommissioning). However, in the early 1990s this mechanism stopped nearly completely, mainly due to the tightening of the licence-schemes in the UK, and caused a major fall in the demand for second-hand vessels.

The role of transferable property rights in changes in fleet-capacity

As noted above, the causes of changes in the capacity of the Dutch demersal North Sea fleet have been complex and it is difficult to assess separately the impact of transferable rights. But it can be stated that the input-rights and output-rights have had several consequences:

i.   Withdrawals from the fleet (apart from decommissionings) realised high earnings from the sale of ITQs.
ii.   Decommissioning of vessels: vessel-owners who left the fishery had to relinquish to the seller their engine-power licence, but they could keep their ITQs. The high earnings from these rights have stimulated decisions to decommission vessels.
iii.   Concentration of rights amongst the owners of the larger beam-trawlers: this has led to the domination of these vessels in the total engine-power of the whole fleet.
iv.   The absence of a high level of construction in the 1990s after a series of profitable years: The effective engine-power 'ceiling' imposed has prevented such an expansion of the fleet. This constraint led in the early 1990s to a shift in investments from vessels towards ITQs. Such investments in ITQs have more-or-less absorbed all the depreciation funds of the companies, so future new vessel construction also will be at a lower level than in the 1980s.
v.   Re-flagging of vessels: the Dutch vessel-owners have acquired much experience in the market for harvesting-rights. High prices for ITQs in the early 1990s prompted them to look at the situation abroad. Low prices for such rights in the UK and other countries have encouraged decisions to buy rights abroad through the purchase of foreign companies. A number of Dutch cutter-vessels have been re-flagged to these foreign subsidiaries becuase they could not operate profitably in Holland at that time.
vi.   Better utilisation of the vessels and a more efficient uptake of quota: in particular the possibility of hiring and renting of ITQs has contributed to a better adaptation of the rules to business practices. The co-management sytem, established in 1993, created an important condition for this improved efficiency.

The Dutch experience demonstrates that co-management can secure the ITQ-right by sound management of group quotas. This includes the monitoring of landings, and measures such as warnings not to land abroad, *etc.*

when a group member has caught almost all his ITQ.[3]. Such group management guarantees that the individual holder can fully take-up his own ITQ. The threat that he will forego catch because of colleagues taking a part of his ITQ by over-fishing their own quota has been removed.[4]

3.3 Consequences of changes in fleet-capacity

The lower capacity of the demersal North Sea fleet, shown in Table 5, has had many consequences:

i.   Improvement of the profitability level of the cutters: the sector has been profitable or at break-even levels since 1991. This represents a long period of good economic returns, in view of developments in fisheries in the 1970s and the 1980s when profitable years were followed by years with adverse results. Fleet expansion through investments in new cutters after good years previuosly used to dissipate such profits, but this is impossible now because of the effective engine-power licence-limitation scheme.

ii.   Decrease of employment from 2750 crew members in 1983, to 1920 by the end of 1997: generally speaking those who have left the fishery could find jobs ashore, particularly in the past years. But now in many cases it is even difficult to find enough capable crew members for the cutters, due to the climate of good economic development in the Netherlands, and the ageing of the labour force.

iii. Decline of fishing communities: the industry fears that the 'critical mass' of some communities may be too small for their sustainability in the longer term.

iv.   A much higher proportion of the larger beam-trawlers in the fleet: this has affected the productivity of the sector. Such vessels show decreasing returns (*i.e.* revenue per kW/day for sole and plaice plus some less important bycatch (Smit *et al.* 1999)), so the fleet-capacity has in fact diminished more than the 'nominal' number of engine-power units indicates. Table 5 gives also the capacity expressed in standard kWs. This measure corrects for lower revenue per kW/day among the larger beam-trawlers so that a better estimate of the real capacity is obtained. In the same way, the real fishing-effort is in fact lower than the index shown in Table 5. When the lower productivity of the bigger vessels is taken into account, the real fishing-effort decreased by some 20% over the period 1983-98, in stead of 7% 'nominal value'.

v.   Difficulties to catch the full quota: apart from the engine-power limitation the cutters are also restricted in their number of days-at-sea. The current MAGP includes further reduction in days-at-sea per vessel, which may make it impossible for the fleet to land all the quota (Smit *et al.* 1999).

4.   CONCENTRATION OF OWNERSHIP

4.1   Status prior to the programme

The demersal North Sea fleet of 610 vessels was owned by 530 enterprises in 1985, the year when transferability of flatfish IQs was officially allowed by the Fisheries Directorate. These firms were family enterprises, in many cases employing several members of the family. Ownership may rest with the father alone or with him and several sons or brothers. In particular the situation of one owner with several sons, has led to expansion of the enterprise, since the aim was that each son should eventually be skipper on his own vessel. This kind of expansion was possible before rights-based fishing became effective, in the second half of the 1980s. It resulted in a number of bigger firms owning more than one vessel. In fact a concentration-process was already going on before 1985, thus leading to more engine-power in the fleet being exploited by fewer enterprises. In 1987 about half of the total engine-power was concentrated in these 'multi-vessel' companies, whereas they only accounted for 13% of the total number of firms (Salz 1987).

The flatfish IQs introduced in 1976 were, up to the mid-1980s, not much more than just 'a piece of paper'. Informal trade of these notes at that time, and the introduction of official transferability, demonstrated their growing importance around 1985. In subsequent years the trade in IQs has led to a concentration of rights with the bigger firms. This accelerated after 1988, because strict enforcement of the ITQ-regime forced the owners of the bigger cutters to acquire enough rights for their vessels. The introduction of ITQs for cod and whiting in 1994, and ITQs for herring and mackerel in 1996, gradually strengthened this concentration-process.

4.2   Restrictions on transfer of ownership

The following more detailed quota regulations were in force in 1998:

i.   A continuous individual-quota regulation, with annual changes in the Dutch part of the EU's TAC and resulting changes in ITQs. These are included in an Appendix to the regulations. This continuity of rights replaced the annual allocation process in 1997.

---

[3] One of the co-management groups expelled three members in October 1999 and held one vessel under arrest because of ITQ over-fishing. Visserijnieuws 29 October 1999.

[4] This advantage of co-management in an ITQ fishery has been emphasized by Dick Langstraat, Chairman of the Dutch Fish Board (pers. comm.). Transfer of some competence from the individual right-holder to the collective of the management group is essential in that case.

ii. Quota for related species are connected, which means that there should always be an ITQ for sole **and** plaice, just like one for cod **and** whiting.
iii. Transfers of the ITQs have to be registered by the Fisheries Directorate of the Ministry of Agriculture, Nature Management and Fisheries.
iv. ITQs should in principle be attached to a vessel, although an exception allows that the rights may be reserved separately during five years at most (from 1 January 1998). This only applies to ITQs that have been included in the total quota of a Group. This allowance for reservation enables right-holders, which have, for instance sold their vessel, to assign their ITQ temporarily to another operator while a new vessel is being built.

Some rules explicitly limit the transfer of ITQs:

i. Selling part of a quota for sole or for plaice to vessels that do not have such ITQs is not allowed.
ii. A quantity of both sole and plaice quota should remain after such a sale; the same applies for cod/whiting ITQs – so as to ensure that the TAC is not exceeded.
iii. ITQ holders are not free to withdraw their ITQ from the group quota during the course of the year, unless the group board agrees and the group quota that has been harvested up to that date does not exceed 90% of the group's total. Sale of the vessel, or bankruptcy, are two other cases in which the ITQ may be separated from the group quota.

A regulation stipulates a time-schedule for a number of requests to the Ministry. This regards mainly:

i. formation of group quota for the main species before 1 February·
ii. requests for transfer of a reduction in an ITQ for sole into one for plaice (or vice versa) before 1 March and
iii. requests for lease/rent transactions of ITQs between groups before 1 December. For fishermen of non-group members this date is 1 March

4.3 Prices received

The prices of ITQs are not publicly recorded, however the co-management groups have a good overview of these prices through their involvement in the trade of rights. Table 7 presents indications of prices for flatfish rights, which are expressed per kg quota right (Davidse *et al.* 1997). These prices are derived from the LEI cost-and-earnings panel, and from interviews with representatives of co-management groups.

Table 7
Price indications of flat fish ITQs

Year	Flat fish ITQ	
	sole/plaice (NLG/kg)[1]	sole only (NLG/kg)
1986	10-15	10-15
1987	70-85	.
1988	100-120	70-80
1989	100-120	70-80
1990	100-120	70-80
1991	130-150	90-95
1992	130-150	90-95
1993	70-95	55-75
1994	65-90	50-70
1995	a)	60-80
1996	a)	75-85
1997	a)	70-90

[1] 1.0 NLG=0.45 EURO or 0.49 US$.
a) Plaice ITQs have been traded more and more separately, at higher prices: NLG 9-13 in 1996 and NLG 10-18 in 1997.
Source: LEI; Co-management groups.

Rather high prices have been paid also for other entitlements: the documents for cod/whiting - NLG 14-17/kg since 1998, and permits for shrimp on the Wadden Sea, some NLG 300 000. Furthermore, engine-power licences were priced at NLG 800-1500/kW in 1998 and 1999.

The prices for flatfish ITQs increased sharply in 1987-88 as shown in Table 7. This reflects the fact that control measures became very stringent in 1988 and therefore effective in conserving the resource. In this year the systematic control of landings was implemented and was carried out by some hundred inspectors. In 1993-94 these prices for sole and plaice quotas dropped, due to the high level of the national sole quota, and diminished catches of plaice. These catches were low in many cases compared with the available plaice quota so that there was no need, generally speaking, to buy plaice quota.

In summary, with respect to price developments of Dutch ITQs, a number of major influences can be distinguished:

i. Enforcement of quotas: a major improvement of enforcement in 1988 caused a sharp price increase.

ii.   Profitability of the fishery: a greater profitability in 1991 also caused higher quota-prices. Formerly, investments in fishing vessels used to increase sharply in such a situation, but in 1991 the investments in vessels were replaced to a major extent by investments in flatfish ITQs.

iii. Availability of fish biomass relative to the quota-level: in 1993/1994 the national plaice quota was rather high in view of the catch-possibilities for this species. This contributed to a downward price trend for plaice quota.

4.4   Effectiveness of regulations governing ownership of rights

The ITQs should normally be attached to a vessel, so that in fact they are owned by the owners of that vessel. Dis-connection from the vessel is allowed for five years (for instance during new vessel construction) provided that the vessel is included in a group quota.

Conditions have been attached to very small boats in order to ensure a distinction between *bona fide* commercial vessels. Conditions for linkage with 'serious' vessels have been strengthened by the requirement that they undertake commercial exploitation. But, the phenomenon of fishermen who primarily remain ashore and try to make a living from leasing-out their ITQs still exists, and this is a matter of concern (although not a major one) for the industry.

The possession of valuable rights has gradually resulted in rather complicated arrangements to facilitate, for example the transfer of ownership of the rights to the son(s) of a rights-holder. Nowadays, ITQs have the same commercial importance as agricultural-rights, such as milk-quota .

4.5   Affects of the programme

The major intensification of enforcement in 1988, accompanied by the decommissioning of vessels, induced more and more transfers of rights. Vessel-owners who had to adjust their flatfish rights to the capacity of their vessel, were prepared to pay high prices for sole and plaice ITQs. The extra revenue from the additional quantities only had to cover the marginal cost of catching and landing the extra fish. Moreover, the desire to avoid heavy fines (through fishing with inadequate quota) was an important condition for this willingness to pay such high prices.

Table 8 shows that the owners of the larger cutters (over 1104kW) possessed 86% of the quota for sole in 1998, compared to 56% in mid-1988[5]. This change in the rights-situation has followed the trend towards larger beam-trawlers in the fleet. But the share of this segment within the total of flatfish-rights has increased somewhat more than its contribution to the total fleet engine-power (54% against 49% in 1988) .

Table 8
Concentration of fishing-rights according to engine-power

Fleet Segment (kW-group)	Mid 1988			January 1998		
	Number of vessels	Total power 1000kW	% sole quota	Number of vessels	Total power 1000kW	% sole quota
0 - 190	141	19	1.0	87	12	0.3
191 - 221	125	27	4.7	143	32	9.6
222 - 1104	201	151	38.5	30	18	4.1
>1104	139	236	55.8	156	264	86.0
Total	606	433	100.0	416	326	100.0

Source: Ministry of Agriculture, Nature Management and Fisheries, Shipping Inspection, LEI.

Apart from these output-rights the input-rights in the form of engine-power licences became more important in the late 1980s and in the 1990s. Owners who wished to increase the engine-power of their vessel, or to build a new cutter, had to buy additional engine-power rights on the market. In this way a trade in engine-power rights has also arisen, in particular during the 1990s.[6]

In Table 9 the distribution of individual quota for sole is considered against the size of the ITQ.[7] This size is expressed as a percentage of the total national sole quota. The values range from the 'mini' ITQs, representing holdings of 0.005% of the total sole-quota (an annual landing of 1.18t of sole on the basis of the 1994 TAC), to 1.5-2.5% (354-590t) for the biggest ITQs.

---

[5] The concentration level has been measured in rights for sole, but the same conclusions can be drawn for the ITQ rights for plaice and cod.

[6] In the first years after the introduction of horsepower licences a quantity of 'floating' licences existed because of extra orders for newbuildings just before this licence scheme was put in place in 1985. Therefore trade of these rights mainly began only after some five years after the introduction.

[7] Analysis from Davidse *et al.* 1997 p.184.

Table 9 shows that some concentration of ITQs for sole occurred during the period 1988-94. In 1994 owners of the bigger holdings of ITQs for sole, having 1% or more of the national sole-quota, represented a higher share (8%) of the total number of ITQ holders, compared with 1988 (4.7%). On the other hand, the percentage of holders owning smaller ITQs (up to 0.5%) decreased since 1988. The ownership-distribution with respect to the ITQs for plaice (not given in the Table) shows the same development, though holders of the biggest ITQs for plaice (1.5-2.5%) were somewhat fewer compared with the ITQs for sole.

Table 9

Distribution of ITQ-holders according to size of the ITQ, expressed as percentage share within the total allocated Dutch quotas for sole in 1988, 1994 and 1997

Share of ITQ within total quota for sole	Percentage of ITQ-holders		
	1988 (n=387)	1994 (n=289)	1997 (n=276)
0.005% (mini ITQ)	20.2	17.3	14.9
0.005 - 0.5%	65.3	57.7	59.7
0.5 - 1.0%	9.8	17.0	17.8
1.0 - 1.5%	3.9	4.5	3.6
1.5 - 2.5%	0.8	3.5	3.6
>2.5%	0.0	0.0	0.4
Total	100.0	100.0	100.0

Source: Ministry of Agriculture, Nature Management and Fisheries, LEI

Table 9 also records the ITQ-distribution against size for the 1997 allocations: it shows that the concentration tendency did not continue clearly during the period 1994-1997. The number of the smallest ITQ-holdings decreased on the one hand, but so also did the number of bigger ITQ-holders (category >1%). The underlying factor seems to be less trade in ITQs, since the number of holders remained rather constant between 1994 and 1997.

The important decrease in the total number of ITQ-holders (by 25%) in the period 1988-94 was caused by selling of sole/plaice ITQs, in combination with decommissioning, or exiting the business for other reasons, or by operating only beam-trawls. The share within the total Dutch sole/plaice quotas, of the holders of the 20% biggest ITQs, is another measure for the level of concentration. In 1994 this group owned almost 60% of total Dutch sole quota and 56% of the plaice quota. In 1997 the corresponding figures were 58% and 56% respectively. Nearly all of these ITQ-holders were companies owning more than one vessel.

The regional concentration of flatfish rights within Holland has also been considered in the property-rights study (Davidse 1997). The conclusion derived from the situation in 1997 was that the share of the main Dutch fishing port (Urk) in the national quota has decreased somewhat, whereas the Den Helder/Texel region has expanded its share. However, there was no major concentration of flatfish ITQs in just a few regions during the period 1988-97.

5.   DISCUSSION
5.1 Reduction in fleet-capacity

The Common Fisheries Policy of the European Union has two main aims: limiting the catches by fixing annual TACs, and reducing fishing-capacity by implementing multi-annual guidance programmes (MAGPs). It must be said that the MAGP objectives for the Dutch demersal North Sea fishery have not been met so far, although an engine-power licence-scheme has been in place since 1985, and subsequent decommissioning schemes have been implemented. The owners of the cutters could not be forced to leave the fishery so that the actual fleet-reduction depends on the profitability of the fishery, and also on the ownership situation, *i.e.* the presence of a successor for the current vessel-owner. In the past eight years most of the cutters have operated on or above the break-even point, so many skipper-owners have not had any compelling reason to stop fishing.

The Dutch Ministry of Agriculture, Nature Management and Fisheries has indeed been successful in preventing expansion of fishing-capacity after several profitable years. This kind of expansion had occurred regularly before the licence-scheme became effective in the early 1990s.

Looking at the developments over the past fifteen years, two reactions by vessel-owners have been contrary to the expectations of the policy-makers:

i.   Extra orders for new vessels were placed just before the engine-power licence-scheme was put into place in 1985. As a consequence the Ministry limited the validity of the 'floating' licences and strengthened the conditions for attachment to a vessel.

ii.   Re-flagging of cutters to other countries, in order to get more harvesting-rights for the remaining vessels in the Dutch fishery. In fact this has not been a major problem for the management of the Dutch fisheries since it contributed to compliance with obligations under the Dutch TAC and the MAGP.

The capacity of the cutter-fleet was 8% above the MAGP requirements at the end of 1997[8]. This target has been expressed in effort (days-at-sea) for the current MAGP which runs from 1998-2002. One goal of this programme is a further reduction by 17% for the segment of the larger beam-trawlers compared with the situation in 1998[9].

The industry strenuously opposes these effort-reductions since the quota allocations cannot be taken up with such a lower fishing effort. The industry representatives assert that the quota should take priority and not the capacity limitations. The co-management system has been able to ensure compliance with the quota since 1993, so that further substantial reductions of capacity are not necessary. In fact now, there are two conflicting kinds of management in the Dutch demersal North Sea fishery:

i.   A management by command-and-control, aiming at reducing the fleet within the framework of the CFP;

ii.   A type of co-management characterized by responsibility for compliance with national quota by groups of quota-holders. Centralized targets for fleet-capacity levels thwart these decentralized quota responsibilities. Therefore, giving priority to fleet-reduction above compliance with national quota would heavily undermine the Dutch co-management system.

This conflict between input-targets and output-targets was discussed in November 1999 between the Dutch government and the European Commission. In the autumn of 2000 a solution was found: with the EC allowing a higher capacity for the cutter fleet. The reason for allowing this correction was a previous wrong assumption regarding the number of vessels that would have a break-even level of revenue, estimated for the first MAGP in 1986. A new analysis of these historical data pointed out that there was 'economic room' in the fleet at that time, for an extra 59 000kW (+17.5%). This correction of the wrong start of the capacity-limitations has resulted in a proportional increase of the current, fourth MAGP, ending in 2001. As a consequence, today there is hardly any over-capacity in the Dutch cutter fleet.

5.2   Concentration of ownership

ITQs for all quota species have now become an important production factor for fishing enterprises. They are major intangible assets on the balance sheet of many fishing firms. The rights can be used as collateral for raising finance, and tax allowances for depreciation are in force, as for other assets. The co-management system facilitates transfers and the lease of rights via the group boards. Thus, co-management of ITQs in the past five years has brought advantages in the form of increases in price, for plaice in particular, through compliance with much lower quota.

Transferability has, since the mid-1980s, led to a continuing concentration of rights (although this has not, at least so far, resulted in just a few companies owning a major part of the rights). It is assumed that this process will continue in future. Newcomers cannot enter the fishery since the prices of rights are too high to make a new firm profitable (Davidse *et al.* 1997), hence in the future the number of enterprises should only diminish.

It is necessary to understand the nature of quota-trade in order to explore future developments. ITQs have mainly been purchased by vessel-owners who already had a vessel with harvesting-rights. Thus, high prices could be paid since marginal revenues only had to cover the marginal costs. Such a high price-level will hamper a major concentration of rights because of financial limitations. Nowadays the economic depreciation of vessels has been absorbed, more or less, by the quota investments, so vessel-replacement may be difficult to achieve in a number of cases. However, the quota-market depends on the future price-levels of rights. In this respect the concentration process is not hampered, apart from some restrictions described above in Section 4.2.

An important aspect of the concentration of rights is the scheduled review of the CFP in 2001. Quota-hopping by Dutch enterprises in the past has made clear that the issue of concentration of rights has gone further than that of just a national situation. The question is whether there will arise from 2001 onwards any new possibilities to acquire harvesting rights in other countries. If so, the concentration of rights may accelerate in a more open, international market. That would bring the fishing sector more in line with other branches of the economy in the common EU market.

---

[8] "Ondernemend vissen", LEI, 1998, p52.
[9] Visserijnieuws 21 May 1999.

6.   LITERATURE CITED

Davidse, W.P., H. Harmsma, M.O., van Wijk, L.V., McEvans, N. Vestergaard. 1997. Property rights in fishing, LEI-DLO report 159. 328pp.

Salz, P. 1996. ITQs in the Netherlands: twenty years of experience, ICES paper. 17pp.

Salz P., A. den Dulk, H. Harsma, W.P. Davidse. 1987. Visserij in cijfers, annual edition for 1987, LEI.

Salz, P., J. Smit, W.P. Davidse. 1988. Vooruitzichten voor de Nederlandse plat- en rondvissector op korte en middellange termijn. LEI-DLO report nr 5.79. 64pp.

Smit, W., P. Salz, W.P. Davidse, M.O. van Wijk, J. de Jager, C. Taal. 1998. Ondernemend vissen, Toekomstperspektief van de kottervisserij. Rep. 1.98.01. LEI. 100pp.

Smit, W. 2001. Dutch demersal North Sea Fisheries: Initial allocation of Flatfish ITQs: 15-23. _In:_ Shotton, R. (Ed.) Case studies on the allocation of transferable quota rights in fisheries. Fish. Tech. Pap. No. 411, FAO, Rome. 373 pp.

Van Wijk, M.O., C. de Ruijter, M.H. Smit, C. Taal. 1999. Visserij in cijfers, annual edition for 1999, LEI

Appendix I
Historical overview of the Dutch fisheries management measures

1975	NEAFC sets TACs and national quota for six species, including plaice and sole.
1976	The Netherlands introduces an IQ system for sole and plaice based on historical catch record. IQs are specified as percentage share of national quota. On the basis of the quota and the share, fishermen are allocated annually a specific weight (in kilograms) of catch of a particular species.
1977	IQ system is revised to take into account heavy investments undertaken in that period. IQ is based a combination of engine-power (50%) and historical catches (50%).
1977-84	IQs are barely implemented. They function mainly as just a 'piece of paper'.
1978	A national investment-stimulation scheme is introduced, including for sea-going vessels. It offers a tax rebate of 12% on new vessel construction.
1981	First codfish entitlements introduced ('k-documents'). They are not transferable.
1984-85	A licensing system is introduced based on the engine-power of the vessels. The ceiling set on the total engine-power of the fleet is made partly ineffective because of large investments made just prior to the introduction of the licenses. All vessels 'under construction' must also be given a license. Licenses are freely tradeable and divisible. Free licenses, not attached to any active fishing vessel, remain valid for 2 years.
1985	Transferability of IQs is officially allowed after a period of substantial informal trade during the previous years.
1987	National reserve is created: part of the national quota is not divided among ITQ holders, but retained by the administration to cover individual over-fishing of ITQs and to allow others to fully utilize their rights. The reserve is about 5% of the national quota.
1987	Uniform prosecution of fishery offences is promoted by creation of regular consultations among the responsible Attorneys-General.
1987	Fiscal subsidy on investment is stopped
1987	Maximum engine-power of new vessels to be built is set at 2000HP. Previously vessels of up to 4500HP were built. Maximum beam is set at 12m.
1987-92	System of obligatory days in port introduced.
1987-89	Legal framework for ITQ management is gradually developed. Responsibilities between the various Ministries are not shared effectively.
1988	Second codfish entitlement introduced (annual and seasonal, 'j-documents' and 's-documents') which are non-transferable.
1988	Strict monitoring of landings is introduced with 120 controllers on 600 vessels. Strict rules are set regarding places, times and conditions for unloading fish.
1988	Several violent encounters occur between fishermen, controllers and riot police.
1988-89	First decommissioning scheme (MAGP II). Industry contributes approximately 10% of the total costs.
1988-90	Creation of first quota-management groups. But these groups failed because of insufficient control and their weak legal position.
1989-90	Administration cracks down on auctions which allow trade in illegal landings. Several individuals are held for questioning. Severe fines are imposed unconditionally.
1990	Fisheries Minister has to resign because of inadequate measures to contain infractions on overfishing of quota.
1990	Overfishing of ITQs is 'punished' by cutting the ITQ of the following year by the nominal amount of overfishing. In this way the government did not have to compensate vessels which were not able to fully utilise their ITQ because of the early closure of the fishery.
1990	Trading in quota is only allowed as long as the vessel has not taken up more than 90% of its ITQ.
1990-92	Legal prosecution of offences becomes more effective. Fishermen give up their previous continuous challenge of legal validity of new regulations after the administration begins to consistently win cases in court.
1992	Days-in-port are changed to 'days-at-sea', based on ITQs.
1993	Upon transfer of any licence which is not attached to an active vessel, its nominal value is reduced by 10%.
1992-93	Second decommissioning scheme implemented (MAGP III).
1993	Validity of 'free licenses' restricted to 6 months (from previous 24 months).
1993	Co-management groups are created, with a mandate to facilitate trade in ITQs (and effort allocations) on behalf of their members. Groups are not (yet) ITQ owners. About 95% of the fleet joins such groups which offer greater facility for trade in ITQs. There is a threat of obligatory decommissioning.

1993	To control the landings, sale through auction becomes obligatory. (Council Reg. 1847/93, par. 9 requires sales-slips to support logbook declarations.)
1993-96	Existing system is maintained. Stress is put on cooperation between the government and sector organizations.
1994	Codfish documents become ITQs for cod and whiting.
1996	Third decommissioning programme implemented (MAGP III).
1996	ITQs for herring and mackerel are introduced.
1996	Validity of free licenses extended to 24 months again.
1997	Fish-quota are treated like milk-quota for succession and gift taxation proceedures, on the basis of the Law 'fiscale structuurversterking' in force from 1 January 1998.
1998	ITQs may be separated from the vessel for a maximum of 5 years, if they belong to a Group quota.
1998	Fishing-effort, expressed in number of days-at-sea, is heavily reduced to meet the MAGP requirements
1998	Gross tonnage included in the fishing-capacity licence. GT-licences are tradable but are reduced by 10% upon transfer.
1998	Within the framework of MAGP IV a new 'segmentation' was applied to the fleet: vessels up-to-10m, vessels up-to-300HP, vessels more-than-300HP, and distant-water trawlers. Transfer of capacity-rights is only allowed within the same segment.

THE EFFECTS OF INTRODUCTING TRANSFERABLE PROPERTY RIGHTS ON FLEET CAPACITY AND OWNERSHIP OF HARVESTING RIGHTS IN ICELAND'S FISHERIES

B. Runolfsson and R. Arnason
Department of Economics, University of Iceland
Oddi V Sturlugotu, Reykjavik, IS-101, Iceland
<bthru@hi.is> <ragnara@rhi.hi.is>

1. INTRODUCTION
1.1 Background

Until the extension of the fisheries jurisdiction to 200 nautical miles in 1976, the Icelandic fisheries were, for all intents and purposes, international and open-access fisheries. Large foreign fishing fleets featured prominently on the fishing grounds taking almost half of the demersal catch. The extension of the fisheries jurisdiction to 200 nautical miles all but eliminated foreign participation in the traditional Icelandic fisheries. Since then international fishing for deep-sea redfish and blue ling outside the 200 nautical miles has developed. However, the initial management measures taken in the demersal fisheries, following the extension of the fisheries jurisdiction in 1976, were inadequate and did not alter the common-property nature of these fisheries as far as domestic fishers were concerned. They were still forced to compete for a share of the catch. Therefore, not surprisingly, the development of the Icelandic fisheries in the post-war era closely followed the path predicted for common-property fisheries exhibiting increasingly excessive investment of fishing capital and fishing effort, compared to the reproductive capacity of the fish stocks. The post-war development of fishing capital and catch values since 1945 is illustrated in Figure 1.

The value of fishing capital employed in the Icelandic fisheries increased by well over 1200% from 1945 to 1983. Real catch-values, on the other hand, increased by only 300% during the same period. Thus, the growth in fishing capital exceeded the increase in catch-values by a factor of more than four, and in 1983 the output-to-capital ratio in the Icelandic fisheries was less than one-third of the output-to-capital ratio in 1945.

Figure 1
Fishing capital and catch values 1945-1997 (index 1945=100).

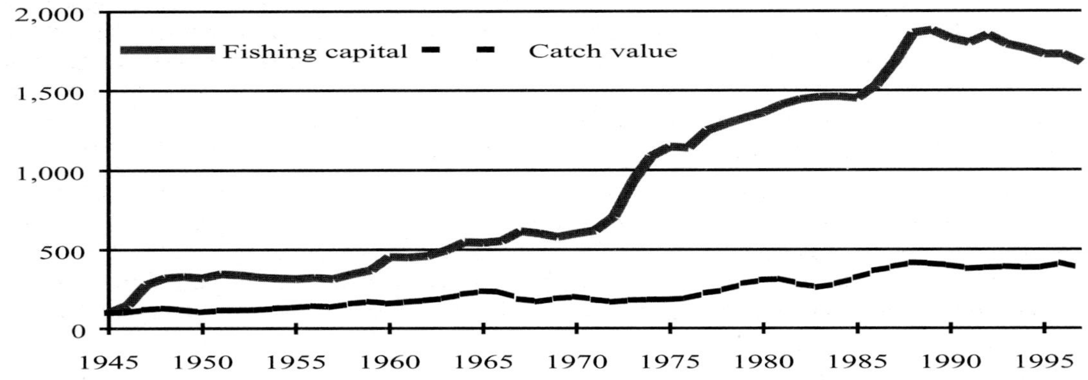

Source: National Economic Institute.

This long-term decline in the economic performance of the Icelandic fisheries did not go unnoticed by the authorities. In fact, over the years, various measures were taken in an attempt to reverse this trend. However, before the extension of the exclusive zone to 200 miles in 1976, effective management of the fisheries, especially the demersal ones, appeared impractical due to the presence of large foreign fleets on the fishing grounds. For this reason, fishery management, subsequent to the extension of the fishing limits to 200 nautical miles, was limited. With the *de facto* recognition of the exclusive 200-mile zone in 1976, the situation changed dramatically: the Icelandic fisheries gradually came under increased management, culminating in a uniform system of individual transferable quotas ITQs) in practically all fisheries since 1991.

1.2 The pelagic fisheries

Due to an alarming decline in the Icelandic herring stocks, an overall quota (total allowable catch, TAC) was imposed on this fishery in 1969. These stocks were fished almost exclusively by Icelanders. Since this did not halt the

28

decline in the stocks, a complete moratorium on herring was introduced in 1972. In 1975, when fishing from the Icelandic herring stocks was partly resumed, it was obvious that the whole fleet could not participate. Hence, an individual vessel quota system with limited eligibility was introduced in 1975. Vessel quotas were small and issued for a single season at a time. The quotas were, therefore, not permanent, but determined annually by dividing the TAC by the total number of eligible vessels applying to participate in the fishery. In 1979, spokesmen for the industry requested fairly unrestricted transfers of quotas between vessels. The Ministry of Fisheries permitted transfers, as it had observed that there were various methods for bypassing the non-transferability of the vessel quotas (Arnason 1996a). The *Fisheries Management Act 1990* made the vessel quota system in the herring fishery part of the general ITQ system.

The capelin fishery, which became a major fishery in the 1970s, was subjected to limited-entry and individual vessel-quotas for licence-holders in 1980 at a time when the stock was seriously threatened with overfishing. Again the arguments in favour of this policy were the same as in the herring fishery discussed above, except in this case the industry asked for regulation. Owners of the bigger purse-seine vessels met in June 1980 and decided that they would ask the Ministry of Fisheries to limit entry into the capelin fishery and allot a quota to each licensed vessel. Only 52 vessels received a licence, but there had been 68 vessels engaged in the capelin fishery during the preceding year.

In 1986, in conjunction with an increasing transferability of demersal vessel quotas, capelin vessel quotas became partly transferable. The capelin vessel quota system became a part of the general ITQ system with the adoption of the *Fisheries Management Act 1990*.

1.3 Demersal fisheries

In connection with the extension of Iceland's exclusive fishing zone to 200 nautical miles in 1976, the major demersal fisheries were subjected to overall catch quotas. These quotas, recommended by the marine biologists, soon proved quite restrictive, and difficult to enforce. As a result, individual effort restrictions, taking the form of limited allowed-fishing-days for each vessel, were introduced in 1977. However, as new entries remained possible, the demersal fleet continued to grow, and the allowable-fishing-days had to be reduced from year to year. In 1977, deep-sea trawlers were allowed to fish for cod for 323 days a year, but in 1981 they were only allowed 215 days of fishing a year. It gradually became obvious to all concerned that this system was economically wasteful.

In 1984, following a sharp drop in the demersal stock and catch levels, a system of individual vessel quotas was introduced. The Fisheries Association of Iceland, an association of vessel owners, processing-plant owners and of fishermen, held its annual meeting on December 2 and 3, 1983. At the end of that meeting, after some heated discussion, it was agreed on to ask the Ministry of Fisheries to institute a system of ITQs in the demersal fisheries for one trial year: 1984. On December 22, 1983, the Parliament passed an amendment to the *Fisheries Act 1976,* basically giving the Minister of Fisheries discretionary power to put a vessel quota system in place. In the upper chamber, the amendment received only the minimum support necessary: 11 out of 20 MPs.

Because of generally favourable results of the system, the quota management system was extended for 1985 and 1986-1987. However, to ensure sufficient political support for the system, an important provision was included in 1985: vessels were allowed to opt for effort-restrictions instead of catch-quotas. On 8 January 1988 the Icelandic Parliament enacted general vessel-quota legislation that applied to all Icelandic demersal fisheries, which became effective between 1988 and 1990. This legislation retained the effort-quota option but made it somewhat less attractive.

In 1990 comprehensive ITQ legislation (individual transferable share quota), the *Fisheries Management Act*, was passed by the Parliament. This legislation abolished the effort-quota option and closed certain other loopholes in the previous legislation, especially as regards the operation of vessels under ten GRT (vessels under six GRT continued to be exempt from the ITQ system). The legislation required licensing of all commercial fishing vessels and a moratorium on the issue of new licences. It also extended the ITQ system indefinitely. Since then, the system has continued to be modified and the Act has been amended on several occasions since 1990.

1.4 The shrimp, lobster and scallop fisheries

The inshore shrimp, lobster and scallop fisheries are relatively recent additions to the Icelandic fisheries. These fisheries were largely developed during the 1960s and 1970s and from the outset have been subject to detailed management, primarily using limited local-entry and overall quotas. An overall TAC was set in the lobster fishery in 1973, with restrictions on the size of vessels that could prosecute the fishery and, subsequently, licensing and vessel quotas in 1984. Legislation regulating the processing and fishing of inshore shrimp and scallop was passed in 1975. This legislation gave the Ministry authority to issue quotas for these fisheries to the fish processors. There are seven inshore shrimp areas, each having specific regulations. In general, the Ministry would set a TAC for each area and allocate shares to each shrimp processing-plant in the area. The Ministry would also decide on the total number of vessels that could catch shrimp in each area and would licence vessels for that fishery. In determining the total number of licences the Ministry would also specify the maximum daily-catch and maximum weekly-catch for each vessel. In addition, the Ministry decided on the allowable size of vessels and the permissable gear. The processing plants would then allocate quotas to vessels that would catch their share of the overall shrimp quota. In 1988, the deep-sea shrimp fishery was also subject to vessel quotas. The management of shrimp and scallop fisheries became part of the general ITQ system with the enactment of the *Fisheries Management Act* in 1990.

1.5   Evolutionary process, not design

As may be inferred from the description in Section 1, the course towards a complete ITQ fisheries management system in Iceland has evolved more by trial and error than by design. In most countries - and Iceland is no exception - there is a strong social opposition to radical changes in the institutional framework of production and employment. A great deal of this opposition derives not from rational arguments, but rather from the desire to protect traditional values and vested interests. From a socio-political view, Iceland probably had to pass through an evolutionary process during which various management methods were tried in different fisheries. The knowledge and understanding gained from these experiments were crucial for the eventual acceptance of a more efficient ITQ system.

It should be noted, however, that the key steps in the evolution of the ITQ system have usually only been taken in response to crises in the respective fisheries due to a sudden reduction in stock biomass levels. Thus, individual vessel quotas were introduced in the herring fishery in 1975, following a collapse in the herring stocks and a prolonged moratorium on herring catches. Similarly, vessel quotas in the capelin fishery and the ITQ system in the demersal fisheries were introduced in 1984, in response to a perceived danger of a corresponding collapse in the stock levels and a serious financial crisis in these fisheries.

This pattern reflects the reluctance of members of the fishing industry to accept changes in the traditional organisation of the fisheries. Only when faced with a disaster in the form of a significant fall in income due to fish stock reductions, or a drop in the world market price for fish products, have interest groups been willing to consider changes in the institutional framework of the fisheries. Rule-changes in fisheries are frequently a response to crises, *i.e.* lower income for fishermen (Libecap 1989). One should bear in mind, though, that even if the adoption of an ITQ system is a rather radical rule-change, it was not new to the Icelandic fisheries, as IQs and ITQs already existed in the herring and capelin fisheries at this time. As early as 1981, the favourable experience with quotas in these fisheries had convinced many vessel-owners that nothing short of an ITQ system was needed. And, despite an increased catch of demersal species, the fishing industry was running with heavy losses in the period from 1981 to 1983.

The passing of the comprehensive ITQ fisheries-management legislation in 1990 constituted a break in this pattern. For the first time, the fishing industry agreed to a significant overhaul in the fisheries-management system without being threatened with the alternative of a financial disaster. This must be attributed to the potentially immense economic benefits of the vessel-quota system, which were now becoming apparent to most of the participants in the fisheries. The modifications of the ITQ system since then have been relatively minor and reflect, on the one hand, the desire of the fishing industry in general to improve the efficiency of the system and, on the other hand, the efforts of special interests (small boats, and regions) to protect their position.

2.   THE NATURE OF THE HARVESTING-RIGHT
2.1   Licences

The current harvesting-right in Icelandic fisheries consists of two parts: a general fishing licence and a catch quota. To carry out any commercial fishing a general licence is needed. To fish species subject to a TAC both a fishing licence and a quota for the particular species are required.

In December 1983 legislation was passed that gave the Minister of Fisheries the power to put a licensing scheme into place for all fisheries as well as those managed using ITQs. The Ministry could now licence vessels in particular fisheries that used particular gear, or vessel sizes, or groups based on gear/vessel-size combinations. The Ministry issued a new regulation in February 1984 that required the licensing of all vessels over ten GRT intending to fish in the Icelandic EEZ. Vessels already in the fishery (for the period November 1982 to October 1983) were issued licences. New vessels could only receive a licence if they had been commissioned before the end of 1983. Other vessels could receive a licence if a comparable ship already in the fishery was decommissioned *i.e.* withdrawn from the fishery. These rules were in effect, with only minor changes, until 1990.

From 1991 regulations on licensing have been issued each year, with small changes or clarification on the definition of a "comparable" vessel. During the first 8 months of 1991 a new vessel could be up to 60% larger than the vessel it replaced, if the older vessel was at least 12 year old. For younger vessels the new replacement had to be comparable in size. From September 1991 to August 1992 this exemption for older vessels was revoked. From 1992 to 1996, up to 3 vessels could be decommissioned for a single new vessel. The size-limit for the new vessel was the sum of the 3 vessels it replaced, with the additional requirement that the largest of the replaced vessels had to be at least 70% of the size of the new vessel. From 1996 to 1997 this 70% threshold was abolished, and any number of vessels could be decommissioned for a new vessel comparable in size to the sum of the replaced vessels. From 1997 to January 1999 a new vessel could be up to 60% larger than the replaced vessel, if the replaced vessel had been within the ITQ system for at least 7 years.

From 1986 there were also regulations on modifying vessels already licensed. Vessels in the fishery before 1986 have had the freedom to be increased in size without limits, although replacing them has mostly been subject to the same rules as for other vessels. Vessels that came into the fishery in 1986 or later, have not been allowed to be increased in size unless other vessels were decommissioned at the same time (such that the now larger-size vessel is really replacing the decommissioned vessel).

The first regulation on the import of small vessels came into effect in 1986. This regulation was not enforced. New legislation, the *Fisheries Management Act* for 1988-1990, required the licensing of all vessels over six GRT, and of gill-net vessels under six GRT. From 1988 a new vessel over six GRT could only receive a licence if a comparable vessel was retired. The *Fisheries Management Act 1990* required all vessels to be licenced. Since 1991, therefore, a new vessel of any size, could only receive a licence if a comparable vessel (or vessels) were decommissioned.

In December 1998 the Supreme-Court reached a decision on a case concerning an application by an individual for a commercial fishing licence and fish quota. The Ministry of Fisheries had declined the application and a lower Court had upheld the Ministry's decision on the basis of the *Fisheries Management Act 1990*. Article 5 of the legislation stated that only vessels already in the fishery at the time of the legislation could receive licences. The Supreme-Court found the article unconstitutional on the grounds that it amounted to unequal treatment of citizens. The Court did not, however, decide on the second issue, the application by the individual for catch quota.

In response to the Supreme-Court decision in January 1999 the Parliament passed legislation to modify the *Fisheries Management Act*. All registered vessels could now apply for commercial fishing licences: access is therefore not restricted any more. Receiving a commercial fishing licence is only one step needed, to fish as harvesting TAC-species also requires a quota. The legislation abolished the restrictions on licensing of new vessels and therefore new vessels can receive a licence without any other (older) vessel being decommissioned. All restrictions on enlarging a vessel have also been abolished.

As of May 1999 17 new licences had been issued to vessels larger than six GRT, and 21 new licences to vessels smaller than six GRT. The 17 larger vessels could participate in the fishery immediately, but had to lease or buy quota to do so. The latter 21 smaller vessels have had to wait until September 2000 to begin fishing, and then they must either lease or buy quota (in a new small-vessel ITQ-system) or be granted a very limited number of fishing-days.

2.2   The ITQ system

Although the ITQ system was instituted at different times and in somewhat different forms in the various fisheries, its application was made uniform through the *Fisheries Management Act 1990*. The fisheries-management system is based on individual transferable share quotas and is therefore appropriately referred to as an ITQ system. The essential features of the current ITQ system are as follows: all fisheries are subject to vessel catch-quotas. The quotas represent shares in the total allowable catch (TAC). They are permanent, perfectly divisible and, with minor restrictions, transferable. They are issued subject to a small annual charge to cover enforcement costs. The ITQ system is fairly uniform across the various fisheries, although slight differences between the fisheries exist, mostly for historical reasons.

It should be noted that the ITQ system was superimposed on an earlier management system designed mainly for the protection of juvenile fish. That system involved restrictions on: certain gears, areas and fish-size, and is still largely in place. The ITQ system has not replaced these components of the more general fisheries-management system.

The Ministry of Fisheries determines the TAC for each species in the fisheries. This decision is made on the basis of recommendations from the Marine Research Institute (MRI). The MRI uses its own vessels to study the state of the fish stocks, but also relies heavily on sampling of catch-landings and operational information from the fishers. In more recent years the Ministry of Fisheries has followed the recommendations of the MRI quite closely. The fishery for cod plays a substantial role in the economy and therefore, not surprisingly, successive governments have been reluctant to curtail the cod TAC in accordance with the recommendations of the MRI. Only in the 1990s has the Ministry, with the general support of the vessel-owners, followed this advice closely, despite some political pressure to the contrary. In 1995 a TAC-rule, which sets the TAC for cod at 25% of the fishable stock, was established.

Currently 19 species are subject to TACs and consequently to ITQs. They include eleven demersal species: cod, haddock, saithe, redfish, Greenland halibut, plaice, wolf-fish, dab, long rough dab, witch and lemon sole; plus two pelagic species: the Icelandic herring and capelin;as well as deep-sea and inshore shrimp, lobster and scallops. Together all these species account for over 95% of the landed value. In addition, Icelandic vessels fishing the deep-sea redfish fishery, the shrimp fishery on the Flemish Cap and the Atlanto-Scandian herring fishery are subject to ITQs[1]. Several species (on which fishing pressure is regarded as slight) are not currently subject to TACs. This means that the corresponding fisheries can be pursued freely, but they are in most cases commercially negligible. Permanent shares are issued for every species for which there is a TAC. These permanent quota shares may be referred to as TAC shares.

The size of each vessel's annual catch entitlement (ACE) in a specific fishery is a simple multiple of the TAC for that fishery and the vessel's TAC share. While the TAC share is a percentage, annual catch entitlements are denominated in terms of volume *i.e.* tonnes.

The Icelandic demersal fishery is a mixed-stock fishery and vessels catch other species than those that they particularly target. The ITQs system permits some leeway in counting catches of one species against the quota of another (5% for demersal species, except cod).

2.3 Transferability

TAC shares are transferable without any restrictions. Any fraction of a given quota may at any time be transferred to another vessel, subject only to registration with the Ministry of Fisheries. The particulars of the exchange, including price, are not registered. Table 1 shows the development of TAC share transfers in the period 1991-1998. As may be seen in the Table, trade has increased during this period, presumably resulting in a more efficient fishery.

Table 1
Transfer of TAC-shares 1991-1998. *Percentage of TAC-shares in each year*

	91/92	92/93	93/94	94/95	95/96	96/97	97/98	98/99
Cod	10.6	13.0	6.7	18.1	18.7	11.8	31.3	12.8
Haddock	11.0	16.6	7.2	18.3	18.1	11.2	27.9	12.2
Saithe	10.3	14.2	9.2	12.8	17.9	10.0	28.8	11.5
Redfish	8.3	12.6	9.7	8.1	16.0	5.9	30.6	4.4
Greenland halibut	3.1	10.3	4.2	9.9	15.4	8.1	34.7	3.5
Plaice	10.7	18.1	10.3	17.1	11.6	11.5	24.8	14.1
Herring	12.0	16.6	12.0	25.0	43.2	16.7	28.8	17.7
Capelin	2.9	6.7	9.4	2.7	11.2	3.8	21.0	18.0
Lobster	22.1	14.1	7.5	30.7	17.2	20.9	19.2	12.1
Deep-sea shrimp	14.7	15.2	13.3	22.6	24.9	20.2	44.4	28.1

Source: Fisheries Directorate.

Transfers of ACEs are subject to some restrictions. First, the Ministry of Fisheries must agree to a transfer of ACE between geographical regions. The rationale for this stipulation is to stabilise local employment in the short-term and to hinder speculation in quotas. In practice, however, it appears that few inter-regional transfers are actually blocked. Second, only up to 50% of ACE are freely transferable between vessels under different ownership. However, exchanges of quotas for different species of equal value are not subject to any such restrictions. Further, as vessel-owners are not allowed to have the crew share in the costs of quota transfers, all ACE transfers as of 1998/99 have to take place through a public body: the Quota Trade Authority. Table 2 shows the transfers of ACE over 1992/93 – 1998/99.

---

[1] In addition Iceland receives shares of the cod TAC in the Norwegian EEZ and the Russian EEZ. These were allocated to Icelandic vessels as ITQs in 1999.

2.4   Exemptions from the ITQ system

There is one minor exemption from the current ITQ system. Under the 1996 amendment to the *Fisheries Management Act*, hook-and-line vessels under six GRT, not already in the ITQ system, must choose between a cod share-quota system and a cod effort-restriction system (maximum number of allowable-fishing-days). As a group, they receive a 13.75% share of the general TAC for cod.

Table 2
Transfers of quota between vessels 1992-1998. *As percentage of total ACE*[1]

Transfer[2]	92/93	93/94	94/95	95/96	96/97	97/98	98/99
Type A	33.0	26.3	41.3	32.5	31.3	38.6	26.7
Type B	20.2	23.9	13.6	18.3	19.4	15.4	-
Type C	12.6	11.3	12.0	7.2	10.1	9.0	8.1
Type D	34.3	38.5	33.1	42.1	39.2	37.0	11.5
Total	66.2	63.7	78.1	71.2	68.1	69.3	46.3

[1] These quotas are measured in cod equivalents and represent temporary annual quota (gross) transfers only.
[2] Type A: Transfers between vessels with the same owner.
 Type B: Transfers between vessels with different owners operated from the same port.
 Type B and Type D are grouped together as of 98/99
 Type C: Offsetting transfers of different species with equal value between vessels with different owners.
 Type D: Transfers between vessels with different owners operated from different ports.
 Source: Fisheries Directorate.

3.   FLEET CAPACITY
3.1   Size

In 1975 the overall fishing fleet was about 97 000 gross registered tonnes (GRT) but had increased to about 111 000 GRT at the introduction of the ITQ system in 1984 (the average vessel age was 18 years). All of the increase in GRT from 1975 to 1984 can be explained by two factors. First, the switch from the herring fishery to the new capelin fishery required larger vessels, and second, a change in fishing technology with the introduction of deep-sea stern-trawlers into the Icelandic fisheries. After the collapse and subsequent moratorium in the herring fishery in 1970, the government encouraged and provided financial incentives for investment in the deep-sea fishery and the capelin fishery (increasing the holding capacity of former herring vessels).

From 1984 to 1998 the increase in the size of the fleet was mainly explained by three factors: (a) the increased number of small vessels, (b) increased size of factory vessels and (c), the increased size of replacement vessels during profitable periods in the fishery since 1984. At the end of 1998 the fishing fleet consisted of 795 decked vessels, and measured about 121 000 GRT,[2] valued at close to US$1 billion. The average age of the fishing fleet is rather high, or about 21 years. The development of the fleet in terms of size and number of vessels is shown in Figures 2 and 3.

Figure 2
Development of the capacity of the Icelandic fishing fleet, in GRT

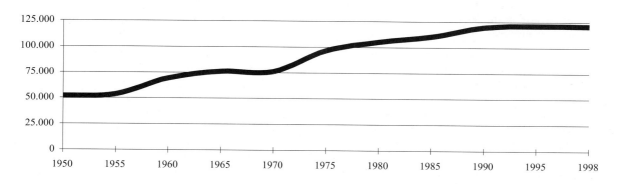

Source: Fisheries Association.

---

[2] A fleet-capacity equivalent to about 10 000 GRT was no longer active in the Icelandic fisheries in 1999 (they did not have quota and/or licences).

Table 3 shows the number of vessels in the fishing fleet in recent years. Note the increase in the number of small vessels (<ten GRT) from 1984 to 1990. The number of vessels larger than ten GRT has decreased since 1985.

Figure 3
Development of the Icelandic fishing fleet, number of decked vessels

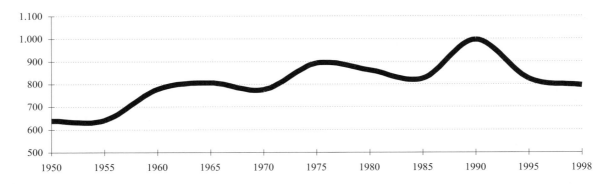

Source: Fisheries Association.

Table 3
Number of vessels in the Icelandic fishing fleet

Year	Vessels <10 GRT	Vessels >10 GRT	Trawlers	Total
1984	1060	573	103	1736
1985	1338	572	106	2016
1986	1357	566	107	2030
1987	1560	551	107	2218
1988	1770	546	108	2424
1989	1894	556	115	2565
1990	2045	542	115	2702
1991	2046	522	112	2680
1992	2001	478	108	2587
1993	1966	437	109	2512
1994	1856	425	109	2390
1995	1721	379	114	2214
1996	1538	360	121	2019
1997	1471	345	115	1931

Source: Fisheries Association.

3.2   Composition of the fleet
3.2.1 Fleet composition

The fleet consists of several vessel-types, and although a particular vessel may actually belong in more than one category it is in some ways convenient to divide the fleet into the following categories: deep-sea trawlers, factory vessels, pelagic-fishery vessels, multi-purpose vessels, and small vessels.

3.2.2 Deep-sea trawlers

The first Icelandic deep-sea stern-trawler started operation in 1970, the numbers increased to 53 in 1975, 106 in 1985, 115 in 1990, 121 in 1996 (of which only 109 had licences), but in 1999 their number was down to 102. They are relatively large vessels usually of between 400 and 1400 GRT (the average size has increased from 490 GRT in 1980 to 615 GRT in 1998) and 130 and 250 feet (40-75 m) in length. There are currently 102 vessels in this group totaling 62 000 GRT, and the average age is 20 years. They are engaged in the demersal fisheries employing bottom- and, occasionally, mid-water trawls. Some are also used in the deep-sea shrimp fisheries. A few also, for a part of the year, catch herring and capelin. Due to their size, the deep-sea trawlers have a wide operating range and are able to exploit practically any fishing ground off Iceland, as well as those in international waters. Each trip on the domestic fishing grounds usually lasts about 5-15 days.

3.2.3 Factory vessels

In the 1980s some of these vessels (mostly deep-sea trawlers, but also some long-liners) were converted so that fish-processing could take place on-board. Often the renovations also involved increasing the size of the vessel. The first factory vessel began operation in 1982: since then there has been a steady increase in their number, and in 1997 there were 54 processor-vessels in operation. In addition a few vessels have freezing and filleting equipment on-board. In the 90s, after the restrictions on ship renewals were relaxed, several new processor-vessels have been built. The fishing trip of a typical freezer-trawler is about 20-30 days, and longer if they go to distant waters (this also resulted in the vessels becoming larger, as crew quarters were made more attractive and comfortable). Currently there are 54 processor-vessels.

3.2.4 Pelagic-fishery vessels

Another vessel-type is the specialised purse-seiner. These vessels - 300 GRT and larger - are primarily engaged in the capelin fishery. These specialised purse-seiners usually follow the capelin schools over great distances and land their catches where it is most convenient. There are 38 vessels in this group (down from 52 in 1980) with a total fleet capcity of 20 000 GRT. The average age is over 27 years. The vessels are renovated periodically and their carrying capacity has usually been enlarged during this process. Most of the purse-seiners participate also in other fisheries, particularly the herring fisheries and, some, in the deep-sea shrimp fishery.

Various other large and small multi-purpose vessels are capable of participating in the pelagic fisheries using purse-seines or pelagic trawls. In recent years between 40 and 60 vessels have participated in such fisheries. There has actually been a steady decrease in the number of vessels fishing the capelin and the Icelandic herring. On the other hand, with the opening of the Atlanto-Scandian herring fishery, other vessels were encouraged to establish a fishing history before the fishery became subject to ITQs in 1997. The same phenomenon is now taking place in the blue whiting fishery.

3.2.5 Multi-purpose vessels

The fourth category comprises the multi-purpose vessels. They cover a wide range in size, from 12 GRT to over 200 GRT. There are 327 vessels in this group (down from 460 in 1980) with a total capacity of 35 000 GRT. The average size being just over 105 GRT, and the average age just over 27 years. The multi-purpose fleet is, for the most part, not specialised with respect to either fishing gear or fishery. Most have been designed as gill-netters or long-liners, although technically capable of employing trawl and purse-seine as well. The geographical range of the smaller multi-purpose vessels is limited and they are normally confined to fishing trips of one to three days, exploiting grounds relatively close to their home port. The fishing trips of the larger vessels can last up to two weeks. A few multi-purpose boats are processing-vessels.

3.2.6 Small vessels

Finally, there is a class of fishing vessels that covers numerous vessels of sizes up to 12 GRT, although most are under ten GRT. There were in 1997 1196 licenced fishing vessels under 12 GRT. Of these, 313 vessels were decked (up from 264 in 1980) with a total of 2500 GRT, and the average vessel age was 12 years. The other 883 are open-decked and the fleet has a combined total of 4400 GRT. These vessels are typically owner-operated on a seasonal basis, employing hand-lines, gill-nets and long-lines. Depending on the gear and fishery, the crew size is one to three persons.

In 1984 all vessels above ten GRT became subject to ITQ restrictions. This restriction led, predictably, to a dramatic increase in the number of vessels smaller than this. In 1984 a total of 978 small vessels were active but by 1990 their number had increased to 1599, an increase of about 63%. By 1991 restrictions on further increases in the number of small vessels had been introduced, and in effect entry into the fisheries became restricted for all vessels. The rule-change required that for every new vessel an older one had to be decommissioned. Furthermore in 1991 all vessels over six GRT were included in the ITQ system. This, along with some change in the fisheries legislation in 1994-96, has resulted in a decrease in the number of small vessels under ten GRT, being down to 1114 in 1997.

3.3 Investment

Investment in new fishing vessels and fishing equipment reached a maximum in 1973-1974. As previously mentioned, the 1980s were a period of renewal of the deep-sea fleet, and 1973-74 was the peak period of investment. The period 1986-90 showed an increased investment in small vessels. Over 1985-89 there was a great deal of investment in factory vessels, mainly through the conversion of older vessels, but also some new vessels in 1988-89. The years 1992 and 1994 saw some replacement of deep-sea trawlers.

3.4 Other measures of fishing-capacity

There is no common method in accessing capacity in fisheries. Capacity is not simply a multiple of vessel number and vessel size. In Iceland, for example, it is common that new ships are larger in size due to the demand for more spacious living quarters on board. Improvement in the handling and storing of the catch on board also effects size, and so do requirements for on-board processing.

Figure 4
Number of small vessels (< 10 GRT) fishing each year

☐ Vessels fishing more than 60 days each year   ☐ Vessels fishing less than 60 days each year

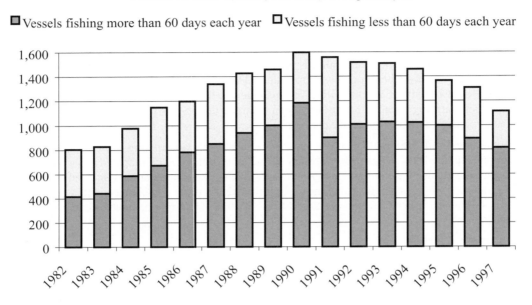

Source: Fisheries Association.

Figure 5
Investment in fishing vessels

☐ Trawlers   ☐ Other boats   ☐ Engines and restoration

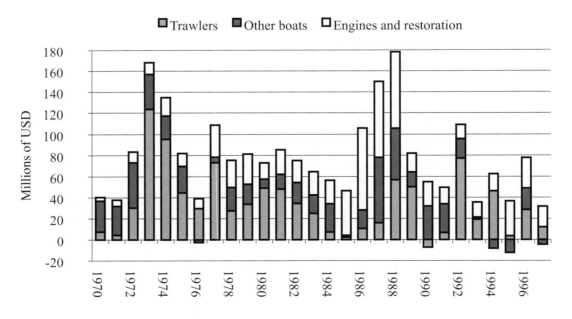

Source: National Economic Institute.

Several other factors therefore effect capacity: vessel age, engine-power, technical equipment, even the skipper's and crew's influence the fishing power of a vessel. Engine power is sometimes used as an indicator of capacity. Figure 6 shows the evolution in recent years of engine-power (in kilowatts) of all vessels larger than ten GRT. It had begun a decrease after 1992, along with the decrease in the number of vessels, but in 1996 and 1997 several powerful deep-sea trawlers were bought for the sole purpose of fishing in international waters. This increased the total engine-power of the fleet again.

3.5 Number of vessels with fishing licences

The number of vessels with commercial permits was 2560 in January 1991; 1433 vessels were in the ITQ system and 1127 had hook-and-line licences. In May 1999 the number of vessels with commercial fishing permits was down to 1688 (1671 if new licences are excluded). Only 740 vessels had TAC shares and 793 vessels under six GRT were active under the small-vessel arrangement. In addition there are 155 vessels with a commercial fishing permit, but no quota. Thus, there has been a total reduction of about 900 vessels since the introduction of the comprehensive ITQ system in 1991.

Figure 6
Engine power for all vessels over 10 GRT

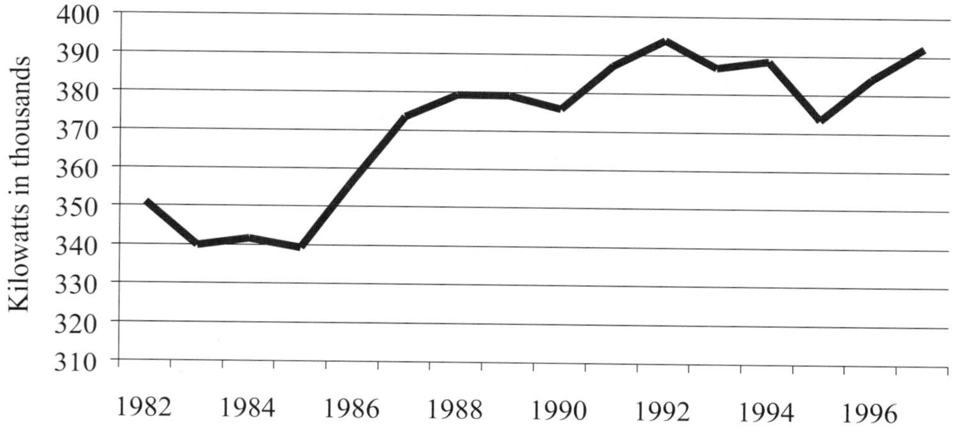

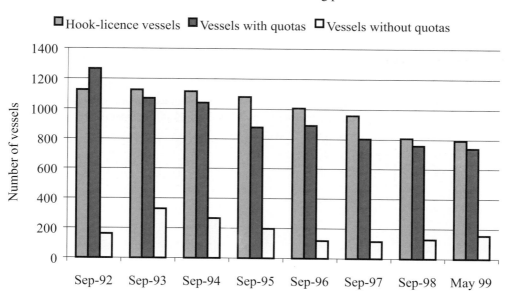

Source: Fisheries Association.

Figure 7
Number of vessels with commercial fishing permits 1992-1999

■ Hook-licence vessels ■ Vessels with quotas □ Vessels without quotas

Source: Fisheries Directorate.

A reduction in the number of vessels engaged in the inshore shrimp fisheries (from 50 to 44), the scallop fishery (from 21 to 15), and the lobster fishery (from 57 in 1992 to 42 in 1998) is included in the numbers above.

4.   OWNERSHIP
4.1   Concentration of ownership

Table 4 shows the development of the quota holdings of the ten largest harvesting-companies. In 1998 these ten largest firms, in terms of demersal quotas (for cod, haddock, saithe, redfish and Greenland halibut), held about 37% of all quotas for these species. Howver in 1991 the ten largest firms had held only about 25% of these quotas. One reason for this recent increase in quota-holdings by the ten largest is the wave of mergers in the fisheries industry. The largest firm in 1991 (Grandi Ltd.) resulted from the merger of three large harvesting-companies in the late 1980s. Their combined share actually decreased from 5.6% of the total in 1984, to 4.8% in 1998. Samherji Ltd., the second largest in 1998, merged with several large and small firms in 1996-97. Haraldur Bodvarsson Ltd., the fourth largest in 1998, merged with three large harvesting firms in the 1990s. Thormodur rammi-Saeberg Ltd., the fifth largest in 1998, merged with three large firms in the 90s. Skagfirdingur Ltd., sixth largest firm in 1998, bought the stock of another large company and they merged. Vinnslustodin Ltd. is the result of the merger of several firms. Snaefell Ltd. is also a combination of several companies, and the same applies for Thorbjorn Ltd. and Basafell Ltd. Only UA Ltd. and Skagstrendingur have been exempt from the wave of mergers, having grown instead by increasing their quota directly[3].

Table 4
The evolution of quota-holdings among the largest harvesting-firms in the demersal fisheries*
Percentage of total quota shares

Harvesting firm	1991/92	1992/93	1993/94	1994/95	1995/96	1996/97	1997/98	1998/99
UA Ltd. (NE)	4.0 (2)	4.6 (2)	4.6 (2)	5.0 (2)	5.4 (2)	5.4 (2)	5.0 (2)	5.5 (1)
Samherji Ltd. (NE)	3.2 (3)	3.4 (3)	3.4 (3)	3.5 (3)	3.6 (3)	4.2 (3)	5.6 (1)	5.5 (2)
Grandi Ltd. (SW)	4.3 (1)	4.9 (1)	4.9 (1)	5.1 (1)	6.1 (1)	5.7 (1)	4.9 (3)	4.8 (3)
Haraldur Bodvarsson Ltd (W)	2.2 (6)	2.3 (5)	2.3 (6)	2.3 (5)	2.6 (5)	3.3 (5)	4.5 (4)	4.3 (4)
Thormodur Rammi Ltd (NW)							4.0 (6)	3.8 (5)
Vinnslustodin Ltd. (S)	2.5 (4)	2.0 (6)	2.9 (4)	2.5 (4)	2.2 (7)	2.0 (8)	4.3 (5)	3.3 (6)
Skagfirdingur Ltd. (NW)	1.5 (9)	1.5 (10)	1.7 (8)	2.2 (7)	2.9 (4)	3.3 (4)	2.8 (8)	3.2 (7)
Snæfell Ltd. (NE)								2.6 (8)
Thorbjorn Ltd. (SW)							2.5 (7)	2.3 (9)
Basafell Ltd. (Wfj)							2.3 (9)	2.3 (10)
Total, ten largest each year	24.9	25.9	27.0	28.2	30.7	31.8	38.1	37.6

*Shares of total cod equivalent values for each year. Quota-holdings in: cod, haddock, saithe, redfish, greenland halibut and plaice as percentage of total allotments of cod, haddock, saithe and redfish
Source: Runolfsson (1999b).

Another reason for the increase and/or decrease in quota-holdings of most harvesting-firms is found in the change in the cod-equivalent values for the various species. The cod-equivalent value for haddock has gone from 0.85 in 1984 to 1.05 in 1998, for saithe from 0.55 to 0.65, for redfish from 0.55 to 0.7 and for Greenland halibut from 0.85 to 2.15. These changes in weighting of the different species may have moved firms up or down the list. Related to this are the changes in the TACs for the various species: the TAC for redfish was 110 000t in 1984 and 65 000t in 1998/9, for saithe from 70 000t to 30 000t, for haddock from 60 000t to 35 000t, for cod from 200 000t to 250 000t, and for Greenland halibut from 30 000t to 15 000t respectively.

---

[3] An analysis of mergers since 1995 shows a slight decrease in concentration in total quota-holdings of the ten largest (see Runolfsson 1999b). It should also be noted that the total sum of ACE has usually been only 75-83% of the combined TAC (in terms of cod-equivalents). The ten largest therefore only hold about 26.7% of the TAC.

Table 5
Percentage distribution of stock in the ten largest demersal harvesting-firms in December 1998

Harvesting firm	Year 1998/99 %	Number of stock-holders	Institutional investors* %	Corporate ** %	Other %	Biggest stockholders			
						One %	Three %	Five %	Ten %
UA Ltd.	5.5	1 720	35	49	16	20	50	64	76
Samherji Ltd.	5.5	3 864	9	1	89	21	62	76	80
Grandi Ltd.	4.8	1 080	18	21	61	26	47	57	71
Haraldur Bodvarsson Ltd.	4.3	1 227	19	37	44	10	24	37	59
Thormodur Rammi Ltd.	3.8	580	18	23	59	19	35	42	61
Vinnslustodin Ltd.	3.3	762	17	35	48	18	38	48	67
Fisk. Skagfirdingur Ltd.	3.2	197	22	8	70	56	74	87	94
Snæfell Ltd.	2.6	119	3	96	1	92	96	98	99
Thorbjorn Ltd.	2.3	368	6	11	83	11	34	51	71
Basafell Ltd.	2.3	332	18	27	55	24	39	48	64
Total	37.6	10 049							

*   Stock owned by municipalities, cooperatives, pension funds, stock funds, etc.
** Corporations and cooperatives listed on the Icelandic stock exchange.
Source: Runolfsson (1999b).

Another view of the issue of concentration is given by the number of share-holders that hold stock in these harvesting firms, since they should really be regarded as the owners of the harvesting-rights. The number of stockholders in these corporations was well over 10 000 in 1998; up from less than 2800 in 1990. Institutional investors, such as retirement funds and investment funds, who represent a majority of the population, are major stock-holders in several of these companies.

Samherji Ltd. is an interesting example of the increased number of share-holders. This firm, although founded in 1972, came under current ownership only in 1983 when its only asset was one (old and rusty) deep-sea trawler. Samherji Ltd. is today the largest quota-holder in Iceland, in terms of overall quotas. The firm has also invested in other firms, domestic and foreign. Samherji Ltd and its subsidiaries operate 20 vessels from five countries. In addition they operate two processing-plants for shrimp, two reduction-plants, one freezing-plant and one marketing firm in England (this is not an exhaustive list of all their investments). Samherji Ltd. became a public company in 1997 and the number of stock-holders in 1998 reached almost 3900.

4.2   Effectiveness of regulations governing ownership rights

There are only three restrictions on ITQ ownership of any significance: (a) ITQs must be attached to a fishing vessel, (b) foreigners cannot own harvesting-companies and therefore cannot own ITQs, and (c) the total ITQ holdings by a single firm may not exceed a certain upper limit.

The first restriction is primarily for bureaucratic convenience and as such it is perfectly adhered to. However, ITQs attached to a given vessel may actually be owned by someone else. The vessel-owner is the registered owner of the TAC share, but through another contract (*e.g.* an option to purchase) the TAC share may actually be owned by someone else.

The second restriction is apparently also well adhered to. However, it is clear that foreign companies can actually be *de facto* owners of fishing-companies and therefore of quotas through a chain of Icelandic companies or by financing and/or catch purchase contracts. So far, however, there have been few signs of this happening.

The third restriction is a recent (1998) addition to the fisheries legislation. No company is actually close to the upper bounds set on ITQ ownership (10% for demersal and 20% for pelagic species). So there has been no reason to thwart these restrictions, although clearly this could be relatively easy by simply forming groups of formally independent companies.

So far cheating in the use of quotas seems to have been relatively limited. Discarding-at-sea appears to be moderate (under 5% for the most discarded species: cod). Landings in excess of quota also appear to be limited, apparently due to relatively few landing places (under 70) and a good system of dockside-monitoring inherited from the previous fisheries management system.

5.   RESULTS OF THE ITQ SYSTEM

The ITQ system in the herring fishery has been very successful. Since 1975 herring catches have increased almost ten-fold. Fishing effort, on the other hand, has not increased: in fact it has declined substantially. The number of vessels in the fishery has decreased from about 65 in 1975, to 30-40 in recent years (their numbers had actually increased up to 145 in 1980)[4]. Catch-per-unit-effort in the herring fishery is now roughly ten times higher than it was at the outset of the vessel-quota system over 20 years ago (Arnason 1996b). The herring-stock biomass is now greater than at any time since the 1950s.

The capelin is a short-lived species and the fishery can be very volatile. Part of the capelin stock migrates seasonally into the jurisdiction of Greenland and Norway. The capelin is therefore a shared stock, but, through an agreement with these two countries, Iceland determines the annual TAC to be shared between the three countries. Iceland's share is 81% of the TAC. In winter, the capelin is fished exclusively in Icelandic waters. The yearly catch of capelin averaged less than 700 000t over 1980-95, but the catch in 1996-98 averaged 1 070 000t. The capelin fleet, on the other hand, has been reduced: the number of specialised purse-seine vessels declined from 68 in 1979 to 38 in 1995 (yet 44 vessels participated in the capelin and Icelandic herring fisheries). The total fleet tonnage (GRT) was reduced by over 25%, and the total days-at-sea for the fleet fell by almost 25%. Thus, there are strong indications that the efficiency of the capelin fishery has increased substantially since the introduction of the vessel-quota system.

In the summer of 1994 the Atlanto-Scandian herring fishery resumed. This herring stock migrates between the waters of Norway, Faroe Islands, and Iceland. ITQs were issued for this fishery in 1998. Icelandic vessels caught 21 000t in 1994, but the catch had increased to 197 000t in 1998. The size of the capelin stock has also been growing and the TAC increasing as a result. This, along with the larger capelin catch, may have induced some vessel-owners to revert to pelagic fishing.

Many of the new, or renovated, large trawlers are multi-purpose vessels, capable of using deep-sea trawls (especially deep-sea shrimp trawls) or also special purse-seines and pelagic trawls for herring and capelin. These larger multi-purpose vessels are therefore not only capable of pursuing pelagic fisheries all year round (capelin in winter and late summer, Atlanto-Scandian herring in early summer, Icelandic herring in the autumn), but can also harvest shrimp (or other species) in between the herring and capelin seasons.

It is appropriate to look at the evolution of the pelagic (purse-seine) fisheries as one, rather than separate herring and capelin fisheries. The changes in the pelagic fishery, in terms of vessel number and size is illustrated in Figure 8.

Figure 8
Evolution of the pelagic fishery.
(maximum number of active purse-seine vessels in any one month)

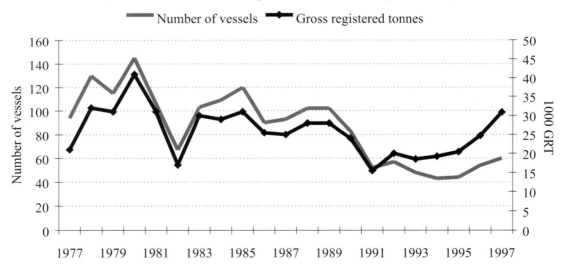

Source: Fisheries Association.

---

[4] The vessel-quota systems in the Icelandic herring fishery only applied to purse-seine vessels. In addition there were another 95 vessels with licences for fishing herring with other gear, and they became subject to vessel quotas in 1985. In 1986 the vessel-quota systems in the herring fishery were abolished, and instead a common ITQ system was instituted.

The trend in the value of fishing capital, and of fishing effort (GRT/days-at-sea) in the demersal fisheries in recent years is illustrated in Figure 9. The previous growth in the value of aggregate harvesting capital halted abruptly in 1984 when the vessel-quota system was introduced. In fact, fishing capital contracted between 1984 and 1985. This was the first time since 1969 that the value of the fishing fleet actually decreased. In the preceding 15 years this capital value had grown at an annual rate of over 6%. Thus, at this point, the vessel-quota system seems to have generated beneficial results, although this halt in investment can hardly be attributed exclusively to the vessel quota system. The years 1982, 1983 and 1984 were periods of heavy losses for the fishing industry. In 1986 investment in fishing capital resumed at a high rate, but this should not, however, be interpreted as a failure of the vessel-quota system as such. After all, the increase in the value of fishing capital since the inception of the ITQ system amounted to just over 2% annually, while during the preceding 15 years this annual increase was over 6%. Moreover, most of the investment since 1986 can be explained by factors extraneous to the ITQ system.

First, a good deal of the investment in fishing capital from 1986 onwards has consisted of the installation of freezing equipment and the corresponding modifications to several deep-sea trawlers. In 1983 there were three processor-vessels, in 1990 they were 26, and in 1997 they were 54. This part of the investment is, in other words, in fish-processing capital employing new and profitable techniques. Second, a part of the investment was in specialised trawlers for the emerging and very valuable deep-sea shrimp fishery, which was not subject to vessel quotas until 1988. Third, by the mid-1980s a significant fraction of the deep-sea trawler fleet was due for replacement. As the years 1986 and 1987 were unusually profitable for the harvesting sector, many firms took the opportunity to replace their ageing vessels.

Fourth, during this period there was a significant investment in vessels under ten GRT (that were not subject to the vessel-quota system). Their numbers increased from 1067 in 1983, to 2023 in 1990. Investment in small vessels accounted for almost 15% of total investment in the fishing fleet over 1984-92. Although the comprehensive *Fisheries Management Act 1990* closed many of the loopholes of the previous ITQ system(s), one loophole did remain: fishing vessels under six GRT were offered the option of remaining outside the ITQ system, provided that they restricted their operations to hook-and-line fishing for demersal species. This exemption, usually referred to as the 'hook licence', was to expire in 1994, but the *Fisheries Management Act* was amended in 1996 so that this group now receives a common share of the TAC for cod, set at 13.75% in 1998. In 1998 there were 807 vessels in this group, 480 of which chose cod-share quotas.

Figure 9
Demersal fishing effort and fishing capital

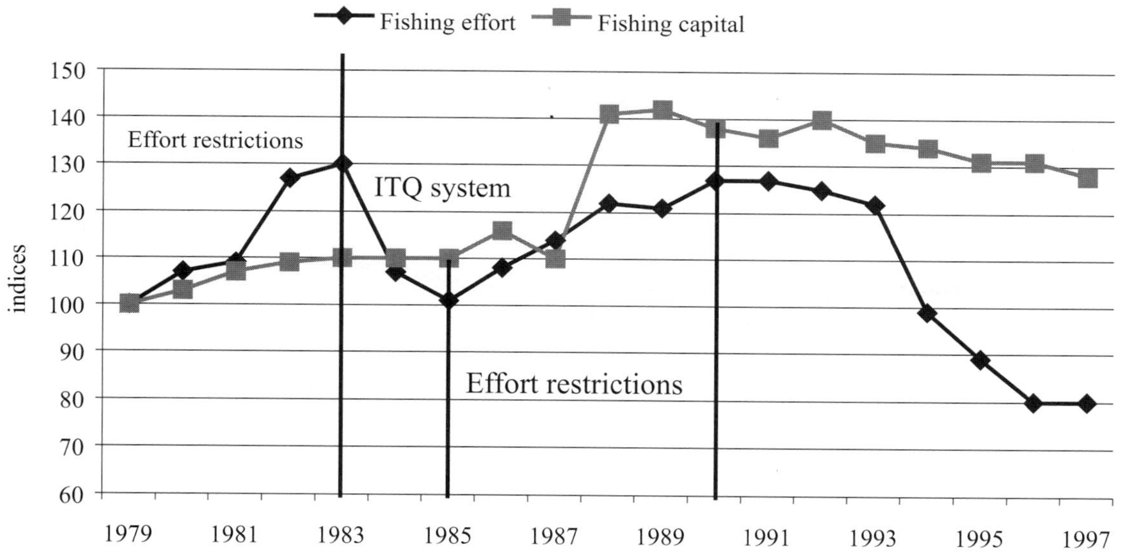

Source: Fisheries Association.

Last, but not least, the effort-quota option in the demersal fisheries, introduced in 1985, undermined the efficiency incentives of the ITQ system, thus inducing many vessel-owners to upgrade or replace their vessels. The effort-quota option was abolished at the end of 1990 and, in fact, a significant reduction in fishing-capital occurred subsequently.

The course of the demersal fishing-effort tells a similar story. As indicated in Figure 7, fishing effort in the demersal fisheries dropped by some 15% in 1984, the first year of the vessel-quota system, and by an additional

6% in 1985. From 1986-1990, on the other hand, fishing effort increased considerably. This is no doubt due to the widespread selection of the effort-quota option within the ITQ system. Another important explanation for the increase in fishing-effort in 1989 and 1990 was the decline in the demersal fish stocks without a commensurate reduction in the TACs. Thus, more fishing-effort was required to fill the catch quotas. Since 1991 demersal fishing-effort has declined substantially.

Over the period 1974-95 as a whole the average annual productivity-growth in the Icelandic fisheries was 2.8%. The extension of the EEZ to 200 nautical miles in 1976 played an important part in the productivity-growth during the early part of this period. However, since 1984, after the introduction of the ITQ fisheries management system, productivity in the fishing industry has grown at an annual rate of well over 4%.

Figure 10

Total factor productivity (index) in the fisheries, at constant prices and adjusted for stock size

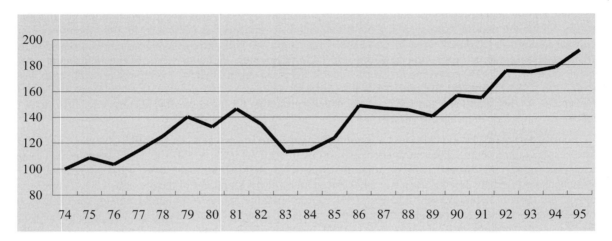

Source: Institute of Economic Studies.

The equity-ratio in the fisheries increased from 5.3% in 1988, to 15% in 1990, and to 26% in 1996. Profitability in the fisheries has also improved, as can be seen in Figure 11. The number of fishermen employed on the fishing vessels actually increased from about 6200 in 1983, to 6500 in 1990, but has since decreased to about 4600 in 1997.

Figure 11

Profitability in the fisheries

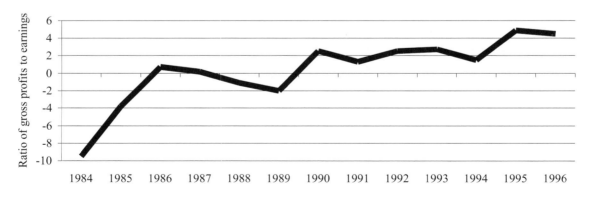

Source: National Economic Institute.

6. CONCLUSION

In Iceland over-capacity and over-capitalisation emerged as issues in the 1970s with the extension of the fisheries jurisdiction to 200 nautical miles and the expansion of the domestic fleet. Over the years, various measures were taken in an attempt to reverse this trend. Most of the early measures were in the form of restrictions on entry and effort, although some financial incentives were also provided to decommission vessels. The most important measure to counter the over-capacity or over-capitalisation problem in the Icelandic fisheries was the introduction of ITQs. Iceland has had vessel-quota systems for two decades now. Since 1991, all major fisheries within the economic exclusive zone have been subject to a uniform system of ITQs, with only minor exceptions. Iceland even subjects its vessels which fish stocks shared with other nations, or in international waters, to an ITQ management system. The evidence on the performance of this system is generally favourable. The system has resulted in increased efficiency of the fishery, and a reduction, albeit a slow one, in the total capacity of the fleet.

In recent years the fishing fleet has decreased drastically in terms of vessel numbers, although in terms of total GRT it seems to have decreased only slightly. But the total GRT given in the official vessel-registry may mask an actual decrease in tonnage, because there are numerous vessels in the registry which cannot participate in the Icelandic fisheries. There has been a reduction of about 900 vessels, since the introduction of the comprehensive ITQ system in 1991, a decrease of close to 35%. The free transfer and trade of quotas has provided incentives for vessel-owners to economise on the number of vessels. Aggregate fishing-effort has decreased by more than 35% since 1992.

7.   LITERATURE CITED

Arnason, R., P. Neher, and N. Mollet 1989. Rights Based Fishing. Boston, Kluwer Academic Press.

Arnason, R. 1990. Minimum information management in fisheries. Canadian Journal of Economics 23:630-653.

Arnason, R. 1994. On Catch Discarding in Fisheries. Marine Resource Economics 9:189-207.

Arnason, R. 1995a. The Icelandic Fisheries: Evolution and management of a fishing industry. Oxford:Fishing News Books.

Arnason, R. 1995b. The ITQ Fisheries Management System: Advantages and Disadvantages.
    In: Johansen, S. T. F. (ed.) Nordiske Fiskerisamfund I Fremtiden. Kobenhave, Nord, p43-70.

Arnason, R. 1996a. On the ITQ fisheries management system in Iceland. Reviews in Fish Biology and Fisheries 6: 63-90.

Arnason, R. 1996b. Property Rights as an Organizational Framework in Fisheries: The Case of Six Fishing Nations, *In:* Crowley, B.L. (ed.) Taking Ownership; Property Rights and Fishery Management on the Atlantic Coast. p99-144. Halifax:Atlantic Institute for Market Studies.

Arnason, R. and H. Gissurarson (Eds.) 1999. Individual Transferable Quotas, in Theory and Practice, University of Iceland Press.

Libecap, G.D. 1989. Contracting for property rights. Cambridge University Press.

Runolfsson, B. 1999a. On Management Measures to Reduce Overcapacity in Icelandic fisheries. A Report to the Ministry of Fisheries, Reykjavik, 1999.

Runolfsson, B. 1999b. ITQs in Iceland: Their Nature and Performance. *In:* Arnason, R. and Gissurarson, H. (Eds.) Individual Transferable Quotas, in Theory and Practice, University of Iceland Press.

CHANGES IN FLEET CAPACITY AND OWNERSHIP OF HARVESTING RIGHTS IN THE UNITED STATES SURF CLAM AND OCEAN QUAHOG FISHERY

B.J. McCay
Rutgers the State University
New Brunswick, New Jersey 08901 USA
<mccay@aesop.rutgers.edu>
and
S. Brandt
University of California, Berkeley
Berkeley, California 94720
<brandt@are.berkeley.edu>

1. INTRODUCTION

The surf clam and ocean quahog fishery of the Mid-Atlantic region was the first federal fishery to be managed with individual transferable quotas (ITQs). We report on its decade of experience with ITQs, focusing mainly on changes in harvesting capacity and in ownership patterns.

The Mid-Atlantic Fishery Management Council (the Mid-Atlantic Council) is one of eight regional councils charged with managing United States fisheries in the 3 - 200 nautical mile zone of federal jurisdiction under the framework of the *Magnuson-Stevens Fishery Conservation and Management Act* (*Public Law 94-265*) which went into effect in 1977. The surf clam (*Spisula solidissima*) fishery, which is under the jurisdiction of the Mid-Atlantic Council, was the first federal fishery subject to restrictions on entry. (The states have jurisdiction from 0 - 3nm; there were some limited-entry systems in state fisheries prior to 1976; in 1977 the State of New Jersey also imposed limited-entry on the surf clam fishery prosecuted in its waters). A moratorium was imposed beginning in 1978, limiting the fishery to the existing vessels, which then numbered 184, a number adjusted to 142 because of vessel inactivity (MAFMC 1990). An annual total allowable catch (TAC) was set and divided into quarterly quotas. Fishing-time limits (per vessel) were also established to encourage and balance distribution of fishing effort throughout the year and stabilize the supply

An example of a hydraulic dredge as used by the Elizabeth C II, a 65' vessel based in New Jersey
Photo credit: J.C. Normant and M. Celestino; N. J. Department of Environmental Protection, Division of Fish and Wildlife, Bureau of Shellfisheries

to processors. The Council eventually agreed upon an explicit policy to set the TAC at a level that allowed for a ten-year supply of surf clams based on the present standing stock. A similar TAC-setting process occurred for the closely related fishery for ocean quahogs (*Arctica islandica*) but without fishing-time restrictions and for a 30-year supply horizon. In September 1989 the Mid-Atlantic Council voted to create individual transferable quotas (ITQs) in both the surf clam and the ocean quahog fisheries. The ITQ system went into effect 1 October 1990 (National Research Council 1999, McCay and Creed 1994).

The commercial fisheries for these species take place in nearshore and offshore waters on the continental shelf of the east coast of the United States, primarily in the Mid-Atlantic region, which stretches from the state of Virginia north to Massachusetts (small ocean quahogs are also found in Maine waters). Surf clams tend to be found closer inshore and in shallower waters than are ocean quahogs. Both species require major investments in technology for commercial harvesting. In the 1940s a system was developed that pumps water into the muddy bottom to raise the molluscs high enough to be caught by a dredge that is run over the bottom. Vessels that had been used for oyster-dredging and otter-trawling for shrimp and fish were converted with hydraulic dredges for the surf clam fishery, which began off the coast of Long Island, New York, but soon moved southwards to New Jersey and on to Virginia. Today, the centre of the fishery has returned to the coast of New Jersey. Ocean quahogs have a broader distribution than surf clams and are found throughout the North Atlantic including Iceland. The commercial fishery for ocean quahogs did not start until the late 1970s and early 1980s, in response to declines in surf clam catches and increased restrictions in the surf clam fishery. Both clams are large and are processed into strips, pieces and broth, canned or frozen, before they reach consumers. The exception is the inshore "mahogany clam" fishery in Maine, based on ocean quahogs that are small enough to compete with other clams on the raw clam market. Although recently made part of the ITQ system depicted below, we will not discuss the Maine component of the fishery.

2.   THE NATURE OF THE HARVESTING RIGHT
2.1   Prior to the introduction of ITQs

Before 1978, the harvesting right in surf clamming was free and open to anyone willing and able to acquire a vessel to prosecute the fishery, which takes place offshore and requires heavy hydraulic dredges as well as access to markets. Market demand is for raw product, which is processed into frozen or canned items before it reaches consumers.

In 1978 the rights were restricted to the owners of vessels then in the surf clam fishery through a moratorium on new vessels in the fishery. At that time there was no significant ocean quahog fishery. Only permitted vessels were allowed to catch and sell surf clams. Entry into the fishery depended on ownership of one of the permitted vessels or their replacements. Replacement of vessels that were severely damaged or lost at sea was allowed with a 10% leeway in their capacity. There were no restrictions on sale or purchase of these vessels and capitalized values of moratorium permits were high, estimated at between $50 000 and $150 000 (MAFMC 1990), and as a consequence many old vessels remained nominally in the fishery, *i.e.* they did not fish. Harvesting rights were conditional upon payment of a modest permit fee, detailed logbook reporting requirements, many restrictions on fishing-time, and for a while, the size of clams. Fishing-time was progressively reduced during this period, as catch per unit of effort increased while the TAC stayed roughly the same. During this vessel moratorium period, 1978-1990, the owners of surf clam boats had rudimentary individual rights, in that each had a right to fish so many days of the year. This had declined to 25 days by the mid-1980s.

Because vessels could not be utilized full-time in the surf clam fishery, many also entered the ocean quahog fishery, which developed in the early 1980s, harvesting rights remained free and open, subject to a modest permit fee and detailed logbook reporting requirements. There were no restrictions on fishing-time or clam size and the TAC was never reached (market demand for ocean quahogs improved during this period but remained lower than demand for surf clams).

2.2   Initial allocation

The initial allocation of ITQs in this fishery, the subject of much controversy and delay in accepting ITQs, is described at greater length in a separate report (McCay 2001). It was divided among owners of all permitted vessels that harvested surf clams or ocean quahogs between 1 January 1979 and 31 December 1988. Logbook data on landings were available for this period of time, which enabled the use of historical landings as well as other criteria in the allocation formula. The ITQ went to the owner of the vessel at the time of the allocation, and that vessel's history and dimensions were factored into the allocation, irrespective of who owned and crewed the vessel in the past. Subsequent to the initial allocation, any person who met the U.S. requirements for owning a fishing vessel could purchase or lease ITQ, whether or not that person owned a fishing vessel or had any other qualifications. Entities with majority foreign ownership are excluded.

The formula finally chosen for surf clam vessels coming from ports in the Mid-Atlantic area – the vast majority of vessels in the fishery – was primarily based on a vessel's average historical catch between 1979 and 1988. The last four years were counted twice and the worst two years were excluded. The resulting figures were summed and divided by the total catch of all harvesters for the period. Eighty percent of a vessel's allocation came from this ratio. A second ratio was computed on the basis of the vessel's cubic capacity (length x width x depth); it accounted for 20% of the vessel's initial allocation. This was in response to complaints by younger and newer participants in the fishery who had invested in larger replacement vessels that did not have strong historical landings and, or, had large vessel mortgages. It was called a "cost factor" and was a key element in enabling an agreement to be reached (Creed 1991).

The method chosen for ocean quahog vessels (which might be surf clam vessels as well) and for surf clam vessels coming from New England ports (a distinct minority) was simpler. The allocation was determined from the average historical catch for years actually fished between 1979 and 1988, excluding the year of the lowest catch (McCay 2001).

2.3 Harvesting-rights after the introduction of ITQs

With the introduction of ITQs in October 1990 the harvesting-right was no longer associated with vessel ownership but rather with ownership or lease rights to shares of the TAC. For both surf clams and ocean quahogs – which are part of the same fishery management plan but managed separately – the ITQ is a percentage of the TAC. The ITQ has two components: (a) the "quota share," expressed in percentages of the TAC, which can be transferred permanently, and (b) the "allocation permit," which takes the physical form of a set of tags that are allocated at the beginning of each calendar year to the ITQ holders. These coded tags must accompany the 32 bushel steel-mesh cages in which the clams and quahogs are moved from the vessel to the processing plants. The tags can be transferred only within a calendar year. The amount of the allocation permit is calculated by multiplying the individual quota share by the TAC in bushels. Bushel allocations are then divided by 32 to yield the number of cages allotted, for which cage tags are issued. Cage tags may be sold to other individuals but are valid for only one calendar year.

The minimum holding of ITQs is five cages (160 bushels); there is no maximum holding and no limit to accumulation except as might be determined by application of U.S. antitrust law. By law the ITQ is not a property right; it is designated a revocable privilege.

3.   CHANGES IN FLEET CAPACITY
3.1   Fleet capacity
3.1.1   Gear; types of boats; age profile, structure in terms of GRT and engine-power and other proxies for capacity

Table 1 documents changes in numbers of vessels engaged in the surf clam and ocean quahog fisheries within the jurisdiction of the Mid-Atlantic Fishery Management Council, that is, between 3 and 200 nautical miles from coastal baselines, the waters inside being under the jurisdiction of the individual states. It shows an increase in the number of vessels during the 1980s despite the vessel moratorium. That was because many permitted vessels were not used in some years, because of low prices, repairs, or for other reasons. Table 1 also shows the major decline in vessel numbers following the introduction of ITQs in October 1990.

The fishing vessels used for surf clam and ocean quahog fishing are highly variable in size, capacity and age (Table 2). Some of the vessels in use in the 1980s were over one hundred years old, having been schooners used in the estuarine oyster fisheries. Some were converted oil-field supply boats while others were ex-shrimp-trawlers. In 1983 their sizes ranged from 57 to 146 feet overall length. Their tonnages ranged from 37 to 297GRT.

The major gear used is a hydraulic dredge, worked either from the side or from the stern of a vessel. Some vessels are equipped with two side dredges, but the standard has become one single large dredge deployed over the the stern. Average dredge-width has increased over time, from about 98 inches to about 110 inches. Another important variable affecting the capacity to harvest these sea clams is the hydraulic system which pumps water into the ocean bottom. The radius of the hoses has increased over time, allowing more water to be pumped. Once released from the bag of the dredge, the surf clams or ocean quahogs go onto a conveyor belt, from which bycatch is removed and the clams are shunted into large steel mesh cages, which hold approximately 32 bushels. In the 1970s and early 1980s some vessels continued to sort and shovel clams by hand, but the process has now become highly mechanized. Boats carry from three to five crew, which includes the captain and mate.

The cages are stowed in the hold of the vessel though some may also be stowed on deck (a practice now closely scrutinized by insurance companies, concerned about vessel stability). At the dock, a crane is used to lift the cages from the vessel and forklift trucks take them to waiting trucks or directly into the storage facilities of processors. The annual allotment based on the ITQ is provided in the form of a coded tag, which is placed on each cage, before or just after, it leaves the vessel. The captains and, or, owners of each vessel record which numbered tags are used in a particular shipment; this is checked at the processor end as well.   The "cage tags" are kept by the processors for eventual inspection by government enforcement agents.

3.1.2   Change in capacity during the surf clam vessel moratorium, 1978-1990

During the moratorium of 1978-1990 there was a major change in the capacity of the surf clam fleet. The fleet was divided into three classes, based on tonnage: Class 1 (≤50GRT), Class 2 (51 - 104GRT) and Class 3 (≥105GRT). The number of boats in Class 1 decreased from 14 to 8 between 1980 and 1987, the size of Class 2 also decreased from 54 to 50 vessels, but the size of Class 3 increased from 59 to 75 in that period (MAFMC 1990) because replacements for moratorium vessels were typically larger, since the replacement policy during the moratorium had been liberal (see Nicholls 1985).   There were other changes, less well documented, but no less significant. For

Table 1
Fishing vessels, surf clam and ocean quahog fisheries, 1983-1999

	Active permitted vessels, federal waters						
Year	**Total surf clam**	**Only surf clam**	**Ocean quahog**	**Only OQ**	**Both fisheries**	**Total FVs**	**Total GRT**
1983	117	89	36	8	28	125	13 992
1984	119	76	57	14	43	133	14 691
1985	130	72	64	6	58	136	14 846
1986	144	74	72	2	70	146	18 006
1987	142	74	71	3	68	145	18 145
1988	134	78	62	6	56	140	17 786
1989	135	72	69	6	63	141	18 150
1990	128	77	54	3	51	131	17 103
1991	75	28	49	2	47	77	10 550
1992	58	25	43	10	33	68	9 464
1993	50	27	36	13	23	63	8 800
1994	45	25	33	13	20	58	7 962
1995	36	26	35	25	10	61	8 420
1996	31	18	34	21	13	52	7 391
1997	33	22	28	17	11	50	7 057
1998	28	22	24	18	6	46	6 507
1999	28	21	21	14	7	42	6 067
Percentage change							
1983-1986	23.1%	-16.9%	100.0%	-75.0%	150.0%	16.8%	28.7%
1983-1999:	-76.1%	-76.4%	-41.7%	75.0%	-75.0%	-66.4%	-56.6%
1989-1999	-79.3%	-70.8%	-69.6%	133.3%	-88.9%	-70.2%	-66.6%
1991-1999	-62.7%	-25.0%	-57.1%	600.0%	-85.1%	-45.5%	-42.5%

Note: The figures do not include state fisheries, which take place in waters 0-3 nautical miles from coastal baselines; New Jersey, New York and Massachusetts have their own fishery management systems for state waters. There is considerable overlap in participation between state and federal (3-200 nm) fisheries. Each of the state fisheries has limited entry, and New Jersey has weekly trip limits and has allowed the consolidation of vessels, but none uses individual transferable quotas.

Table 2
Mean dimensions, all vessels in surf clam and ocean quahog fleet, 1983-1999

Year	Gross tons	Main engine horsepower	Dredge width (inches)	Age (years)
1983	116	468	96	24
1984	120	479	96	24
1985	124	509	98	24
1986	130	534	98	21
1987	130	553	99	22
1988	134	577	99	22
1989	130	577	100	22
1990	131	594	100	22
1991	136	622	105	20
1992	143	646	109	19
1993	143	652	115	18
1994	138	643	115	18
1995	143	661	115	19
1996	148	694	118	21
1997	143	677	116	22
1998	145	676	115	22
1999	149	679	108	24

Note: Dredge width data were incomplete for most vessels.

example, vessels that used one dredge converted to the use of two dredges, and the radius of the hose used for hydraulic water pressure was increased, as did the average width of the dredges themselves. The moratorium did not include any constraints on technological change beyond the vessel itself.

The increase in capacity during the moratorium (1978-90) is also shown in Table 1: fleet tonnage increased from 13 992 in 1983 (already a significant increase over the tonnage of 1979) to 18 145GRT in 1987, an increase of 30% (see also Figure 1). This accompanied a 17% increase in active fishing vessels, from 125 in 1983 to 146 in 1986, much of which was due to increase in ocean quahog fishing.

Figure 1
Gross tonnage, surf clam and ocean quahog fleet, 1983-1999

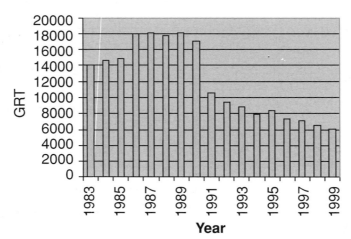

The ocean quahog fishery was open-access until October 1990 when it became part of the ITQ regime. The number of vessels involved in ocean quahog dredging doubled from 36 in 1983 to a peak of 72 in 1986. Virtually all of them were also involved in surf clamming, but new vessels did enter the ocean quahog fishery.

There was also an increase in surf clamming activity. In 1986 virtually the entire "moratorium-permitted" fleet was active in surf clamming. In most other years, many permitted vessels did not actually land clams due to lack of markets, need for repairs, etc. The short times allowed for fishing (as little as 6 hours every three weeks) contributed to the deployment of all available vessels, in order to meet the demand for clams. Another factor that probably led to the large numbers of surf clam vessels in the period 1985-1990, was anticipation of some kind of individual allocation based on landings history. Although the decision to adopt ITQs was not made until the fall of 1989, by 1985 and 1986 it was clear that some kind of individual allocation, with or without transferability, would take place. The "cut-off" date was 1988; landings after that date were not part of the allocation formula. However, the allocation formula for surf clams double-weighted the years 1983-1988.

The over-capitalized fishery that developed during the pre-1978 open-access period had become even more bloated during the moratorium. The only way to participate in the fishery was to own one of the permitted vessels or its replacement vessel. Restrictions on fishing-time were used to try to keep catches below the TAC; these restrictions were assigned to individual vessels, and their owners were not allowed to combine allowable fishing-time from two or more vessels onto one. Consequently, the only way to spend more time catching surf clams was to operate more permitted vessels. Some chose to fish ocean quahogs, and the state fisheries provided additional opportunities, but the general strategy was to acquire more vessels. Several clammers put together large fleets, one with as many as 28 vessels at one point in time.

The incentives to increase vessel-capacity were obvious to many. The management system had quarterly quotas and time-limits for vessels but no limit on how much each vessel could bring in. Each vessel competed with the others and against time to catch as much as possible. As time went on, and some kind of allocation based on the vessel's catch history seemed likely, the motive for maximizing catch came to include increasing one's share of impending property rights. The rise in catch per unit effort (MAFMC 1990) during this moratorium period, was important, reflecting both changes in harvesting-capacity and changes in the resource. Vessel-owners achieved this in two ways – larger catches per tow and per hour spent at sea, countered by shorter periods of time that they were allowed to be on the water. Frustration with the latter was important in fueling the social and political process that led to ITQs. In addition, larger vessels were needed to prosecute the ocean quahog fishery, which typically takes place much farther from port and in deeper waters, than the surf clam fishery. Diversifying into ocean quahog fishing was an important strategy as time limits became shorter in the surf clam fisheries (smaller vessels were used in the state fisheries, particularly New Jersey's state surf clam fishery, which was restricted to the winter months, a season when smaller vessels had difficulty in offshore waters).

The occupational health and safety issues associated with the system were large, indeed tragic. Vessels frequently sank and men's lives were lost each year in Mid-Atlantic waters during the moratorium. A study of mortality rates in New Jersey showed that fishing was one of the most dangerous occupations in the state, a result, almost entirely, of the surf clam and ocean quahog fisheries (pers. comm. P. Guarnaccia, Department of Human Ecology, Rutgers the State University, New Brunswick, N. J., USA). For example, five clam vessels capsized in New Jersey waters in 1989. A study of fishermen's perspectives on marine safety (McCay 1992) showed that surf clamming and ocean quahogging were widely seen by commercial fishermen as the most dangerous fishery, partly because of the technology and partly because of the regulatory system (for surf clams). These created incentives to harvest as much as possible in a short time, often in bad weather. This too contributed to the decision to implement ITQs. However, in January 1999, almost a decade after ITQs were implemented, another four vessels capsized and ten lives were lost.

3.2 Changes in fleet-capacity arising from the introduction of transferable property rights
3.2.1 Numbers of active vessels

This period included that of the moratorium on the entry of new vessels to the surf clam fishery, which began in 1978 and lasted until September 1990, and the ITQ-based management regime, which began October 1990 and still continues. The harvesting sector of the fishery has become smaller, with fewer vessels, even though quotas and catches are larger than in pre-ITQ times (Table 3).

Table 3
Clam and ocean quahog catches and quota (TAC), 1979-1999

Year	Surf clams		Ocean quahogs	
	Catch (bushells)	Catch (bushells)	Catch (bushells)	Catch (bushells)
1979	1 674 209	1 800 000	3 034 696	3 000 000
1980	1 924 033	1 825 000	2 961 789	3 500 000
1981	1 976 438	1 825 000	2 888 287	4 000 000
1982	2 002 830	2 400 000	3 240 775	4 000 000
1983	2 411 940	2 450 000	3 215 640	4 000 000
1984	2 967 026	2 750 000	3 962 967	4 000 000
1985	2 909 330	3 150 000	4 569 509	4 900 000
1986	3 180 642	3 225 000	4 167 205	6 000 000
1987	2 819 819	3 120 000	4 743 025	6 000 000
1988	3 031 681	3 385 000	4 469 373	6 000 000
1989	2 838 408	3 266 000	4 930 280	5 200 000
1990	3 113 976	2 850 000	4 622 417	5 300 000
1991	2 673 413	2 850 000	4 839 824	5 300 000
1992	2 812 270	2 850 000	4 938 700	5 300 000
1993	2 834 717	2 850 000	4 811 941	5 400 000
1994	2 846 670	2 850 000	4 611 395	5 400 000
1995	2 545 305	2 565 000	4 628 323	4 900 000
1996	2 569 319	2 565 000	4 391 428	4 450 000
1997	2 413 575	2 565 000	4 279 059	4 317 000
1998	2 365 374	2 565 000	3 897 487	4 000 000
1999		2 565 000		4 500 000

Source: C.E. Heaton and T.B. Hoff August 1999. Overview of the Surfclam and Ocean Quahog Fisheries and Quota Recommendations for 2000. Mid-Atlantic Fishery Management Council, Dover, Delaware.

In 1978 (the onset of the vessel-entry moratorium) there were 142 vessels active in the clam fishery (the ocean quahog entry fishery had not yet started). By 1983, 125 vessels were active; 89 were fishing only for surf clams, eight for ocean quahogs, and 28 for both species (Table 1). In 1989, on the eve of the ITQ system (which began in October 1990), there were even more active vessels: 141, of which 63 were fishing for both surf clams and ocean quahogs, 72 fished only for surf clams, and six fished only for ocean quahogs.

After the implementation of ITQs, the numbers of vessels dropped dramatically as owners took advantage of the system to divest themselves of the older vessels they had accumulated during the moratorium period; some owners sold out or chose to lease their ITQ to others. The single vessel owner-operator was rare by the eve of introduction of the ITQs management system, and thus many owners were in a position to withdraw one or more of their vessels from the fishery in order to economize. Both consolidation within firms, and exit by firms significantly changed the fleet. By 1999 the total had declined 66.4% from the 1983 level (and 70% from the 1989 level!) to 42 vessels. Twenty-one were surf clamming, 14 were ocean quahogging, and seven were doing both.

The numbers of vessels fishing both surf clam and ocean quahog continued to decline during 1991-1999, after the ITQ management system was put into force and consolidation and down-sizing continued. Overall, there were 45.5% fewer vessels in 1999 than in 1991, the first year of ITQs (Table 1). There were even fewer in 2000 because of the loss of four vessels in a series of storms in January 1999.

Weninger and Just (1997) have analyzed these data for the period up to the 3rd quarter of 1994. They showed that after the initial major vessel exodus (about 35 vessels, or 27% of the active Mid-Atlantic surf clam and ocean quahog fleet, as soon as ITQs began on 1 October 1990), the rate of exit became slow, although steady, with an accumulated exit of 38 vessels by the 3rd quarter of 1994. Those that left were relatively inefficient.

3.2.2 Vessel specialization

One of the striking structural consequences of ITQs appears to be greater specialization, in contrast to the pre-1990 pattern of either surf clamming or combining surf clamming with ocean quahogging. Prior to the introduction of ITQs, most of the larger vessels fished for both surf clams and ocean quahogs because of the strong incentive to use capital assets that would otherwise be idle due to the constrained surf clam fishing-time. This situation changed with the advent of ITQs. In the 1991-1999 period, particularly from 1995 on, there was significant specialization in the vessels fishing either surf clams, or ocean quahogs. More vessels fished only for ocean quahogs rather than combining that fishery with surf clamming over the course of the fishing year (Table 1).

Changes in vessel tonnage and engine power

There has been a substantial increase of tonnage and engine-power in the fleet since 1983. For all-vessels-combined, the trend has been steadily upward for both average tonnage and average engine-power (Tables 1, 2). Following the implementation of ITQs in late 1990, tonnage and engine power declined in the ocean quahog fleet, as some large vessels were retired. Eventually, during the 1990s, average tonnage reached previous levels though average engine-power was considerably higher, probably because during this period the vessels tended to go farther from port in search of productive quahog beds, as a result of low catch-per-unit effort in customary ocean quahog grounds and the decision of one major processor to move to New Bedford, Massachusetts.

For vessels fishing surf clams (some of which were also used for ocean quahog fishing), their dimensions and engine power increased under ITQ management as older and smaller vessels were retired, but there was not much further increase during the 1990s.

3.2.3 Changes in aggregate fleet tonnage 1983-1999

During the mid-1980s there was a major increase in the number of vessels, largely due to the entry of ocean quahog vessels, but also because of the increased use of vessels that had moratorium permits, possibly in anticipation of a quota-allocation system based on historical landings. As shown there was also a structural change, whereby larger vessels replaced smaller ones. This translated into a dramatic increase in aggregate capacity, as measured by gross registered tonnage (GRT). Figure 1 shows this change and also the dramatic reduction in GRT after 1990 when ITQs were implemented and both the fleet and industry down-sized.

3.2.4 Age of vessels

The average age of vessels in the fleet also increased in the 1980s – not only as existing vessels aged, but also because of the increased use of the older "moratorium" fleet of surf clam vessels in the ocean quahog fishery. Older vessels left the fleet soon after the introduction of ITQs, but the subsequent structure remained virtually unchnaged thereafter, with few replacements, thus the surf clam fleet continued to age. The average age of the surf clam vessels was initially much higher than that of the ocean quahog vessels because the moratorium had resulted in the retention of an old and aging fleet, whereas new vessels could enter the ocean quahog fishery. In 1983 the average age of an ocean quahog vessel was about 17 years; for surf clammers it was 25 years. With ITQs the average age of the surf clam vessels declined considerably, to 16 years in 1993, as older vessels were retired from the fleet. In 1993 the average age of the ocean quahog vessel was also relatively low: 17 years. By 1999 the difference had narrowed and ocean quahog vessels were older; the average surf clam vessel was 22 years old and the average ocean quahog vessel over 27 years old.

An analysis of the characteristics of the vessels shows that those that left the fleet were on the average considerably older (Weninger and Just 1997). Their data show that the average year of construction of the 35 vessels that left in the third quarter of 1990, when ITQs were implemented, was 1953 (over 35 years old); the average year of construction of the 38 boats that left the fishery from the 4th quarter of 1990 to the third quarter of 1994 was 1963 (Weninger and Just 1997). In contrast, the average year of construction of the vessels still operating in the surf clam or quahog fisheries in the 4th quarter of 1994 was 1976 (or 18 years old).

3.3   Consequences of changes in fleet capacity
3.3.1   Economic and impacts

Appraisals of the surf clam and ocean quahog fisheries have shown that since the introduction of ITQs in October 1990, economic efficiency in clam harvesting has increased and excess harvesting capacity has declined (Adelaja *et al.* 1998a, 1998b, McCay and Creed 1994, Wang 1995, Brandt 1999). The rapid reduction in the number of vessels in the fleet encouraged organizational changes that allowed more efficient use of production inputs (Menzo *et al.* 1997, Adelaja *et al.* 1998). The effects were more noticeable in the surf clam fishery, which had a much greater problem of over-capitalization than in the ocean quahog fishery. However, it is argued that down-sizing and economic restructuring should have occurred faster and had greater effects than it did (Weninger and Just 1997). This ITQ system was designed with few restrictions on the trading of harvest-rights in contrast with many others that have accumulation-limits on ownership, and rules restricting ownership to certain classes of people.

Several analyses of the economic effects of ITQs have been undertaken. Brandt (1999) estimated Tornqvist Productivity Indices[1] for the surf clam industry for three periods: from 1980 through 1984, when the fishery was managed by limits on the allowable number of fishing-hours per week, he showed that total factor productivity averaged 0.84. Between 1985 and 1900, during the period of increasing restrictions on allowable fishing-hours and ongoing negotiations regarding allocations of exclusive quota shares, it fell to an average of 0.70. Four of those years (1985-90) had a negative growth-rate of total factor productivity, a result consistent with the strategic behavior of industry participants. During 1991-1995, the initial years of ITQs, the index averaged 0.85. The largest growth-rates were in 1991 and 1992, the first two years after the transition in policy. Brandt used regression analysis, controlling for variables such as changes in the surf clam population, effective fishing-hours, the number of processing-plants, and the price of alternative clam harvest, to show that the mean Index of Productivity under ITQs was 39.8% higher than it was under the "command-and-control" regime of the previous period.

Adelaja *et al.* (1998a, 1988b) and Menzo *et al.* (1997) developed an econometric model that showed the effects of ITQs in the fishery. They showed that the use of ITQs in the surf clam fishery accentuated the effects of other variables on how many clams were caught. The rapid reduction in the number of vessels encouraged organizational changes that allowed more efficient use of production inputs. The effects were less noticeable in the ocean quahog fishery, which was not over-capitalized to the same degree. The major effect of ITQs in the ocean quahog fishery came from the initial period of reduction in vessel numbers; those remaining after the initial round of ITQ allocation transfers had greater catch and market share than before. They also found that during the period 1991-1994 catches (as measured by average monthly landings) by different sized firms responded differently to changes in price, suggesting that industrial reorganization was taking place. Their results support the theory that large firms were relatively buffered against price changes whereas small- and medium-size firms are either more vulnerable to changes in price, or more flexible in responding to them.

---

[1] A Tornqvist Productivity Index provides a measure of technical progress in the change of aggregate outputs produced per unit of aggregate input. It does this by aggregating inputs and outputs based on observed costs of inputs, price of outputs and quantities of outputs. The productivity of all inputs, in year t (TFP$_t$) is calculated as:

$$TFP_t = Y_t / X_t$$

The aggregate variable input is written as a weighted sum of inputs with the weights equal to the cost shares of the inputs

$$X = S_i ((W^i X^i)/C) X^I$$

where: W^i is the cost of input i, X^i is the quantity of input i, and the total cost is $C = S_i W^i X^i$.

Likewise, the aggregate output is written as a weighted sum of outputs with the weights equal to the revenue shares of the outputs
$$Y = S_i ((P^i Y^i)/R) Y^I$$

where: P^i is the price of output i, Y^i is the quantity of output i, and the total revenue is $R = S_i P^i Y^i$.

For further information on this estimator, see Applied Production Analysis: A Dual Approach by Robert Chambers 1988, Cambridge University Press. Pages 233, 243 and 248-249.

Weisman (1997) used a hazard rate model to examine patterns of exit in the surf clam and ocean quahog fisheries from 1990 to 1994, finding that exit (from fishing, not necessarily from owning ITQs) was greatest at first, decreased during the first two years, and then increased, reaching an apparent equilibrium in the fourth year. In their analysis of the surf clam and ocean quahog fishery, Weninger and Just (1997) show similar evidence of a delay in the fleet restructuring expected from ITQs. They argue that the cost-minimizing, long-run equilibrium, (optimum) fleet-structure would consist of approximately 20 vessels each of 1485GRT with per vessel harvest-levels of 35 625 bushels of surf clams and 66 250 bushels of ocean quahogs per quarter. The estimated aggregate GRT was 2970. However, as of the 4th quarter of 1994, the fleet had approximately 2.5 times more GRT and vessels than the model predicted (Weninger and Just 1997). Using our data (which may differ slightly, given some discrepancies in counts of active vessels), as of the 4th quarter of 1999, there were slightly more than two times the number of vessels (42), and the aggregate GRT was 5943, also about 2 times Weninger and Just's optimal equilibrium. Consequently, 10 years later the surf clam and ocean quahog fishery still appears far from the theoretical economic equilibrium.

Weninger and Just (1997) ask why so many owners of ITQ rights continue when their operations from a normative perspective are inefficient. For example, why didn't they sell to the more efficient firms right away? These authors developed a game-theoretic analysis, which predicts a strategy of wait-and-see in a situation of high uncertainty about the value of ITQ and marginal inefficiency. The political decision not to charge for ITQs at the initial-allocation contributed by reducing the cost of holding ITQ and by not discriminating against the continuation of less efficient operators. In addition, fishers who wish to purchase ITQs to become more cost-efficient may also benefit from delaying investment. This industry, however, is one of minimal uncertainty, and there are other possible reasons for delays in operators leaving the fishery. One factor concerns the information and transaction costs involved in quota-trading. Another factor not yet developed analytically, is the "job satisfaction" factor, or the value that owners may place on remaining in the fishery despite the opportunity costs (Gatewood and McCay 1990). For example, one family firm we interviewed has tried to keep as many boats operating as possible in order to provide employment for themselves, plus the relatives and neighbors they have long worked with. There are others remaining in the business who, as some have said "know nothing else and can't imagine doing anything else". Such behaviour is often described as "stickiness"; a more accurate one is commitment and tenacity. Also it takes time for the market to clear vessels. Rather than eliminating all of one's boats on the market at once, it makes sense to gradually sell-off boats over a period of years.

3.3.2 Social impacts

The restructuring of the industry that has taken place may be considered as a social impact, but in this analysis we follow custom by restricting discussion to employment and community impacts. The section on ownership that follows describes in detail the social dimensions of the fleet-restructuring.

As might be expected from the down-sizing of the surf clam and ocean quahog fleets, employment opportunities in the fisheries have declined. We estimated that within two years of the introduction of ITQs the number of jobs in the fleet was reduced by one-third. This was less than would be expected by the reduction in the number of boats. But, employment had already been reduced by the time ITQs came along. During the 1980s, when fishing time was severely restricted, many vessel-owners moved their captains and crew members among two or more boats, reducing employment accordingly (McCay and Creed 1987). Nonetheless, when 35 vessels left the fishery on 1 October 1990, as well as another 38 over the next five years, the unemployment that resulted was considerable. Some owners tried to mitigate the impacts by keeping boats fishing even when not needed, but the reduction of crew, as well as boats, still remained a serious problem.

A major social effect of the fleet reduction was the lower bargaining power of crew members and captains, which is symbolized, and to some degree exacerbated, by changes in the share system of returns (McCay *et al.* 1990, McCay and Creed 1994). A common practice adopted with the introduction of ITQs was for the owners of vessels to deduct from the amount that would be shared out, the cost of leasing quota. (Almost all remaining active vessels leased more cage tags than their owners held; in some cases the leases were among corporations owned by the same persons or company). The lease price for cage-tags or "allocation" became an operating expense similar to the cost of fuel and food. The price paid to the crew by the vessel-owner also might be reduced. For example, an owner might receive $8.00 per bushel for surf clams from a processor, but only pay the crew $4.00 per bushel – to be shared out – because he deducted the cost of leasing allocation from the processor buying the clams or quahogs. This might be done even if the vessel-owner actually owned the allocation, or the owner might transfer the allocation to the processor and lease it back, to create a legitimate paper trail for tax purposes (Ross 1992). Thus, the share of catch revenues for crew-members declined sharply. The negative effects on crew incomes have been compensated, it is alleged, by the increased catches of the vessels on which they work – but they also must work much longer than before. This is a sensitive area and few accurate data exist on what has happened. There are reports of difficulty finding qualified crew, suggesting that the monetary and other returns may not compete with alternatives.

The social distance between vessel-owners and the crew (including hired captains) has increased with the introduction of ITQs. Many owners had been crew-members and captains themselves, working their way to vessel-ownership from similar regional and social backgrounds. With the advent of ITQs, the owners became holders of harvest-rights as well as owners of technology. Some of them, particularly those who had established sizeable fleets prior to the initial allocation of ITQs, became wealthy from their new assets. Prospects for working one's way up to become a vessel-owner now appear bleak, given the large cost of owning or leasing the ITQ needed to participate in the surf clam or ocean quahog fisheries.

There are no data on the effects of the ITQ schemes on local ports and communities. Down-sizing of the fleet has clearly affected employment opportunities for welders, vessel-supply companies and other ancillary fishing industries. And, it is difficult to disentangle the effects of down-sizing from the effects of shifts in industry activity subsequent to 1990. Changes in clam and quahog abundance, and other non-ITQ factors, led to the movement of many vessels towards the north during the 1990s. One large processing firm relocated to New Bedford, Massachusetts, and it was able to bring to work there many ocean quahog vessels from its earlier location. The surf clam fleet concentrated its activities on the grounds off New York and northern New Jersey. Ports in Virginia and Maryland were hurt by these moves, but ports to the north were helped. The processing component of the fishery has direct community impacts, too, particularly for the low-income and minority rural and urban communities that supply most of the processing labor. This sector has undergone many changes since 1990, including consolidation and down-sizing, some of which might be related to ITQs, but it is difficult to demonstrate the linkages.

Another important impact on communities has been health and safety. Improved safety was a major selling point for the ITQ system. During the 1980s the system of restricted fishing-time had promoted unsafe races to and from the clam beds. Overloaded boats and other unsafe practices contributed to a series of vessel sinkings and loss of life. In the Mid-Atlantic region an average of one vessel a year was lost at sea. In the early period of ITQs, 1990-1992, three more vessels sunk, throwing doubt on the safety value of ITQs. In interviews, people said that ITQs did not help because the processors still demanded that vessels dredge for shellfish when the product was needed, regardless of weather conditions (Beal 1992, McCay and Creed 1994). For five years there were no vessel losses, but between 1997 and early 1999 six clam boats and eleven lives were lost out of a much smaller fleet. As noted earlier, sea clamming remains a dangerous occupation, with or without ITQs, and the impacts on communities are memorialized and given cultural significance in the services for the dead, the benefits held for the families left behind, and in statues created to honor fishermen lost at sea. One was erected at Point Pleasant Beach, N.J. in the year 2000 in memory of all fishermen from the area lost at sea; the process was in response to the tragic losses in the surf clam fleet in the January 1999 storms.

3.3.3 Management effects

Prior to ITQs, the surf clam and ocean quahog fisheries were problematic to federal, Council, and state managers. Deciding on a management-regime to replace the vessel-moratorium became one of the most time- and labor-intensive tasks of the Mid-Atlantic Council. Enforcement of the various restrictions of that period, particularly the limit on fishing-time (plus a minimum-size limit and a closed-area, in place for some years to protect young clams), was very costly. Another major cost was the staff time devoted to running alternative scenarios for determining the initial allocation and rules of the ITQ system. Although ITQs are very data-demanding, the staff involved in enforcement, data-management, and management report that demands and costs are much lighter than during the moratorium period.

4.   CONCENTRATION OF OWNERSHIP
4.1   Ownership structure before and after the introduction of transferable property rights
4.1.1 Number of owners

Prior to the introduction of ITQs in the surf clam and ocean quahog fisheries, the number of owners increased in the latter 1980s, with the rise in fishing activity for ocean quahogs and the entry of new firms from New England. There were 43 individuals, families or businesses who owned active surf clam and/or ocean quahog vessels in 1983; in 1990 there were 49 (the peak was 55 in 1988). Interestingly, in 1999 there were 51 owners of ITQ allocations, a slight increase, but only 21 owners of vessels actually being used for fishing (Table 4).

4.1.2 Fewer "true" owners after ITQs were introduced

The harvesting sector of the fishery has become smaller, with fewer vessels, even though quotas and catches are in fact larger than in pre-ITQ times (prior to 1990). There are also fewer individuals and firms who own the vessels. The numbers of owners of ITQs themselves has also declined but at a lower rate (Table 4).

We worked closely with industry members and Mid-Atlantic Council staff to develop a record of "true ownership".[2] Table 5 shows changes in the numbers of owners of vessels active in the surf clam and, or, ocean quahog fisheries (there were other owners of permitted vessels that were not active). Evident is a rise in ownership during the latter 1980s, with new entrants in the ocean quahog fishery (and also in the New England part of the fishery). Following the initiation of ITQs in 1990, the number of owners among those holding surf clam ITQ declined, from 54 in 1990 to 47 in 1991, and among those holding ocean quahog ITQ from 43 to 36, with a decline from 57 to 52 overall. Some vessel owners who had been allocated ITQ in 1990 sold out entirely, while some just sold their vessels and chose to lease out their ITQs to others.

Between 1989 and 1999 the number of vessels used in these fisheries declined by 70% (from 141 to 42), and the number of owners of vessels (rather than owners of ITQs–although almost all owners of vessels also owned some ITQ) declined by almost 60% (from 52 to 21). The highest rate of decline in the number of vessel owners occurred between 1990 and 1992, when 15 had given up vessel ownership. Another period of rapid decline was between 1995 (33 owners) and 1999 (21 owners). In contrast, overall the number of owners of ITQ allocation has remained very stable since 1991 (Table 4).

Table 4

Owners of active vessels, 1983-1999 and owners of ITQ allocations, 1990-1999

Year	Active vessel Owners	Vessels	All	Allocation owners Surf clam	Ocean quahogs
1983	43	125			
1984	43	133			
1985	44	136			
1986	51	146			
1987	54	145			
1988	55	140			
1989	52	141			
1990	49	131	57	54	43
1991	37	77	52	47	36
1992	34	68	53	48	37
1993	35	63	53	47	35
1994	32	58	51	45	32
1995	33	61	51	45	31
1996	29	52	51	45	31
1997	26	50	50	45	30
1998	23	46	53	47	31
1999	21	42	51	45	30
Percent change					
1983-1999	-51%	-66%	na	na	na
1990-1999	-57%	-68%	-11%	-17%	-30%

Note: "Owners" are the best estimate of the number of true owners or beneficial owners, i.e., individuals, families or firms widely known to be the owners of vessels or ITQ; they are not necessarily the same as the recorded owners on the permit and transfer files. In many cases ownership of record is a vessel corporation. One "true owner" may have several corporations.

4.1.3 Stable ownership of ITQ allocations

The "true owners" are making a transition from owning vessels to owning allocation, hence the growing discrepancy between the numbers of vessel-owners and the numbers of allocation-owners. For example, in 1999 there were 21 owners of vessels and 51 owners of ITQ. Although there has been marked decline in the number of individuals, families, or firms owning surf clam or ocean quahog fishing vessels, some former owners of vessels continue to own ITQs, choosing to lease them out. And a few owners of vessels currently used for surf clamming or ocean quahog fishing do not own ITQs, instead leasing them from others.

---

[2] True ownership is difficult to determine. The official record of ownership in government permit files is often for a corporation that was created for a particular vessel. It may be one of several or many owned by an individual, family, or business. In addition, with the advent of ITQs many vessel owners signed over ownership to banks or other financial institutions. Because the ITQs are explicitly defined as revocable privileges rather than property rights, they cannot be used as collateral for loans. It has become common for lending institutions to own the ITQs themselves, letting the original owners use them until the loan is repaid. The listing of banks in the ownership record of quota is an artefact of the ITQ system.

The number of owners of surf clam and ocean quahog ITQ (as reported at the beginning of each calendar year) have remained remarkably stable since the first year of ITQs. In 1990, at the outset of the program, there were 57 entities that owned quota allocations (54 surf clam and 43 ocean quahog). In 1991 there were only 52. However, between 1992 and 1999 the number varied at the same overall level: between 50 and 53 owners. A similar pattern is evident for owners of surf clam ITQ: remaining at 45-47, after an initial reduction from 54 to 47. The number of owners of ocean quahog ITQ has shown more consolidation. There were 43 "true owners" in 1990, 36 the following year, then a continued decrease but which stabilized at between 30 and 31 during the period 1995-1999 (Table 4).

The market for quota apparently has not developed as much as simple economic theory might predict. Most market activity concerns the yearly quota-shares, which change hands through leasing, and much of that leasing is reportedly through long-term contracts, which in effect sustain and replicate the institutional structure.

4.1.4 Average sizes of fleets

An important industry variable is the size of a business. Prior to the introduction of ITQs a major indicator of size in the harvesting sector was the number of vessels that an individual or business owned and operated. During the moratorium period of 1979-1990, the system of management through time-limits gave strong incentives to buying more boats for surf clam fishing ,as each boat was allowed only so much time to dredge for clams. Moreover, as the advent of ITQs approached, some entrepreneurs realized the potential advantage of purchasing old boats that would confer them the rights to part of the TAC allocation. A few former deckhands and hired captains managed to put together sizeable fleets of vessels. There were also large fleets owned by vertically-integrated processing firms.

The large independent vessel-owners and the vertically-integrated processors drastically reduced the numbers in fleets between 1983 and 1999 (Table 5). The average number of boats in 1999 was 4.0 for the independent fleet owners, a 43% reduction from the 1983 total (7.0), and for the processors it was 2.2, a 68% decline from the 1983 average number (6.9). The major decrease occurred, as expected, at the onset of ITQs. The decline in number of boats per owner was higher for the ocean quahog fleets than the surf clam fleets (but be reminded that many of the fleets were essentially the same).

4.1.5 Changes in the proportion of small-scale "independents" to large-scale "independents" and processors

Throughout the moratorium period, the ownership structure was highly skewed, with few large and powerful players and numerous smaller ones. The long period of debate about ITQs, particularly about the method of initial allocation and about regulations concerning industry consolidation, was fueled by concerns about the power relations among different kinds of owners. Within the industry, people distinguish between "independents" and "processors" as the major contrast in types of owners; the "processors" being vertically-integrated firms. Throughout the history of these fisheries, vertically-integrated firms have been involved. Some of these are subsidiaries of multinational food corporations with fleets of a dozen or so boats; others a family businesses with large fleets; and yet others were small rural clam-processing operations with one or two boats of their own. Their ability to rely on their own vessels to supply raw product for their plants gave them bargaining power *vis-à-vis* the "independents." The independents include small-scale owner-operators, individuals and families with one or two vessels. Independent operators also included sizeable numbers of vessels owned by individuals, families, or businesses which, by virtue of their magnitude, exercise more "market power" *vis-à-vis* the processors. Some of the owners of these fleets have from time to time developed processing operations of their own.

The number of "independent" owners increased during the late 1980s, but declined dramatically with the advent of ITQs, as many withdrew from the fishery (Figure 2). The decline is marked for both the small-scale and large-scale "independents." Granted, some of the larger-scale independents became small-scale ones by consolidating their efforts from three or more vessels to one or two vessels, which explains the relative stability of the number of small-scale owners between 1992 and 1996. The general and striking pattern though is one of severe decline in the owner-operator nature of this fishery. By the end of the 1990s proportionately more of the owners were vertically-integrated firms or owners of large fleets of boats: 43% in 1999 compared with 25% in 1989. Moreover, even where the owner had only one or two vessels, it had become rare that he worked on the vessel.

4.1.6 Change in the proportion of vessels owned by firms of different ownership types

The structural results of down-sizing, in terms of ownership of active surf clamming and ocean dredging vessels, are not much different from that which existed in the past (Figure 2). Although there are relatively few vertically-integrated processor firms, they own just about as many vessels as either the large fleet owners (three or more vessels) or the small-scale independents (one or two vessels). The situation was similar in the early 1980s, albeit at a much higher level of capacity. Figure 2 shows the pre-ITQ expansion of the large fleet class. It also shows the rise in numbers in the small-scale independent class between 1993 and 1996. This largely happened because several independent owners of fleets down-sized to only one or two vessels. Within a short time, by 1997, this class declined in numbers of boats. A couple of large fleet owners increased their ownership. Data for 2000 are not yet available, but further consolidation occurred in 1999 and 2000, as one of the largest independent fleet owners bought a controlling share of a large processing corporation.

Table 5
Mean number of boats per owner, 1983-1999, by species, or by independent fleets versus processors

Year	Both	Surf clam	Ocean quahog	Independent fleets: both	Processors both
1983	5.1	5.1	5.7	7.0	6.9
1984	5.0	6.4	5.3	7.1	6.6
1985	4.9	5.2	5.2	6.8	6.7
1986	4.6	4.6	4.9	5.6	7.1
1987	4.4	4.5	4.8	5.3	6.9
1988	4.5	4.5	4.7	5.2	7.0
1989	5.2	5.2	5.7	6.1	8.3
1990	5.0	5.0	5.4	6.0	7.8
1991	2.9	2.9	3.0	4.0	3.7
1992	2.8	3.0	2.9	4.2	3.2
1993	2.7	2.9	2.9	4.2	2.7
1994	3.0	3.1	3.2	5.0	2.8
1995	3.0	3.1	3.1	5.3	2.5
1996	2.8	2.8	2.9	5.0	2.3
1997	2.7	2.7	2.8	4.5	2.4
1998	2.6	2.5	2.5	4.0	2.4
1999	2.5	2.5	2.4	4.0	2.2
Change					
'83-'99	-51.0%	-49.9%	-58.2%	-42.9%	-67.9%
'89-'99	-52.6%	-51.5%	-58.0%	-34.9%	-73.6%

Note: "Independent Fleets" are owners with three or more active vessels; Processors are vertically integrated owners with one or more active vessels

4.1.7   Ownership and concentration of harvesting

With the introduction of ITQs, ownership of ITQs began to become more important in the structure of the industry, than ownership of vessels. Accordingly, relative market power is indicated by holdings of ITQ rather than numbers of vessels. In this report we focus on the proportion of the overall TAC that is held by the top owners (continuing to use "true owners" rather than "recorded-owners" – the latter include large financial institutions, which have become owners as part of lending arrangements). Where possible we have identified the "true owner" of these and other holdings).

The structure of the industry was asymmetrical before ITQs, dominated by owners of large fleets and by vertically-integrated firms. After ITQs, the structure as indicated by ownership of ITQs is clearly asymmetrical too. The top 10 owners of ITQ allocation represent about 20% of the total number of owners, but account for between 67 and 75% of the total surf clam quota and between 70 and 80% of the total ocean quahog quota (Tables 6 and 7). There is a discernable but small tendency for these proportions to increase over time. The number of surf clam ITQ owners was reduced by 21% between 1990 and 1999, but the percentage of the surf clam allocation held by the top-ten owners was increased by 12% (Table 6). The number of ocean quahog ITQ owners was reduced even more, by 30%, and the percentage of the top-ten owners also increased by 12%. However, in that case the proportion held by the top-three, top-four, or top-five owners actually declined or stayed the same (Table 7).

4.1.8 Changes on the buying/processor side

We were not asked to address changes in the processing sector of the fishery but it should be noted that in the post-ITQ period (since 1990) the number of buyers of surf clams and ocean quahogs – whether middlemen or processors – has declined. Some small processors (that did not get substantial initial allocation of ITQs because they had few or no vessels at the time) suffered difficulties getting financing, compared with other processors, and essentially left the business. The role of ITQs in these and other changes is hard to disentangle from other influences. For example, consumer market, labor, and environmental problems have worked against these and other processors, contributing to consolidation and reduced participation in the industry. In addition, the situation is very dynamic, including major changes among the processors with or without consolidation.

Figure 2
Number of true owners by ownership type, 1983-1999

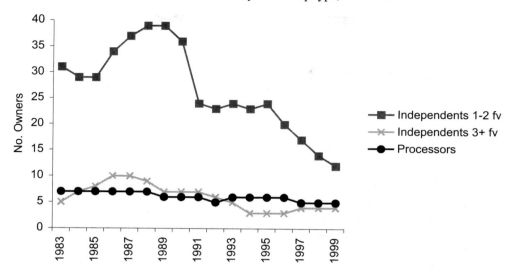

Table 6
Percent of surf clam quota held by the top-three, top-four, top-five, or top-ten owners, 1990-1999

Year	Top-3	Top-4	Top-5	Top-10	Number
1990	47.5	51.8	55.0	67.1	57
1991	44.0	49.8	54.8	70.1	47
1992	44.0	49.8	54.8	70.1	48
1993	40.7	48.2	53.6	70.1	47
1994	39.4	48.7	55.2	71.8	45
1995	38.2	47.6	54.0	70.8	45
1996	41.9	49.6	56.0	72.6	45
1997	44.4	50.9	56.1	72.6	45
1998	47.7	53.4	58.6	72.5	47
1999	50.3	55.9	61.1	74.9	45
Percent change 1990-1999	6	8	11	12	-21

Table 7
Percent of ocean quahog quota held by the top-three, top-four, top-five, or top-ten owners, 1990-1999

Year	Top-3	Top-4	Top-5	Top-10	Number
1990	46.1	50.3	54.4	70.7	43
1991	42.8	48.7	53.8	75.4	36
1992	42.8	48.7	53.8	75.4	37
1993	42.2	47.8	53.2	75.5	35
1994	44.6	51.3	56.9	80.2	32
1995	36.1	44.2	51.0	75.6	31
1996	33.3	41.4	48.2	73.8	31
1997	33.8	41.9	48.8	75.9	30
1998	40.6	47.4	53.1	77.7	31
1999	42.4	48.8	54.4	79.1	30
Percent change 1990-1999	-8	-3	0	12	-30

Reflecting consolidation on the processing side of the business, the owners of surf clam and ocean quahog fishing vessels are likely to sell to fewer different buyers than before (Table 8). In 1983 the overall mean was about 5.0 buyers per vessel; in 1999 it was 2.4, overall. Median data are even lower in 1999, closer to 2.0 for surf clams and 1.0 for ocean quahogs. This trend is also a reflection of increased reliance on long-term contracts.

4.2 Restrictions on transfer of ownership

There are few restrictions on the transfer of ownership of either ITQs or the annual allocations of cage tags. The initial owners were owners of permitted vessels in the surf clam and/or ocean quahog fisheries. Thereafter, any person or business that meets U.S. legal requirements for owning a fishing vessel in the United States is eligible to own ITQ. The minimum holding of annual allocations is five cages (160 bushels). There is no maximum. There is also no limit to how quota tags can be held by one person or business except as might be determined by

Table 8
Number of buyers per vessel owner, 1983-1999

Year	Mean number of buyers per owner		
	Both	Surf clam	Ocean quahog
1983	5.0	5.0	5.5
1984	4.8	5.5	5.1
1985	4.3	4.5	4.4
1986	4.3	4.3	4.7
1987	3.8	3.9	4.2
1988	3.7	3.7	3.9
1989	3.7	3.7	3.9
1990	4.0	4.0	4.3
1991	4.2	4.2	4.4
1992	3.4	3.5	3.7
1993	3.0	3.3	3.0
1994	3.4	3.3	3.6
1995	3.4	3.3	3.6
1996	2.6	2.7	2.5
1997	2.8	3.2	2.9
1998	2.3	2.5	1.8
1999	2.4	2.7	1.8

application of U.S. antitrust law. Cage tags are transferred only within a given year and cannot be transferred between 15 October and 31 December of any year. All transfers, whether permanent sales of ITQs ("quota shares") or temporary transfers of the annual allocations, must be approved by the regional director of the Northeast Region of the National Marine Fisheries Service. This is primarily for monitoring and enforcement purposes.

4.3 Prices received

In the surf clam and ocean quahog management-system, prices received for quota shares or for the annual allotments of cage tags are private, not public, knowledge. The administrative system of the NMFS does not require provision of price information even though all transfers must be approved. Consequently, there is no problem about biased information. Based on what we learnt through interviews with vessel-owners and processors during the Spring of 2000, the industry is almost totally dependent on long-term contracts, including long-term leases of ITQs.

4.4 Effectiveness of regulations governing ownership of rights

Because there are no effective restrictions on ownership, the problem of loopholes, false accounting, *etc.* is not relevant to the surf clam and ocean quahog fishery. Nonetheless, the issue of ownership concentration is extremely important because it is an issue of much public discussion of fisheries policy.

Obtaining information about concentration of ownership is complicated by the widespread practice in the fishing industry of this region (and especially of the surf clam and ocean quahog industry) of creating a legal corporation for each vessel. The result is that the official government records of ownership only list such corporations. Much of our work has been directed toward the task of identifying functional ownership.

5. DISCUSSION

5.1 Reduction in fleet capacity and policy objectives

Until 1988 most industry representatives and Council staff assumed there would be some kind of constraint on the rate at which owners of large numbers of boats could reduce the size of their fleets. However, the final design of the ITQ system had no limits on the rate of "consolidation." Rapid reduction in fleet capacity was an objective of the design of this transferable property-rights system, and in a general sense the results were expected. As noted above, the only major "surprise" is that down-sizing of the surf clam and ocean quahog fleets has not been faster (Weninger and Just 1997). A related issue is the unexpected event that ownership of ITQs has remained fairly stable, at least in terms of the numbers of quota-owners, since the mid 1990s. This was apparently enabled by the proliferation of long-term contracts among ITQ owners, vessel-owners, and buyers (mainly processing firms).

5.2   Concentration of ownership and policy objectives

The ITQ system was designed with no limits on accumulation or trade, and thus with the tacit recognition that concentration could take place. There was an expressed intent that no one could accumulate an "excessive share," but this has proved of little value in controlling the rate, or degree, of quota-holding concentration.

In the decade-long debates leading up to the choice of ITQs in 1989, the issue of ownership concentration was central. The industry has always had one or more large players with concentrated harvesting- or market-power and the future relationship of these large players to the smaller players, particularly the owner-operator and small "independents," was a major question framing negotiations over initial allocations, transfer rules and other matters. In the preamble to the final ITQ system agreed upon was a provision that the government would periodically monitor the number of quota-shares owned by each person, and advise the Department of Justice if any one had an "excessive share" (MacDonald 1992). This provision was intended to compensate for the lack of limits on accumulation in the plan. Attorneys are uncertain about how, if at all, this could be applied - which raises questions such as to whether or not the surf clam and ocean quahog quota market is a "market" within the meaning of the U.S. anti-trust act (MacDonald 1992). The "excessive share" provision has no definition, and the courts thus far have not been concerned unless concentrations approached monopoly levels, which appeared to not be the case in this fishery, at least during the first half of the 1990s (Milliken 1994: Sea Watch International *v.* Mosbacher, 762 F.Supp. 370 [D.D.C. 1992]). The decline in the number of owners of ocean quahog allocations by the latter 1990s, and the overall decline in the number of owners of allocation by almost half (from 39 in 1992, to 21 in 1999) may result in renewed inspection of the "excessive share" provision. However, this had not occurred by early 2001, despite even fewer owners in 2000.

In conclusion, the surf clam and ocean quahog ITQ-system for federal waters of the Mid-Atlantic region of the United States has met the objective of rapidly down-sizing an over-capitalized fishery. It was designed with few restrictions on market forces, and market forces have led to increased concentration of ownership, as well as reliance on long-term contracts.

The concentration of ownership and related changes observed in the surf clam and ocean quahog fisheries have contributed to resistance to the use of ITQs in other fisheries in the region and the nation. On the other hand, the administrative and enforcement burdens of the agencies involved (primarily the National Marine Fisheries Service) have also been reduced from the pre-ITQ level, which involved a complicated and costly regulatory structure. Administration also has been eased by the far smaller number of key actors in the fishery following consolidation. Moreover, ITQ- and vessel-owners in the industry have shown the capacity to organize far better than in the past to influence decisions at the Mid-Atlantic Council, possibly related to the rights secured through ITQs and the reduction in the number of players involved. They have also been at the forefront of a recent surge in "collaborative research" in the region, contributing money, vessel- and crew-time, and expertise to at-sea research in order to test and calibrate the gear used in government stock-assessment surveys, and to expand the surveys.

6.   ACKNOWLEDGEMENTS

The research on which this report is based was supported in part by the New Jersey Agricultural Experiment Station, the National Science Foundation, the New Jersey Sea Grant College Program, and the CMER (Cooperative Marine Education and Research) Cooperative Agreement between Rutgers the State University and the Northeast Fisheries Science Centre of the National Marine Fisheries Service, NOAA, Department of Commerce. Particular thanks must be given to people in the surf clam and ocean quahog industry, including consultants Dave Wallace and Ricks Savage, as well as vessel-owners: Barney Truex, Warren Alexander, the Marriner family; the Osmundsons and Joe Garvilla, for trying to explain the industry and allowing us to go out clamming; and to people in the agencies who cooperated, especially Clay Heaton, Tom Hoff, and David Keifer (ret.), at the Mid-Atlantic Fishery Management Council. Thanks are also due to Dr. Barbara Grandin, for her assistance in developing the database, and to graduate students who worked on aspects of the project: Carolyn Creed, Jonathan O'Neil, David Weisman, Julia Menzo, all formerly at Rutgers University.

7.   LITERATURE CITED

Adelaja, A., B.J. McCay and J. Menzo 1998a. Market Share, Capacity Utilization, Resource Conservation and Tradeable Quotas. Marine Resource Economics **13**(2): 115-134.

Adelaja, A., J. Menzo and B.J. McCay. 1998b. Market Power, Industrial Organization and Tradeable Quotas. Review of Industrial Organization **12**(2): 589-601.

Beal, K.L., 1992. Surf Clam/Ocean Quahog ITQ Evaluation Based on Interviews with Captains, Owners, and Crews. Exhibit 2. *In*: MacLeod, E. Memorandum, Review of the Effectiveness of Our Administrative and Enforcement

Obligations Under the Surf Clam/Quahog ITQ Plan. February 25, 1992. Northeast Region, National Marine Fisheries Service, Gloucester, Massachusetts.

Brandt, S., 1999. Productivity and Industrial Structure Under Market Incentives and Traditional Regulation; A Case Study of Tradable Property Rights in the Middle Atlantic Surf Clam Fishery. Working Paper No. 900, Department of Agricultural and Resource Economics and Policy, University of California at Berkeley.

Creed, C.F., 1991. Cutting Up the Pie: Private Moves and Public Debates in the Social Construction of a Fishery. Unpub. Ph.D. dissertation, Department of Anthropology, Rutgers the State University of New Jersey.

Gatewood, J.B. and B. McCay 1990. Comparison of Job Satisfaction in Six New Jersey Fisheries: Implications for Management, Human Organization **49**(1): 14-25.

MacDonald, G. 1992. Subject: Individual Transferable Quota (ITQ) Management system. Exhibit 3. In E. MacLeod. Memorandum, Review of the Effectiveness of Our Administrative and Enforcement Obligations Under the Surf Clam/Quahog ITQ Plan. February 25, 1992. Northeast Region, National Marine Fisheries Service, Gloucester, Massachusetts.

MAFMC - Mid-Atlantic Fishery Management Council, 1990. Amendment #8, Fishery Management Plan for the Atlantic Surf Clam and Ocean Quahog Fishery. June 20, 1990. Dover, Delaware: Mid-Atlantic Fishery Management Council in cooperation with the National Marine Fisheries Service and the New England Fishery Management Council.

McCay, B.J., 1992. From the Waterfront: Report on Interviews with New Jersey Commercial Fishermen about Marine Safety and Training. February, 1992. Report to the National Marine Fisheries Service, Saltonstall-Kennedy Program. Fort Hancock, N.J.: New Jersey Marine Sciences Consortium. 112 pp.

McCay, B.J., 2001. Initial Allocation of Individual Transferable Quotas in the U.S. Surf Clam and Ocean Quahog Fishery. 86-90. In Shotton, R. (ed.). Case studies on the allocation of transferable quota rights in fisheries. *FAO Fisheries Technical Paper*. No. 411. Rome, FAO. 2001. 373p.

McCay, B.J. and C.F. Creed, 1987. Crews and Labor in the Surf Clam and Ocean Quahog Fleet of the Mid-Atlantic Region. A Report to the Mid-Atlantic Fisheries Management Council, October 1987. Dover, Delaware: Mid-Atlantic Fisheries Management Council.

McCay, B.J. and C.F. Creed, 1990. Social Structure and Debates on Fisheries Management in the Mid-Atlantic Surf Clam Fishery. Ocean & Shoreline Management 13: 199-229.

McCay, B.J. and C.F. Creed, 1994. Social Impacts of ITQs in the Sea Clam Fishery. Final Report to the New Jersey Sea Grant College Program, New Jersey Marine Sciences consortium. February, 1994.

McCay, B.J., J.B. Gatewood and C.F. Creed, 1990. Labor and the Labor Process in a Limited Entry Fishery Marine Resource Economics 6: 311-330.

Menzo, J., A. Adelaja and B.J. McCay, 1997. Supply Response Behavior Under a Tradeable Quota System: The Case of the Mid-Atlantic Surf Clam and Ocean Quahog Fishery. Paper presented to the annual meetings of the International Atlantic Economics Society, London, England, March, 1997.

Milliken, W.J., 1994. Individual Transferable Fishing Quotas and Antitrust Law. Ocean and Coastal Law Journal 1:35-58.

National Research Council, 1999. Sharing the Fish; Toward a National Policy on Individual Fishing Quotas. Washington, D.C.: National Academy Press.

Nicholls, B., 1985. Management of the Atlantic Surf Clam Fishery Under the Magnuson Act, 1977 to 1982. *In:* FAO, 1985 Papers Presented at the Expert Consultation on the Regulation of Fishing Effort (Fishing Mortality), Rome, 17-26 January 1983. pp17-26. FAO Fisheries Report No. 289, Supplement 3. FAO, Rome.

Ross, R., 1992. Summary: Surf Clam ITQ Implementation-Processor Evaluation. Exhibit 1. *In:* MacLeod, E. Memorandum, Review of the Effectiveness of Our Administrative and Enforcement Obligations Under the Surf Clam/Quahog ITQ Plan. February 25, 1992. Northeast Region, National Marine Fisheries Service, Gloucester, Massachusetts.

Weisman, D., 1997. An Economic Analysis of the Mid-Atlantic Surf Clam and Ocean Quahog Fishery Using Logit, Hazard and Survival Rate Functions. Unpublished M.S. thesis, Department of Agricultural Economics and Marketing, Rutgers University, New Brunswick, New Jersey.

Wang, S., 1995. The Surf Clam ITQ Management: An Evaluation. Marine Resource Economics 10:93-98.

Weninger, Q.R. and R.E. Just, 1997. An Analysis of Transition from Limited Entry to Transferable Quota: Non-Marshallian Principles for Fisheries Management. Natural Resource Modeling. **10**(1): 53-83.

CHANGES IN FLEET CAPACITY AND OWNERSHIP OF HARVESTING RIGHTS IN THE UNITED STATES WRECKFISH FISHERY

J.R. Gauvin
Director, Groundfish Forum, Inc.
4215 21st Avenue West, Suite 201, Seattle, Washington 98199, USA
<gauvin@seanet.com>

1. INTRODUCTION TO THE WRECKFISH FISHERY AND ITS MANAGEMENT HISTORY

During the mid-1980s, swordfish and tilefish fishermen discovered commercial concentrations of wreckfish (*Polyprion americanus*) on the Blake Plateau, a deep fishing-ground located about 120 nautical miles due east of Savannah, Georgia, on the Atlantic coast of the United States. Although wreckfish resemble grouper in appearance, wreckfish are members of the temperate bass family found in the Atlantic Ocean. Found only at considerable depth in northwestern Atlantic waters, they are thought to be closely related to striped bass (*Morone saxatilis*) also found off the east coast of the United States. Wreckfish are known to occur in commercial landings in fisheries off mainland Portugal, Madeira and the Azores Islands.

The initial development of the U.S. fishery for wreckfish is described in detail by Sedberry *et al.* (1993), and the wreckfish fishery management plan (FMP) by the South Atlantic Fishery Management Council (Snapper/Grouper FMP Amendment 3), being one of the eight regional management councils in the United States (SAFMC 1990).

Wreckfish are caught in the northwest Atlantic at depths from 450-600m over benthic structures characterized by rock ridges and relief features extending vertically more than 30m (Sedberry *et al.* 1993). Although wreckfish can exceed 100kg in weight, most taken from the Blake Plateau weigh approximately 15 kilos (Vaughan *et al.* 1993).

The fishery was developed by harvesters using hook-and-line gear employing hydraulic reels spooled with steel wire, and terminal rigs consisting of a monofilament spine with eight to twelve monofilament leaders deploying a single circle hook per leader. Frozen squid is the primary bait used for wreckfish and 10 to 20kg lead weights are deployed to bring gear and baited hooks down to the ocean floor in a semi-vertical fashion. Typically, four to six hydraulic reels are used per vessel, spaced evenly across both sides of the fishing deck. Vessels fishing wreckfish range in length from 15 to 25m (SAFMC 1990).

In some respects, the initial phase of management in the wreckfish fishery was similar to other U.S. efforts to regulate commercial fisheries. The fishery developed rapidly with little or no long-range planning or control, expanding from fewer than five vessels in 1988 to more than 40 vessels in 1990 (Sedberry *et al.* 1993). In the spring of 1990, a rapid influx of refitted shrimp trawlers, and the concurrent introduction of bottom long-lines by some fishermen, resulted in increased landings and harvester conflicts (SAFMC 1990). With the addition of vessels from the shrimp fishery, managers estimated that as many as 60 to 70 vessels were geared to fish for wreckfish in 1990-1991, two years prior to ITQ management (SAFMC 1991a). There were approximately 90 vessels permitted to fish for wreckfish during the first year that permits for the fishery were required (Gauvin *et al.* 1994).

To establish control over the burgeoning fishery and attempt to resolve user-conflicts, the first of a series of fishery management plan amendments was approved in June of 1990. These initial measures were "fast tracked" by federal Emergency Rule to "prevent a fishery that would seriously interfere with the necessary protection of the resource" (United States Department of Commerce 1990a, b). The fishery management plan and emergency rules established several management measures including: an April-to-April fishing year, a two million pound total allowable catch (TAC of 0.9 million kg) for the 1990-1991 fishing year, a prohibition on the use of bottom long-line gear for wreckfish, and a vessel catch-per-trip limit of 10 000lb (4,100 kg). Even with these emergency measures, the fishery effectively exceeded the TAC by landing roughly 3.6 million pounds (1.6 million kg) during 1990-1991, effectively catching fish faster than the new regulations could be implemented. From that point in time forward, fishermen continued to press managers for measures to address declining earnings and increasing user-conflicts (SAFMC 1991a).

Responding to the rapid expansion, managers moved expeditiously to request public comment and develop measures for different proposals for limited-entry and individual quota systems during 1991. The end result, achieved approximately one year later, was the development of an individual transferable quota (ITQ) system for wreckfish, only the second such programme implemented for a fishery under federal management in the United States.

This paper evaluates the changes in catch-capacity attributable to the implementation of a rights-based management regime for the fishery, although attribution of all changes in capacity to ITQs cannot be made unequivocally. The paper also reviews the degree of share-concentration in the fishery under ITQs, and evaluates what is known of the effects of concentration on such issues as: prices received by fishermen; and the functioning of markets for permanent and annual rights in the fishery. A very cursory evaluation of effects on prices to consumers, and a conceptual evaluation of social effects of rationalization of the fishery are also provided. Information used for this paper was drawn principally from three sources: the fishery management plan amendments developed by the management council (SAFMC) responsible during the development of the wreckfish ITQ programme, and two separate follow-up studies of the programme (Gauvin *et al.* 1994, Richardson 1994).

The focus of this analysis is limited to the first five years (1990-1994) of the fishery, because roughly four years after the ITQ system became effective, the fishery, in effect, began to collapse. Because this failure of the fishery is not likely directly related to ITQ management, the fishery still serves as a reasonable case-study for the issue of share-concentration and effects on fishing capacity. One has to recognize, however, that the collapse probably influenced incentives for share-concentration although the latter is evaluated for the early period of ITQ management, and the factors contributing to the eventual decline of the fishery are not thought to have been of major consequence during the early period.

The reason for the failure of the fishery is not known exactly, but fishermen and managers cite several possible explanations. Below, I have reported, without prejudice, some of the leading opinions to explain the fishery's demise. These opinions represent the current thinking by fishermen and managers as conveyed by Mr. Robert Mahood (pers. comm., Executive Director, SAFMC).

One explanation is that the fishery suffered from an initial over-estimation of sustainable yield from the "stock" (assuming wreckfish off the southeastern United States is a separate stock, a question that has never been resolved). Hence, according to some managers and fishermen, wreckfish are no longer found in concentrations that support economically feasible harvests. Another explanation is that former wreckfish fishermen are taking advantage of more profitable inshore fishing opportunities in lieu of exercising their options to fish wreckfish. Still others contend that the ex-vessel price does not adequately compensate fishing operations. Whatever the reason or combination of reasons that might best explain the situation, I will focus on the first few years of the ITQ fishery because participation in the fishery today is apparently limited to one or two full-time vessels per year. Although the annual total allowable catch has been maintained at the 2 million pound level, only approximately one-tenth of that harvest has been achieved in recent years (1999 and 2000).

2.   NATURE OF RIGHTS PRIOR TO AND UNDER ITQ MANAGEMENT

Before ITQs, the fishery was virtually unconstrained, since the aggregate annual catch-limit for the fishery first constrained the fishery only one year prior to the formation of the ITQ programme. As the fishery developed, the rapid expansion in participation and catch-rates prompted fishery managers to implement control measures to stem harvest-rates and user-conflicts. These measures generally failed, however, to address declines in economic returns to fishermen. Trip-limits probably contributed to the erosion of profits, especially for the larger vessels, which according to public comment, would operate more profitably if allowed to catch more than the allowed amount of fish per trip (SAFMC 1991a and 1991b). The erosion of economic performance as a result of the fishing-derby itself, and the manner in which the trip-limit quantity of harvest affected vessels of different sizes, are identified as key considerations in the selection of objectives for the ITQ plan (SAFMC 1991a). Related objectives of the plan are economic efficiency, long-term incentives for resource conservation, reductions in user-conflicts, and lower management and regulatory costs.

The documents summarizing public comment during the development of the ITQ plan's amendment elaborate on the concerns of fishermen regarding the erosion of earnings in the fishery. Of note are comments submitted by fishermen claiming to have developed the fishery. These comments state that the lack of real restrictions on entry and effort, in conjunction with rules established to make the fishery more manageable under open-access, contributed to the loss of economic viability on an individual-firm basis (SAFMC 1991b).

The requirement for harvester permits in the fishery (implemented in 1990) also failed to create an effective barrier to entry or exclusivity of rights, because virtually anyone could obtain a permit simply by filling out and submitting an application to the Southeast Regional Office of the National Marine Fisheries Service (NMFS), and paying what amounted to a nominal fee for the processing and handling of the permit application. In fact, the *Magnuson Fishery Conservation and Management Act* as it stood when the wreckfish ITQ programme was developed, expressly prohibited the NMFS from collecting fees in excess of the costs of processing and handling permit applications (SAFMC 1991a).

The prohibition on collecting fees even extended to the collection of fees or resource royalty/rent recaptures once ITQs were in place, a limitation that according to the management-plan amendment establishing the ITQ system, troubled some managers and temporarily served to persuade some council members not to vote in favor of an ITQ system (SAFMC 1991a).

Once the ITQ programme was established, however, the rights provided to harvesters were relatively free from restrictions. Rights were assigned in perpetuity and were allowed to be traded freely to anyone, regardless of whether or not that person or business entity was able to document a vessel in the United States (SAFMC 1991a).

ITQ rights were granted to a percentage share of the annual TAC. Recipients received a percentage share-certificate and paperwork entitling them to the quantity of wreckfish (annual individual quota) for a given year amounting to the share-holder's percentage share of the annual total allowable catch.

The only limitation placed on the rights distributed by the initial allocation was that recipients had to be qualified wreckfish fishermen (see below), and that no single entity could receive an initial allocation of more than 10 of the 100 (10%) of the shares initially distributed (see below). An additional minor requisite was that leases of annual rights could only be between permitted wreckfish fishermen. This leasing restriction amounts to little more than a requirement that any owner of a vessel not granted an initial share, simply had to complete the appropriate paperwork in order to fish for leased wreckfish quota.

There are several statements of intent in the management plan that appear to be aimed at further limitations on the rights granted to qualified wreckfish fishermen. These are, ostensibly, guidance to NMFS rather than actual elements of the management plan. For instance, in several places the plan notes that it is the council's intent that major violations of the ITQ regulations be met with forfeiture of the shareholder's permanent rights in the fishery (percentage shares) (SAFMC 1991a). Another similar statement is that management fees and royalty or economic-rent extractions, should be considered for the wreckfish fishery under ITQs if, and when, such collections become allowable under the applicable laws governing fishery management (SAFMC 1991a). While possibly helpful for NMFS' interpretation of the intent of the plan, in reality, the council members recommending the plan for final approval by the Secretary of Commerce actually have no purview over the matters of enforcement and collection of fees. Those areas of management are the responsibility of the NMFS Enforcement Branch and NMFS respectively.

Although the programme established a relatively unimpeded market for wreckfish harvesting rights, it did require that transactions of annual or permanent rights to the fishery be recorded by the National Marine Fisheries Service, Southeast Regional Office. Paperwork requirements established that transactions of permanent shares of the fishery were not official until recorded by the agency, although the "coupon" paperwork for transfers of annual rights allowed for the recording of the transfer on the coupon itself, provided that the permit numbers of the parties were recorded on the forms and the coupons were signed by both parties.

The management plan also established that the NMFS would make public each year the percentage shares of all holders of permanent rights to the fishery, through its regular public information activities.

NMFS collects nominal fees for transactions of percentage shares, and annual rights can be transferred by filling-out and signing transfer paperwork attached to the coupons denominating quantities of catch. No fees are collected for transfers of annual rights.

3.   MEASURING CHANGES IN FLEET CAPACITY
3.1   Wreckfish fleet capacity prior to and since ITQ management (1990-1994)

Reliable indices of fleet capacity prior to ITQ management are not available. Estimates of the number of vessels fishing wreckfish during the open-access period are thought to be of limited accuracy, because the state fish-ticket data-systems in some south Atlantic, states probably overlooked marginal or infrequent participants. More critically, the data-collection systems were generally not set up to handle the situation where a vessel was permitted in one state, but landed its wreckfish catches in other states. The practice of landing out-of-state was apparently common for wreckfish because many of the buyers were located in northern Florida, while many wreckfish boats had South Carolina home ports (SAFMC 1991b). To further complicate matters, the requirement to possess a permit only became effective one year prior to the ITQ programme, and thus a comprehensive list of vessels that potentially landed wreckfish was not available for most of the years during which the fishery was expanding.

Given the data-limitations, the generally accepted estimate of the minimum number of vessels equipped to fish wreckfish during the two years before implementation of ITQ management, is approximately 60-70 (SAFMC 1991a, Gauvin et al. 1994). Table 1 below lists the number of vessels with documented wreckfish landings during the last year of open-access management and the first two fishing years of ITQ management.

The table is divided into two vessel-classes: "fish boats" and "shrimp boats". This division is meaningful because the transferable rights programme appears to have reduced the number of larger "shrimp boat" vessels more profoundly than for the smaller "fish boat" vessels.

Table 1

Indirect measures of fleet capacity

Vessel type	Fishing year	Number of vessels making wreckfish trips	Fishing trips	Mean vessel length (feet)
Fish boat	1991-1992	26	281	49
	1992-1993	17	206	48
	1993-1994	18	150	47
Shrimp boat	1991-1992	12	26	69
	1992-1993	4	10	65
	1993-1994	1	1	73

Source: E.J. Richardson and Associates 1994

Given the significant decline in the number of vessels with documented landings, the wreckfish ITQ plan appears to have markedly reduced fleet capacity during the first two years the programme was in place. If the data-system overlooked any vessels that fished in the early years of the fishery, then the reduction could even have been larger than the official numbers suggest.

The original participants in the fishery were mostly vessels belonging to the "fish boat" category, and the larger shrimp vessels came into the fishery in anticipation of obtaining some of the profits that were apparently being garnered by the early participants (Richardson 1994). Vessels principally used for shrimp trawling installed hydraulic reels and started fishing wreckfish in the waning days of open-access, mostly in 1991. While these vessels could fish more reels at once than fish boats, this did not necessarily provide an advantage to the larger fishing platforms, because the "fish boats" were apparently more maneuverable and thus possibly more effective at the vertical fishing technique (Richardson 1994). The fishing technique that was apparently easier for smaller vessels, involved positioning lead weights suspended by cables so that the baited hook rigs came close to, but did not snag on, the extreme vertical relief of the bottom where wreckfish are found.

The early years of the wreckfish ITQ programme likely brought about the expected effect in terms of fleet-capacity reduction, such has been observed in many cases where an ITQ system replaces open-access. Data in Table 2 strongly suggest that the smaller, possibly more efficient, "fish boats" purchased rights from the larger vessels. This occurred despite the fact that the SAFMC had lifted the catch-per-trip limit which it had imposed in order to control the fishery in its rapid development phase. Commenting on the proposed plan, the public apparently still believed that the trip-limit had protected smaller vessels and if lifted, even under ITQs, the larger vessels would regain an advantage (SAFMC 1991b). Despite this lingering concern over domination by larger vessels under ITQs, data on share-transfers suggest smaller vessels were more efficient fishing platforms, particularly given the removal of incentives to race for fish (SAFMC 1991b; Richardson 1994).

Table 2

Landings, landed price, and total annual revenue

Year	Landings quantity (pounds x1000)	Price ($US/lb)	Total revenue ($US)
1990-91	3300	1.35	$4.45 million
1991-92	2000	1.50	$3.00 million
1992-93	1300	1.70	$2.20 million

While the rights-based system appears to have functioned to remove redundant fishing-capacity according to expectation, it is important to recognize that wreckfish catch fell short of the total allowable catch during the first year of the ITQ system (and from then on as well) (Table 2). While catch undershot the TAC by a only a relatively small amount in the first year of the ITQ (0.7 million pounds of the 2 million pound TAC, or 35%), that was the year of the most precipitous decline in the number of vessels active in the fishery, particularly the larger shrimp boats. Overall, however, for the first year of the ITQ programme, the decrease in the number of vessels is most probably attributable to economic rationalization of fishing capacity, not attrition caused by whatever brought about the eventual collapse of the fishery. Most of the annual and permanent share-exchanges happened right at the outset of the first fishing year under ITQs, before the shortfall in harvest of the TAC took place (Gauvin *et al.* 1994). For later years, it is probably more difficult to distinguish economic rationalization from other incentives for egress.

Using the available data on changes in the overall number of annual trips taken for wreckfish, is even more problematic for measuring changes in fleet-capacity. These data suggest that the number of trips declined from 307 for fishing year 1991-1992 (the year prior to ITQs) to 216 in 1992-1993 (Table 1). This is roughly a 30%

decrease, which appears to be almost directly proportional to the quantity of TAC not harvested, and so one might be tempted to conclude that the same amount of fishing effort occurred per unit of catch. This would seem to dismiss the possibility of fishing-efficiency gains with the ITQ system, but several considerations prevent this type of conclusion. First, no data are available to compare trip-length, so changes in the number of trips could be misleading. Secondly, trip-length was artificially constrained during the last year of open-access because a limit on the quantity of catch per trip was enforced. This limit was dropped under ITQ management. Lastly, as will be discussed below, changes in allowable-gear may not have become effective until ITQ management was in place, and this could have greatly affected trip-length. In summary, the change in the number of trips based on available data, needs to be viewed with some caution.

3.2 ITQ incentives as a possible reason for adjustments in gear and techniques

With the elimination of the race for fish afforded by the assigned-rights system of management, one could possibly expect gear and other fishing practices, to be modified so as to increase efficiency over what was achievable before. This may have occurred but is largely impossible to establish from available data.

The fishing gear that was the most likely candidate for no longer being employed under ITQ management was bottom longlines. This is because although long-lines caught wreckfish faster than vertically-fished rigs, the gear-loss rate was apparently quite high (SAFMC 1991b). It was also frequently reported that parted long-line cable 'soured' fishing grounds, making them practically un-fishable for the "traditional" vertical gear (SAFMC 1990; SAFMC 1991b). In the absence of a race-for-fish, fishermen could avoid some of the expense of lost gear, and the problem of 'souring' of fishing grounds due to the "hangs" created from parted cable, by using only vertical fishing gear (SAFMC 1991b).

The reason that the change in gear-type is not attributable to a change in incentives imparted by ITQ management, is that managers actually mandated changes in gear just prior to the advent of the ITQ system. Regulations prohibited the use of bottom long-line fishing for wreckfish, essentially allowing only the vertical-fishing method, and this became effective just prior to the ITQ system (SAFMC 1991a). Although it was often reported that despite the ban, some fishermen continued to use long-lines up until the ITQ system was in place, there is no way of establishing that ITQ incentives curtailed the use of long-lines (Mr. Robert Mahood, pers. comm.).

In addition, fishermen commenting on the proposed ITQ management-system claimed that they would consider reducing the number of hydraulic reels per boat and the number of deck hands operating the reels, once the premium for catching fish faster was removed (SAFMC 1991b). Unfortunately, a follow-up study to collect data on these types of adjustments was never undertaken.

3.3 Longer term social and economic consequences of changes in fleet capacity

While the ITQ system likely induced considerable egress of fishing capital from the wreckfish fishery, as well as fishing and service-industry jobs soon after the programme went into effect, there are several considerations for evaluating social and economic effects of the programme. The wreckfish management plan itself provides a theoretical framework for evaluation of expected economic and social effects of "shake out" from economic rationalization of the wreckfish fishery (SAFMC 1991a). The expectations under ITQ-management were: fewer fishing jobs for wreckfish, fewer processor jobs in packing houses, and a decrease in gross economic activity in the related service sector. The plan states, however, that the remaining jobs and economic activity under ITQs, were expected to be more viable and profitable than before.

According to the plan, potential negative social effects of economic rationalization and consolidation were considered by the SAFMC prior to its approval. It is interesting that the plan actually concluded that decreases in gross employment and support-industry activity from ITQs would probably occur under continued open-access. This is because managers assumed that an over-capitalized wreckfish fishery could lead to an inability to manage the fishery in a biologically or economically sound manner. The plan presumed that political or economic pressure would not allow managers to limit catches, because the same outcome had occurred in the southeast region with the tilefish fishery. Managers reportedly had allowed that fishery to develop unbridled, and the end result was overfishing and a collapse of the stock (SAFMC 1991a).

Ironically, it appears that the same outcome may have occurred for wreckfish (either for biological or economic forces), in spite of the existence of a rights-based management system. Nevertheless, this outcome is unlikely to be attributable to the ITQ programme. The failure of the fishery under ITQs does point out that if biological management is inherently unsound and risks are not appropriately assessed, many of the same risks exist even with a rational economic management system. In the case where an ITQ system is in place, an advantage is probably that an efficient market for fishing-rights facilitates the economic transition.

4.   CHANGES IN OWNERSHIP CONCENTRATION

From the record presented in the management plan, it is clear that managers considered constructing limitations on concentration of ownership, but their final decision was to allow the unfettered accumulation of shares if that is what the market would eventually dictate. Essentially, the SAFMC opted to let existing anti-trust laws sort out issues of market-concentration or price-control (SAFMC 1991a). Written public comment during development of the plan certainly addressed problems stemming from share-concentration, and several letters expressed concern for the specter of monopolistic pricing or control if the council did not construct direct controls on share-accumulation (SAFMC 1991b). One letter even predicted that foreign control of the fishery would be imminent.

In 1994, a preliminary evaluation of share-concentration in the fishery was undertaken (Gauvin *et al.* 1994). Focus on the wreckfish fishery was largely due to the fact that it was one of the first ITQ systems in the United States, and the decision not to construct limits on share-accumulation had been controversial for the SAFMC. That decision had also received considerable attention in the trade press (Biro 1992). Outside the south Atlantic region, the question of restricting ownership accumulation was proving to be pivotal for many managers considering ITQs at the time. Because the SAFMC had established a data system to track share-concentration that was accessible to the public, wreckfish appeared to be an ideal case study for the question of concentration. It is also important to note that one and one-half years after implementation of the ITQ for wreckfish, there was no obvious sign that the fishery was experiencing problems that might be affecting incentives for egress and concentration, outside of the normal incentives for increasing efficiency and profitability.

Data made available for the study by the SAFMC indicated that the number of vessels landing wreckfish from 1992 to 1993 had declined from 44 to 14. Three points in time were selected to evaluate shareholder-concentration: share-holdings at the time of the initial allocation, share-holdings in October of 1992 (which amounted to roughly the midpoint in the time series of data), and in June 1993 - the latest point in time for which information was available.

The study relied mostly on the Herfindahl index, an index of concentration that was selected because its application is possible in cases where individual share-information is readily available. That index was also appealing because it was thought to be one of the easiest indices for interpretation of concentration-changes. The Herfindahl index is thought to provide an unbiased assessment of concentration-changes over time for an industrial structure that is made up of a relatively small number of firms with small nominal differences in market share (Scherer 1970). The index essentially normalizes share-holdings by summing squared individual shares, thus creating a theoretical range from close to the zero for "perfect competition", to approaching one, for "perfect monopoly".

Applying this approach to wreckfish shareholder-information, the study noted an increase in the index from 0.027 to 0.042 from share-holdings in the programme from the outset of the programme to October of 1992, and 0.042 to 0.064 from October of 1992 to June of 1993 (Gauvin *et al.* 1994).

While this amounted to a more-than-doubling of the index over the period, which indicated that share-consolidation had certainly taken place, the end point still suggested a rather competitive structure for a fishery. The question of the number of boats remaining in the fishery was evaluated both in the context of the number prior to ITQs, and the practical number of vessels that the fishery could economically support. It was concluded that, in practical terms, the fishery was only likely to be able to support less than twenty vessels of the scale needed for the fishery (Gauvin et *al.* 1994).

Given that managers had elected not to construct limitations on consolidation, it is unlikely that the shareholder data failed to accurately reflect ownership and control in the fishery. Limits on ownership often induce incentives for structuring layers of ownership such that it is veiled. This can be accomplished by creative corporate structures, or "proxy" ownership, through placing title to ownership in the names of individuals who effectively carry out the decisions of the actual owner. The array of possible ways of structuring ownership and control, so as to thwart ownership restrictions, can be impressive and can make assessment of share-accumulation impossible in some cases. This was likely not the case for wreckfish because the programme placed no restrictions on ownership after the initial allocation.

Regarding the possibility of price-setting power, the study noted that wreckfish was essentially functioning as a substitute for grouper and other grouper-like fish that are popular in the southeastern United States (Gauvin *et al.* 1994). Annual production of grouper and grouper-like fish in the region was estimated to be more than 10 times greater than wreckfish production, and annual consumption (including imports) was at least 20 times greater (United States Department of Commerce 1992). From this information, a conclusion was made that it was highly unlikely that any entity in the wreckfish fishery could exert much control over wreckfish prices.

5. EFFECTS OF ITQS ON EX-VESSEL PRICES

5.1 Financial performance of the fishery

Ex-vessel prices increased over the period just prior to ITQ management and every year thereafter (Table 1). The reason for price increases is not easily determined, however, because quantity supplied over the same period varied considerably, and an adequate measure of price flexibility for wreckfish or the "grouper and grouper-like fish" is not available. Total ex-vessel revenue from the wreckfish fishery decreased over the period, indicating that the change in quantity supplied was not fully compensated by any price response, either through changes in the bargaining power of fishermen, or price response to the decrease in quantity supplied (Table 1).

From this cursory information, it is virtually impossible to evaluate whether the ITQ system had the theoretically-expected effect on price to fishermen, although one cannot eliminate the possibility that fishermen made price and total-revenue gains compared to what would have occurred under a significant decrease in landings over time, in the absence of the ITQ programme.

Richardson's cost and earnings study of the wreckfish fishery evaluates the question of annual economic-rent formation for wreckfish quota-holders, in the context of performance from wreckfish fishing and from other fisheries in which wreckfish fishermen participated during the years studied. While he came to the conclusion that "highliner" rents were achieved by some operators in the fishery (and the other fisheries in which they participated in those years), from an overall perspective the study found that no rents were created (Richardson 1994). This, however, may have been driven by one single dominating factor: most dedicated wreckfish fishermen paid what turned out to be fairly high prices to regain the amount of catch they had had prior to the outset of the ITQ programme (see Gauvin 2001).

In the end, most of their purchases of annual and long-term rights amounted to complete net losses for the year for which the *pro forma* analysis was undertaken. As it turns out, the year analyzed was one where most fishermen did not land all the fish for which they had acquired rights (whether through additional purchases or their initial allocation shares). Expenses incurred to increase rights to wreckfish were, in many cases dead losses, and this clearly affected the potential for rent-generation (Richardson 1994).

5.2 Fleet capacity and ownership concentration: actual changes compared to theoretical expectations

As can be gleaned from the above discussion of share concentration, capacity, price, and revenue changes in the fishery, the effects of the wreckfish ITQ are confounded to a large degree by the reduction in catches that occurred in the years following the formation of the ITQ programme. Because such a great decrease in capacity occurred during the first year of the programme, it is likely that the ITQ system was achieving the expected effect of elimination of excess fishing-capacity. One can speculate that dedicated wreckfish fishermen must have believed that they would be able to achieve higher margins on the quota-shares they were acquiring, than the initial recipients from whom they purchased these rights. One has to believe that their purchasing decisions reflected rational behavior, but that the information upon which those decisions were based was rather imperfect.

On the issue of concentration of shares, the wreckfish programme did not reflect any obvious signs of concentration that observers would conclude was problematic, to the market or even to the local economies of fishing communities in the southeastern United States. Had the stock or the market for wreckfish been capable of sustaining harvests of the two million pound TAC for a period of time after the ITQ became effective, then a later evaluation of the fishery would have revealed whether concentration would have increased, and whether economic-rent generation was achieved.

Assuming the fishery was able to sustain the two million pound harvest, would it have been problematic if the number of vessels had decreased to ten or fewer? Given the variable- and fixed-cost structures of the firms owning fishing vessels detailed in the cost and earnings study, it is actually hard to imagine the fishery being able to economically sustain more than ten full time vessels (Richardson 1994). This is because in the absence of large increases in wreckfish ex-vessel prices, which seems unlikely given that wreckfish is a substitute for grouper and prices are probably affected largely by quantity of grouper supplied to the market, the wreckfish fishery was in reality a fishery capable of producing between $3 million and $4 million on an annual basis (assuming an annual sustainable yield of 2 million pounds). There seems to be no way such a fishery could sustain the 60 to 70 vessels that had apparently geared up to fish for wreckfish, or the 40 or so boats that made wreckfish landings just prior to the formation of the ITQ.

In this sense, the ITQ probably brought about a more rapid, and perhaps more thorough, economic rationalization of the fishery than would otherwise have occurred once the fishery was over-capitalized and the boom of the initial development turned to a period of economic "shake-out". What is unfortunate is that managers and fishermen invested considerably in the development of a management programme that might have worked well for the fishery, had a stock collapse or market erosion not occurred.

Even if one writes-off the wreckfish experience as a pilot programme to learn about ITQs in United States fisheries, the problem is that the "crash" of the fishery that occurred for reasons other than the ITQ programme, occurred too early to tell how well the programme would have worked. For this reason, most of the SAFDC's efforts to craft compromises and proactively address expected problems for ITQ management were, for all intents and purposes, squandered.

6. ACKNOWLEDGEMENTS

The author thanks Mr. Rober 'Bob' Mahood, Executive Director of the South Atlantic Fishery Management Council (SAFMC) for the information provided and insights on reasons for the decline of the wreckfish fishery. The author also wishes to thank Mr. Ed Richardson for information provided to supplement his 1994 study of the wreckfish ITQ programme.

7. LITERATURE CITED

Biro, E. 1992. Wide support for wreckfish ITQs. National Fisherman 73 (7): 16-1992.

Gauvin, J.R., J.M Ward. and E.E. Burgess 1994. A description and evaluation of the wreckfish (*Polyprion americanus*) fishery under individual transferable quotas. Marine Resource Economics 9 (2) 99-118.

Gauvin, J.R. 2001. Initial allocation of Individual Transferable Quotas in the U.S. wreckfish fishery. 91-98. *In:* Shotton, R. (Ed.) Case studies on the allocation of Transferable Quota Rights in fisheries. Fish. Tech. Pap. No 411, FAO, Rome.

Richardson, E.J. 1994. Wreckfish economic and resource information collection with analysis for management. A report pursuant to National Oceanic and Atmospheric Administration Award No. NA37FF0047-01.

SAFMC - South Atlantic Fishery Management Council 1990. Wreckfish amendment number 3, regulatory impact review, initial regulatory flexibility determination, and environmental assessment for the snapper/ grouper fishery of the South Atlantic region. Fishery management document, South Atlantic Fishery Management Council, 1 Southpark Circle, Charleston, South Carolina, USA.

SAFMC 1991a. Final Amendment 5 (wreckfish ITQs), regulatory impact review, initial regulatory flexibility determination, and environmental assessment for the snapper/ grouper fishery of the South Atlantic region. Fishery management document, South Atlantic Fishery Management Council, 1 Southpark Circle, Charleston, South Carolina, USA.

SAFMC 1991b. Snapper/grouper FMP amendment 5 (wreckfish ITQs) informal review comments. Compilation of letters from individuals, organizations, and agencies, summaries of scoping meetings and public hearing minutes by the South Atlantic Fishery Management Council, 1 Southpark Circle, Charleston, South Carolina, USA.

Scherer, F.M. 1970. Industrial Market Structures and Economic Performance. Rand McNally Academic Publishers, Chicago.

Sedberry, G.R., G.F. Ulrich and A.J. Applegate 1993. Development and status of the fishery for wreckfish (*Polyprion americanus*) in the southeastern United States. Proceedings of the Gulf and Caribbean Fisheries Institute, Vol. 58, 118-129.

United States Department of Commerce 1990a. Emergency Rule. "Snapper/grouper fishery of the South Atlantic" Federal Register 55, no. 153, 8 August 32257-59.

United States Department of Commerce 1990b. Notice of Closure. "Snapper/grouper fishery of the South Atlantic" Federal Register 55, no. 155, 10 August 32635-36.

United States Department of Commerce 1992. Fisheries of the United States 1992. Silver Spring , MD.

Vaughan, D.S., C.S. Manooch, J. Potts and J.V. Merriner 1993. Assessment of South Atlantic wreckfish stock for the fishing years 1988-1992. Manuscript of the Beaufort Laboratory, National Marine Fisheries Service, Beaufort, N.C. United States.

EFFECT OF TRADEABLE PROPERTY RIGHTS ON FLEET CAPACITY AND LICENCE CONCENTRATION IN THE WESTERN AUSTRALIAN PILBARA TRAP FISHERY

N.J. Borg and R. Metzner
Fisheries Western Australia
Locked Bag 39, Chifley Square Post Office, Perth, 6850, Australia
<nborg@fish.wa.gov.au>
<rmetzner@fish.wa.gov.au>

1. INTRODUCTION

1.1 Current spectrum of commercial fisheries access-rights in Western Australia

In Western Australia (WA), there are a number of different commercial management categories which confer a spectrum of commercial fishing-rights:

i.   managed fisheries
ii. interim managed fisheries
iii. condition-based fisheries
iv. exemption-based fisheries and
v.   "unmanaged" fisheries.

"Managed Fisheries" are formally 'declared' under the *Fish Resources Management Act 1994*[1] and are governed through a management plan. There are 30 "managed fisheries" [2] in Western Australia at present. The property rights conferred by, or associated with, licences in these fisheries are the strongest under the fisheries legislation, because the licences are automatically renewed annually unless the licence owner commits three serious breaches of the fisheries legislation; under such circumstances, the licences may be suspended or cancelled. The majority of managed fishery licences are transferable, but they are not divisible because the licence is issued in respect of a boat. However, in some instances, the licences may be the platform for holding transferable units, which may be able to be transferred individually and, thus, are more divisible.

"Interim Managed Fisheries" are fisheries for which there is little or no long-term management direction. These fisheries are also declared under the *Fish Resources Management Act 1994* and are governed through a management plan. Currently, there are three "interim managed fisheries" in Western Australia. Permits for these are renewed annually for the life of the plan and are not divisible. Although permits for such fisheries are often non-transferable, this is not the case for the current three interim managed fisheries, all of which serve as platforms for management systems based on individual transferable effort (ITE) units.

"Condition-based fisheries" are those where access is granted through a permissive licence condition which allows the licensee relief from a current prohibition. These fisheries are gradually being reviewed by the fisheries management agency and decisions are being made about the most appropriate form of long-term management for each fishery. Some may become managed fisheries, some may continue with licence conditions or have management regulations placed on the fishery licences under the *Fish Resources Management Regulations 1995*[3]. The rights associated with access to these fisheries are usually lower-level property-rights because there may not be a guarantee of long-term access to the fish resources within these fisheries.

A final category of access-rights results from a specific *exemption* from a specific *prohibition*, *i.e.* specific permission to engage in a clearly defined activity. The access-rights are granted under "exceptional" circumstances and are not given via a 'licence'. Instead, these rights are conferred via an 'exemption' certificate that sets out the details of the allowed fishing activity (gear, area, *etc.*), and the time within which the fishing activity may take place. The exemptions and the associated rights have a limited life, are non-divisible, and non-transferable.

Outside these categories there are a small number of ways in which licensed fishers can engage in open-access fisheries (generally referred to as "unmanaged" fisheries). Fishing activities can be undertaken just by virtue of having a Western Australian Fishing Boat Licence, which entitles the registered boat to be used within

---

[1] The enabling legislation for fisheries management in Western Australia is the *Fish Resources Management Act 1994*.

[2] These fisheries were previously referred to as Limited Entry Fisheries under the *Fisheries Act 1905*.

[3] A variety of administrative- and management-related regulations are promulgated under these Regulations.

a commercial fishery; this licence is generally transferable. These activities are not governed under formal management, that is, there is open-access to any fish resources not covered by other forms of management[4].

There is nothing static about either the different categories of commercial fishing management or the various rights they confer. Historically, a fishery has been able to - and typically will - move from relatively unmanaged status to increasingly formalized management arrangements, with the associated rights evolving along the way in response to the pressures and needs of various stake-holder groups[5].

1.2   Fisheries management in Western Australia

During the 1960s and 1970s, some of the major invertebrate fisheries within the State were brought under management plans, but the majority of fishing (mainly finfish) activities in the State were still open-access in nature. By 1985, there was concern that competing pressures and uncontrolled development would erode the sustainability and economic viability of the remaining open-access fisheries.

It was at this time that the Western Australian Government, with extensive involvement from the commercial fishing industry, undertook a complete review of fisheries management in Western Australia with the view to determining how the remaining open access fisheries should be developed. This review consisted of consultative stages that are still the basis for consultation in Western Australian fisheries today.

When considering changes to management arrangements in a fishery, in most instances the first stage is for the Government (through the fisheries management agency) to write a management discussion paper setting out the issues and a number of possible management options to address these issues. The next phase involves some level of consultation. Where there is a working group or a Ministerial advisory committee[6], the issues are thoroughly discussed in this forum before seeking comments from the wider industry, and in some cases, the general public. If there are broader issues to be considered, the Agency will often hold industry and public meetings to allow for wider discussion on the management issues. If there are any submissions gathered as part of this phase, they are analysed by the relevant industry/government group advising the Minister for Fisheries on the particular matter. Finally, as a result of Ministerial prerogative and discretion, the process may involve a number of rounds of consultation until the Minister is satisfied that the recommendations resolve the management issues before him.

In 1985/1986 the Agency conducted a major review of fisheries using such a process: a government management position paper was developed and released, the industry convened a major workshop in 1986 to consider its position, and the report from the workshop made a series of recommendations to the Minister about future directions for management of the remaining open-access fisheries in Western Australia. The majority of the recommendations were accepted by the Minister, including a recommendation to manage five types of open-access fishing activities in order to ensure their sustainable development.

The five fishing activities to be managed included the trap-and-line activities off the State's northwest Pilbara coast, and it is the management and rights-related changes in the Pilbara trap-fishery which are the subject of this case study.

1.3   The Pilbara trap-fishery

It was during the early to mid-1980s that trapping for finfish gradually moved north up the Western Australian coast into the Pilbara area. The only requirement for access to the fishery was a Western Australian Fishing Boat Licence. Thus, fishing activities were virtually unmanaged and had minimal associated property-rights. Although the total number of available state fishing licences had been frozen in 1983 there were nearly 2000 fishing units in the state and these licences were freely available on the open market at a price within the reach of most crew.

---

[4] Future management of these open-access finfish fisheries is a subject of the State's Coastal Finfish Initiative, which will address ongoing resource sharing and security of access issues.

[5] More recently, Fisheries Western Australia developed and implemented a "Developing Fisheries Strategy" which moves new fisheries through an increasingly defined and explicit set of management systems of associated rights. However, this strategy does not apply to fisheries which were already in existence.

[6] In most major fisheries in Western Australia there is a Ministerial advisory committee (MAC) appointed to provide advice to the Minister on the management of a particular fishery. Where such a MAC does not exist the Minister may establish, on an *ad hoc* basis, a working group to provide management advice on a particular matter. Both a MAC and working group comprise members of key interest groups in the fishery, such as commercial fishers, recreational fishers, processors, community representatives, and government fisheries staff.

The waters off the Pilbara coast were fished extensively for finfish by foreign trawl fleets until the late 1980s. Very little fishing effort was exerted by Australian boats off north-west Australia at this time, with the exception of prawn (shrimp) trawlers which fished in adjacent shallow inshore waters.

It was not until the mid-1990s that the fishery came under increasing fishing effort and hence management scrutiny as a result of: adjustment in adjacent fisheries, the appearance in the fishery of experienced trap-fishers, the desire to rebuild stocks after the Taiwanese fleets had left, and increased consumer demand for deep-water tropical fish.

As a result of the 1985/1986 review and the decision to manage the Pilbara trap-and-line fishery, extensive consultations continued throughout the Pilbara in 1986 and 1987, including the release of a management paper by the Agency. In lieu of any formal management advisory committee for the Pilbara fisheries, an industry/government working group was established to put recommendations to the Minister on management arrangements for a suite of Pilbara fisheries, including the trap-and-line fishery. This working group met in early 1988, submitted a draft interim report in April and, after further consultation in the region, submitted its final report in August of that year.

Concurrently, and as an interim measure to curb the development of the fishery pending the outcome of the working group, the government gazetted Notice No. 313 in 1988 prohibiting the use of fish-traps in WA waters unless authorised to do so via a licence endorsement[7]. Then, in March 1990, the then Minister for Fisheries approved a series of management recommendations, which included establishing a limited-entry fishery (now re-named a "managed fishery") for trapping[8] of finfish off the northwest Pilbara coast of Western Australia, roughly between the longitudes of 114°E and 120°E, landward of the 200 metre isobath.

2.   THE NATURE OF HARVESTING-RIGHTS
2.1   Phase One: The Allocation of Non-transferable Licences

The Pilbara Trap Limited-Entry Fishery Notice (ostensibly, the management plan) was gazetted on 24 March 1992 with a commencement date for the fishery of 1 May 1992. Initially, the rights issued with the commencement of the limited-entry fishery were non-transferable. The intention of non-transferability was that the number of participants would decrease over time through natural attrition, to a level where latent effort was removed and the remaining participants were taking the available sustainable catch (which had been estimated to be 250-300t of the target species of deep-water finfish). The non-transferability provision was to be reviewed three years after the introduction of the management plan with the view to removing the provision if latent effort had been removed.

The initial allocation of these non-transferable licences was based on:

i.   a benchmark (or cutoff) date, which was the date after which fishing activity may not be considered when deciding access under future management arrangements and
ii.   fishing history, which took into account catches not only for the three years before the benchmark date of 11 February 1988, but also continued performance in the fishery based on catches in 1988 and 1989 of at least one tonne per year.

Other management elements included:

i.   the creation of two zones[9] within the scope of the fishery
ii.   implementation of performance criteria (minimum tonnage and days-fished)
iii.   closure of the inshore portion of the fishery in recognition of the importance of recreational fishing in the region
iv.   no limit on the number of traps used by each operator at the commencement of the fishery[10] and
v.   "supplementary-access"[11], mostly for those fishermen in the adjacent prawn trawl-fisheries who had also trapped in the area of the fishery.

---

[7] The number of these licence-endorsements peaked at 31. The endorsements were not transferable in themselves, but could be transferred with the licences to which they were attached.

[8] Line fishing-effort was so low that it was not considered necessary to manage this component of the fishery and it was left open-access.

[9] Until the plan was amended in late 1999, there were two zones in the fishery: Zone-1 access covered the entire area of the fishery, while Zone-2 covered only waters east of 116°45'E. The two zones afforded extra protection to the waters west of 116°45'E, which were more exploited than waters in the east of the fishery. Access-criteria for Zone-1 were more stringent than those for Zone-2.

[10] After consultation with industry, this entitlement was modified in 1994, again through legislation, to limit the number of traps that could be used by each licensee, to 20.

[11] "supplementary access" was for those fishers who had participated in the trap-fishery but who could not meet the entry criteria because their primary fishery was the adjacent prawn-fishery. It was considered at the time that this access was an important component of their fishing business. They were not subject to performance criteria, however the management plan for the fishery allowed the Minister to revoke this access if necessary.

The then Fisheries Department of Western Australia assessed against the entry criteria the monthly fishing returns[12] of all fishermen with current trap endorsements. Those who met the entry criteria were granted a limited-entry licence for the Pilbara Trap Fishery. Those who did not qualify were given the opportunity to appeal to the then Minister for Fisheries, who was the final decision-maker on who gained access to the commercial fishery. After the appeals process was concluded a total of 20 licences were issued: eight for Zone-1, six for Zone-2 and six supplementary licences.

2.2 Moving towards more flexible rights

By 1995, a number of licensees had left the fishery as a result of: voluntarily relinquishing their licences, the transfer of fishing-boat licences (whereby any attached Pilbara Trap licences were forfeited), or licence cancellations due to the failure of participants to meet performance criteria based on minimum catch requirements. As a result, given the level of remaining effort that was being exerted at the time, catch-levels in the fishery appeared to be sustainable. However, it was becoming obvious to both the government and the industry that the fishery was in a process of change.

Adjustment in adjacent fisheries, market demand for quality deep-water tropical fish, and the financial incentives for the new fishermen who were fishing on leased licences in this fishery[13], meant that effort in the fishery was likely to increase and that, as a result, the current arrangements would no longer be adequate to ensure sustainability in this fishery. In addition, the changing dynamics of the fishery were placing demands on the management system to allow restructuring, and with that, to increase the strength of access-rights held by the fishermen.

A major concern was that, given that the fishery was estimated to be only capable of sustaining a catch of 250-300t of the key target-species of red emperor, Rankin cod and job fish, the fishery would not be able to sustain the increased catches, which could happen if transferability was introduced. New entrants would likely want to catch more fish, but the existing management rules would not prevent this.

This obvious need to reassess the suitability of the Pilbara Trap Fishery management arrangements coincided with the previous undertaking that the fishery would be reviewed three years after the introduction of the management plan. The review commenced with a government and industry meeting in June 1995. The main subjects of the review were: the provisions on non-transferability, the number of traps permitted for each licensee, and the retention of performance criteria. The review used scientific recommendations from Agency research scientists, economic parameters provided by the fishers, and offered management options suggested by the Agency based on this information.

In order to gain ownership from industry, the Agency presented a picture of the status of the fishery at that time, assisted fishers in identifying problems within the fishery, and discussed the possible solutions with them. The result was a request from industry for consideration of an individual transferable effort (ITE) system.

Following the meeting, the Agency put a proposal to industry for a limited-access management-system based on transferable trap units within the system of transferable fishery licences[14]. The proposition was that, to maintain a sustainable catch below 300t at current fishing-effort and trap-utilization levels, no more than 80 traps could to be allocated within the fishery. Should such a system be implemented, each of the 7 fishermen in the full-time fishery would receive 11 traps. There would be full transferability allowing for flexibility of fishing operations, yet total effort would be capped at 77 traps, and for consistency with management arrangements in an adjacent trap-fishery, each licence-holder would continue to hold no more than 20 traps. In addition, as the licence was only a platform for the trap-units, if all trap-units were removed from a licence then that licence would be cancelled.

The holders of supplementary access licences were not included as part of the 80-trap fishery-limit. The Agency proposed that, because these participants were applying little fishing effort in the fishery at that time, each should receive a non-transferable allocation of 11 traps.

---

[12] Under Western Australia's fisheries Regulations, fishermen are required to submit monthly returns of catch and effort for all commercial fishing activities.

[13] See Section 4.2 for further discussion of this subject

[14] The fisheries legislation requires licences to be issued for fishing activities and these form the basis for any transferable fishing units, however, it was a matter of management culture within the State at the time that licences remain limited once transferable units are attached. This may change over time.

The proposal was supported by the trap fishermen but not by the Minister for Fisheries. In particular, he was not satisfied that the proposal adequately dealt with the potentially high level of effort that supplementary-access holders could exert if they chose to do so, prompting him to request that the issue of supplementary-access be better addressed. Eventually the Minister invoked the clause under the Pilbara Trap Fishery management plan that allowed him to remove supplementary-access. In doing this he provided supplementary-access holders with the opportunity to 'explain' why their access to this fishery should not be revoked.

The process of removing supplementary-access took considerable time because of a number of factors:

i.   There was no pre-specified process for showing cause. This resulted in fishers being given a number of opportunities to show cause, that is, to argue their case to the Minister why their licence should not be cancelled.

ii.   Before he made a final decision, the Minister wanted a thorough discussion of the matter with all licence-holders at the 1995 Annual Management meeting[15] for the fishery.

iii. The introduction of the *Fish Resources Management Act 1994* (FRMA), part way through this process in October 1995, moved the decision-making power for access to fisheries from the Minister to the Executive Director of Fisheries. This change of jurisdictions and responsibilities required legal direction on how to proceed.

After one final opportunity for review, the Minister moved to cancel all supplementary-access to the Pilbara Trap Fishery in April 1996. However, the new FRMA provided a Right to Object which gave holders of supplementary-access the ability to seek an independent tribunal's decision on any decision affecting access with which a licensee did not agree. One of the supplementary-access holders submitted an objection and, given that the initial allocation of trap-units was a function of the number of licensees, the Pilbara Trap Managed Fishery Management Plan could not be changed until the objection was resolved. The Tribunal, which did not report until October 1996[16], found in favour of Fisheries WA. This opened the door for the management changes to be legislated.

Finally, in March 1997, amendments to the management plan introduced measures allowing the full transferability of licences and an individual transferable effort (ITE) system that was denominated in traps. The traps were to be allocated equally amongst the licence-holders, thus the six licensees in the fishery received an allocation of 13 traps each[17].

2.3 Phase Two: Tradeable input-controls and changing harvest-rights

The gradual introduction of licences, transferable traps, and then transferable trap-units opened the way for the fishery to change. Licences could be traded on the open market and individual trap-allocations could be traded between licence-holders; or they could continue to be leased. Financial institutions recognized the status of these fully transferable assets and this enabled participants to finance further expansion in the fishery and, as a matter of course, provided the impetus for additional management changes.

The pressure for these changes came from two sectors. The commercial participants wanted to improve their economic efficiency, gain market advantage and maximise individual profits. The fisheries management agency, although interested in economic efficiency, was primarily concerned with the sustainability of the targeted fish stocks.

With increased transferability, the predicted increase in fishing effort occurred. At the October 1998 Annual Management Meeting for the Pilbara Trap Fishery, the scientists reported that trap numbers needed to be reduced by 50% if the catch was to be constrained to 300t in 1999, and that there might need to be further reductions in the numbers of traps in the future. If this reduction were to occur, each fisherman would receive only 6 traps[18]. Despite the size of the proposed reduction, the scientists also noted that the reduced number of traps would not cut the catch in the fishery; the cutback would only remove a significant amount of the existing latent effort, thereby making it more difficult for catch levels to increase.

---

[15] The main consultation and reporting mechanism for the Pilbara Trap Fishery is an Annual Management Meeting. These meetings have been held each year since the fishery was established in 1992. The small number of fishermen involved in this fishery has meant that these meetings, supplemented by extra meetings or correspondence, has been the most efficient and cost-effective method of consultation.

[16] The Tribunal process can be a lengthy one. It involves establishing a new Tribunal for each objection, setting hearing dates, preparing and presenting cases, and receiving a decision.

[17] The total number of 78 traps in this fishery was based on scientific advice and reflected the belief that such a number would permit a sustainable catch.

[18] The management plan in force at that time allowed the Executive Director of Fisheries to determine the maximum number of traps which could be used in the fishery and to make such adjustments to the units of effort within the fishery.

This recommendation was based on the following facts:

i.   increasing catch - even with the input-controls which were in place, the average trap-catch from the fishery over the five years to 1998 had been 250t, with a range of 179t to 302t and

ii.   increasing efficiency - in the 1990s it took about half the number of fishing-days to catch the same amount caught in the 1980s, and it was clear that the Agency's desired target catch of 300t could in theory be caught by 2-3 full-time boats instead of the six that were in the fishery.

Although the industry recognised that there was considerable latent effort in the fishery, they resisted the proposed cuts in trap numbers on two grounds. First, the nature of the cuts severely limited their ability to respond to the business challenges in an equitable way. Second, the trap fishers felt that there was no equity in the size of the cuts they were being asked to make, compared with the cuts being required from a finfish trawl-fishery that overlaps the trap fishery in both area and certain of the species that were exploited.

After further industry representation and a meeting with the Executive Director, the latter determined that there would be no reduction in trap numbers for 1999, but that new management arrangements were to be formulated for 2000. On 1 January 2000, eight years after the gazettal of the Pilbara Trap Limited-Entry Fishery Notice, the amended management arrangements for the Pilbara Trap Fishery's unitized time/gear-based access-system came into effect under a management plan which clearly spelt-out participants' entitlements, how entitlements could be changed, and the operational conditions for participating in the fishery.

There is a *de facto* minimum holding of units, because if all the traps on a Pilbara Trap managed fishery licence are leased or sold, the licence is automatically cancelled. Fishermen have fully tradeable, annually-renewable harvest rights, which can only be revoked as a result of serious breaches of fisheries legislation. There is no maximum to the number of units a licensee may hold, thus enabling participants to pursue their respective business requirements with minimal involvement of the Agency in such changes. The Government maintains a full register of entitlements for this and other managed fisheries, and this register is accessible to the public. It includes a register of interest in entitlements where this is required by third party interests.

An example of a typical vessel in the Pilbara trap fishery

3.   MEASUREMENT OF FLEET CAPACITY

3.1   Characterising fleet composition and capacity

In 1987 the average trap-boat operating off the Pilbara coast was about 14-15 metres in length. These vessels, although dedicated in their fishing activity, had come from other fisheries and were not purpose-built trap-boats and were not necessarily configured for trap-fishing. On average, the fishers worked 6 traps for about 9 months of the year, making 3 trips per month. During the summer months of November to February, boats would be repaired and refitted.

By the early 1990s, the majority of dedicated trap-fishers had moved north out of the Pilbara fishery, following the availability of target species. The remaining small number of boats used a variety of gear types including traps, and pursued a variety of fish stocks. Boats generally ranged from 11 to 15 metres in length, used a small number of traps, and mostly only operated 9 months of the year in total, mainly because of cyclones.

In 1999 the average boat in the fishery was still not a purpose-built trap-boat although a couple of larger boats were now participating in the fishery. The average boat still did not work more than 11 traps, due to its size, despite the fact that provisions in the management plan now allow participants to work more traps than this.

There are inherent difficulties in providing a useful measure of fishing-capacity in this fishery. One of the difficulties is accessing data about the vessels themselves, but the more significant point is that the type and size of vessel has had little impact on the fleet's harvesting-capacity. Capacity is more a function of:

i.   the number of traps used
ii.   the number of times they are pulled each day and
iii.   the experience of the skipper and crew.

Thus, for the purposes of this analysis, capacity has been measured in terms of:

i.   the number of licences in the fishery at each stage
ii.   the number of traps available for use in the fishery and
iii.   the number of months these traps were employed.

Since 1 January 2000, the fishery has operated on a system of time/gear units which, although an additional factor in determining fishing-capacity, has not been in force long enough to be able to analyse its long-term effect. Finally, it is important to remember that fishing experience, although difficult to quantify, is a critical factor because the highliners (*i.e.* the most skilled skippers and crew) can and do affect the level of catch in this fishery independent of the characteristics of the boat.

3.2   Changes in fleet-composition and capacity arising from the introduction of transferable property rights

The composition of the fleet began to change around 1996 once it became likely that transferability would be introduced into the fishery. Prior to this time the changes in the fleet were mostly due to the cancellation of licences that occurred due to fishermen not meeting the performance criteria in the fishery, or through sale of licence packages containing a Pilbara trap licence. Participants' activity-levels were relatively stable and there was little, if any, leasing of licences.

Table 1 shows the change in licence numbers over time, but it does not tell the story of what was happening with actual fishing activity over the past five years. Its does, however, show the reduction in boats that has occurred since the introduction of the time/gear access-system in January 2000. Fishermen have bought or leased time/gear units and concentrated their use on a smaller number of boats.

Table 1
Number of boats/fishermen licensed for full-time access (Zones -1 and -2)

	Year	No. of managed fishery licences[a]	No. of boats licensed	No. of boats active
At initial issue of licences	1992	14 (8+6)	14	8 (7+1)
After transferability introduced	1997	6	6	6
After transferability and unitisation	2000	6	5[b]	3

(a) Only one boat allowed per licence.
(b) Currently one owner has two licences but only one boat.

The season, or the number of months fishermen were active each year, has also changed with the introduction of transferability. Nine months of fishing had been the norm for the dedicated operators until 1997, however, with the introduction of transferability, dedicated trap-fishermen increased the number of pulls per day and began fishing up to 11 months of the year. Some participants also began using alternating crews in order to keep the vessels working more days per month.

The introduction of unitised time/gear access in 2000 has changed fishing patterns once again. The numbers of traps is no longer the basic unit of harvesting-capacity - units of time/gear are now the critical operating factor.

Fishing is now undertaken at times when fishermen expect to maximise the catch in the time they have available to them, and leasing or purchasing of additional time/gear units also occurs.

In reality, the fleet's harvesting capacity has more to do with the fishermen than the characteristics of the vessels, because the type of boat used has not changed significantly during the development of the fishery even though the harvesting capacity of some of the fishing units has changed. As mentioned previously, the size and style of the boats currently used limits the number of traps that can be carried, so the difference amongst the participants is a function of the fishermen's trapping experience and the number of times per day the traps are pulled.

Fishermen retrieving a fish trap containing Red Emperor ("Red Things") - *Lutjanus sebae*

3.3 Consequences of changes in fleet-capacity

It appears that major changes within this fishery are only just commencing and hence many of the consequences can only be speculated at this stage. At present, effort is increasing in response to declining catch-rates, with fishermen having to travel further from the original grounds in order to maintain catch-rates acceptable to their level of operation. The fishery appears to be evolving into one consisting of two to three dedicated, full-time operators, with diversified operators opting to move out of the fishery by selling their units or trading their licences. Because it does appear that there are returns to scale, the consolidation of units onto two or three licences should maintain or even increase the profitability of the remaining operations.

Fishers now have the option to exit the fishery with a financial gain, rather than losing any benefit attached to the right to the Pilbara Trap Fishery, as was the case prior to transferability. This money provides fishers with the option either to remove themselves entirely from fishing or to reinvest in alternative fisheries.

Because of the small size of the fishery, the employment impacts coming from such changes are likely to be small and will be the result of the new operating strategies that some participants have developed (such as alternating crews so that the boat can operate on a continuous basis for a large part of the year). Thus, regional employment is unlikely to be significantly affected.

4. CONCENTRATION OF OWNERSHIP
4.1 Status prior to the unitised time/gear programme

The concentration of ownership has been triggered by three stages of management changes:

i. the introduction of limited entry in 1992
ii. the introduction of transferability in 1997 and
iii. the unitisation of access in 2000.

The use of the 1988 benchmark-date plus qualifying criteria meant that 14 full-licences and six "supplementary-licences" were initially issued in 1992. Over the following four years, 1992-1996, these numbers were reduced to six full-licences and no supplementary-licences, through the various means mentioned previously. Hence, immediately prior to transferability, there were six licences operated by six licensees, who operated at various levels of effort.

Access to the fishery had already been decided through the introduction of limited-access in 1992. The introduction of unitisation resulted in the only change in the allocation-method. In itself, unitisation did not change the distribution of licences among participants; it only changed the nature of trap-units held by each licensee, because trap-allocations were expressed as trap-units rather than actual number of traps[19]. However, unitisation did provide a market-based mechanism that participants could use to alter their unit holdings and hence, the distribution of units.

4.2   Restrictions on transfers of ownership

Prior to the introduction of transferability in 1997, licensees could only sell or lease traps to other holders of Pilbara Trap Fishery Managed Fishery Licences. However, even though these licences were formally non-transferable prior to 1997, there was *de facto* change in ownership through what is known as a "one-dollar lease". The licence would remain in the name of the original licensee who would still own a small portion of the licence. The lessee would pay most of the purchase price for the licence and then lease the total package for a minimal amount. With the introduction of transferability, the buyer would pay the "last dollar" and the contract would be filled.

After transferability was introduced there were no restrictions on who could buy a Pilbara Trap Fishery Managed Fishery Licence, but it is not possible to buy trap-units unless one held a Pilbara Trap Managed Fishery Licence, as there is no "platform" on which to place the units unless a licence is held. The other restriction on ownership transfers is that if all trap-units are transferred off a licence, the licence is cancelled.

4.3   Prices received

There is currently no specific mechanism in place to require the reporting of actual ex-vessel or other prices of the fish landed in this fishery. Any prices that are used for calculations by the Agency are obtained through spot checks at fish markets and by surveying a small number of processors. The total value of the fishery is based on a weighted average according to the species mix taken in the fishery. The average prices used by Fisheries WA are public information and are published in various reports, but are biased by the extent processors desire to give this information to the government, and a reluctance of fishers to divulge where fish are sold or prices received for those fish. Additionally, and more recently, because fishing access-fees are currently based on gross value of product from the fishery, fishers are loath to provide information that may result in an increase in access-fees.

4.4   Effectiveness of regulations governing ownership of rights

The regulations governing ownership of rights in this fishery prior to 1997 did little to restrict effective ownership. New players moved into the fishery in anticipation of the change in management arrangements, by skippering existing boats and, if satisfied that there was potential, by leasing the licences under long term leases, effectively changing the ownership of the licence and boat. This practice undermined the use of non-transferability as a management tool to a large extent, but not totally. Non-transferability did help to reduce the number of licences from the original 14 to six immediately prior to the change in rights-structure.

The actual restructuring of the fishery, in this case, trading of trap-allocations and licences, did not commence until the transferability provisions were legislated. The introduction of time/gear units has allowed further restructuring, which appears likely to continue until only two or three licensed boats remain in the fishery.

4.5   Effects of introducing transferability

The effects of introducing transferable licences and trap-allocations can be followed through licence numbers, trap numbers, and the number of months these traps were employed each year. For example, with the introduction of transferability, the effects of management changes could be seen in the movement of traps, that is, the change in the number of traps held by each licensee. Although this movement of traps was not directly caused by the introduction of transferability, it was facilitated and made more explicit by it.

Some of the participants who had previously been leasing licences bought them, but the total number of licences was not affected by the change in 'rights' status brought about by the introduction of transferability (a couple of licences changed hands, but there was no aggregation of licences and, hence, no change in the number of licences). Concentration of licences did not commence until the introduction of the unitised time/gear access system in January 2000.

Another effect of introducing transferability provisions was that it enabled operators to more accurately determine their level of activity in the fishery. One group of participants in the fishery has not altered their behaviour greatly. These are the multi-fishery fishers, or diversified operators, who trap when convenient and

---

[19] For example, instead of six licensees being licensed to use 13 traps each and the fishery consisting of 78 traps, the capacity of the fishery was determined for the year 2000 at a maximum of 5867 trap-days. Because one trap unit was defined as one trap-day at the commencement of the amendment on 1 January 2000, for every trap owned, 75 units were allocated. (With 78 traps in the fishery, the initial conversion meant that every trap could be fished for 75 days.)

who undertake other fishing activities at other times. These fishers have not changed their season or increased fishing effort within the existing fishing season. Thus, the programme has had minimal impact on these fishermen, except that they now have the option to sell or lease trap-allocations that are not being used within their fishing operation, to their obvious financial advantage. A second group who wanted to exit the fishery could sell their licences to those who wanted to increase their level of participation, and this third group of dedicated trap fishers could then extend their fishing activities from nine to eleven months, being primarily limited only by bad weather, particularly cyclones. These fishers also increased the number of times that traps were pulled each day and the number of trips undertaken each month.

Because of transferability of trap-units, there was little reason for a licensee to invest in multiple boats in the Pilbara Trap Fishery. There was no advantage in accumulating licences to gain access to the trap-allocations because traps could be traded separately. Although there was a natural limit on the traps that could be used due to the size of the boat, it has not yet been worth investing in a larger purpose-built trap-boat. This may change if consolidation results in sufficient accumulation of units to warrant operating a larger boat.

The effects of introducing time/gear units, although early in their implementation, are already evident through further trading of licences and the first movement towards eventual concentration of licences. Although there are still six licences, there are only five licensees (one now owns two licences) and four operators in the trap-fishery. Another owns one licence, and leases all but one trap from another who is not active in the fishery. The other two licensees retain their units but do not fish extensively within the fishery at this point in time.

5. DISCUSSION
5.1 Reduction in fleet-capacity

Within the broad objective of maintaining sustainable fisheries, the more specific policy objectives for management of this fishery have varied at each stage of its development. Initially, the objective was to limit the number of boats with access to trapping off the Pilbara coast and to reduce the numbers of those boats over time.

The objective set out through the 1995 review, which was the basis of all future management decisions in this fishery, was to maintain fishing effort levels such that the total catch did not exceed the estimated sustainable catch of 300t.

In both these stages of development, the objective has been reached. The initial system of non-transferability and licence cancellations assisted in successfully reducing the number of participants/ boats to six which, with fishing practices at the time, was considered a reasonable number to take the available catch. When fishing practices changed through the introduction of more experienced trap-fishers into the fishery, management arrangements were modified to allow the management agency to limit trap numbers as a proxy for limiting the total catch to a maximum of 300t.

Later, the introduction of transferable licences and transferable trap-allocations allowed more flexibility among different operators in the fleet. It kept the mechanism for reducing fleet trap-allocations through across-the-board trap reductions, but allowed fishers to trade their traps and to more closely hold the number they needed for the nature of their particular operations. The next stage of introducing transferable unitised trap time/gear access has further increased the flexibility of fishermen to be able to maximise their return and fishing times within the constraint of the estimated sustainable catch-level.

The success of the gradual changes in management can be attributed to the fact that participants and managers have not had to address both catch-reduction issues, at the same time that they were making the transition from non-transferable to transferable access, largely due to:

i.   early intervention by government to limit access to the fishery off the Pilbara coast, to issue trap-licence endorsements that identified the pool of trap fishermen in Western Australia, then to analyse catch data to identify those that fished off the Pilbara coast

ii.   a small participating pool of fishers, which facilitated consultation and the discussion on management issues and

iii. early scientific advice which estimated a sustainable catch limit of 300t and was accepted at an early stage of management and which still serves as the recommended maximum for the fishery. This means that there have not been catch-reduction issues to be faced concurrent with the transition of the fishery from non-transferable- to transferable-access.

The fishery is still in a process of change. The opportunity now exists for a change in fleet-capacity, but this will take time. The system now provides the business environment and incentives for further investment and the concentration of ownership; whether this will happen remains to be seen.

5.2 Concentration of ownership

Consolidation in the fishery began with the introduction of individual trap-allocations and has been facilitated by the increased divisibility offered by the transferable time/gear unit system. To date the movement of transferable time/gear units suggests that the fishery may be moving towards one based on two to three dedicated boats, but as mentioned previously, the concentration of ownership of licences has yet to occur to any great extent.

6. ACKNOWLEDGEMENTS

I would like to acknowledge the assistance of staff of Fisheries Western Australia, particularly John Nicholls, Peter Millington, Dr Mike Moran, Heather Brayford, Peter Stephenson, Christine Mary Keys, and David Donatti.

7. LITERATURE CONSULTED

Chairman's summaries from Annual management meetings of the Pilbara Trap Fishery 1993 - 1998 (unpublished).

Fisheries Department of Western Australia, 1985. Arrangements for Entry to all fisheries off and along the WA Coast. Fisheries Department of Western Australia. Perth, WA.

Fisheries Department of Western Australia, 1986. Fishing access arrangements FINS Vol 19 No 4: pp 12-14.

Malone, F. (unpublished). Report by the Chairman Mr F. Malone to the Executive Committee. 1986. Australian Fishing Industry Council (Western Australia Branch).

Stephenson, P. and J. Mant, 1998. Fisheries Status and Stock Assessment for the Pilbara Demersal Scalefish Fishery in relation to the Pilbara Trap Managed Fishery. Fisheries Western Australia (unpublished).

CHANGES IN FISHING PRACTICE, FLEET CAPACITY AND OWNERSHIP OF HARVESTING RIGHTS IN THE ROCK LOBSTER FISHERY OF WESTERN AUSTRALIA

G. Morgan
106 Barton Tce. West, North Adelaide, SA 5006, Australia
<garymorg@ozemail.com.au>

1. INTRODUCTION

Western Australia has for the past 40 years managed many of its fisheries through a system of limited-entry, with the first of these fisheries (the western rock lobster fishery for *Panulirus cygnus*) being made a limited-entry fishery in 1963. This fishery, and the way in which the initial allocation of the quota was managed, has been described by Morgan (2001a).

The Western Rock Lobster Fishery began in the early 1900s supplying a small local market. However, it was not until the 1940s, when an export market developed to supply first of all the armed forces and later the American market, that catches and fishing capacity began to increase rapidly (Sheard 1962).

Western Australia lobster boat at idling speed

With increasing prices being paid for rock lobster tails, the fishery rapidly expanded until by the late 1950s, it supported over 1000 fishermen who took approximately 8600t of rock lobsters from coastal areas as well as the Abrolhos Islands. As the number of boats and fishermen continued to increase and the fleet became more efficient, it became apparent that restrictions would have to be placed on the unbridled expansion of the industry if the stocks of rock lobsters were to be managed for long term sustainability.

On 1 March 1963, the first restrictions were put into place in the Western Australian rock lobster fishery with the number of rock lobster vessels in the industry being limited to those already engaged in the industry and, to the present time, no new additional licences have been issued.

The management system, which has evolved over the years within the context of a limited entry philosophy, has essentially stabilized catches and fishing-capacity. However, because the fishery remains an input-controlled fishery, the problem of containing fishing-capacity in the face of technological change remains an issue that needs continual management attention. The fishery today is a vigorous, wealthy and stable fishery with the catch in 1999/2000, which is taken by 594 licence-holders, being approximately 14 450t valued at some $A380 million. The fishery is, therefore, the largest single rock lobster fishery in the world, accounting for approximately 24% of global production.

The history of the allocation of access-rights in the western rock lobster has previously been documented as part of this case study series (Morgan 2001a). This paper examines the changes that have taken place in fishing practices, fleet capacity and ownership of harvesting-rights subsequent to the introduction of transferable access-rights in this fishery.

The data for this examination relies on occasional economic performance reports of the fishery. No social impact data have been collected on these effects and hence comments on social impacts are based on *ad hoc* observations. Provision of catch and fishing-effort data is a mandatory reporting requirement in all Western Australian fisheries and have been collected as part of the monitoring programme for the western rock lobster fishery since the earliest days of the fishery.

2.   THE NATURE OF THE HARVESTING RIGHT

Access to the fishery was restricted in 1963 to those licence-holders who were part of the fishery on 1 March of that year. The access-right entitles the holder to operate a fixed number of pots (*i.e.* a retrievable fish-trap), with that number being restricted by the length of the vessel from which they are fished. Additional pots may be purchased (or, as is common practice, leased) but the number that can be operated is still restricted by the length of the vessel. Although restrictions on vessel replacement have been part of the fishery since 1965, these have recently been abolished so that there are currently no impediments to the building of larger vessels, thereby enabling a larger pot-entitlement. However, because the number of pots permitted in the fishery is fixed at 69 288, additional pots to operate from a larger vessel must be acquired in the open market from other licence-holders.

Both the access-right to the fishery (formalized by a commercial fishing licence) and the pot-entitlement are freely tradable and, because of the highly profitable nature of the fishery, these currently command significant prices on the open market. Pots (*i.e.* the right to operate a single pot) currently change hands for approximately $A27 000. Therefore, the free market price for an access-right with the average pot-holding of 116 pots would be approximately $A3.13 million. This does not include the cost of the vessel and other equipment.

The tradable entity is therefore the pot and transfers occur freely in an open market. The market, however, is restricted to those who have a rock lobster licence to fish in this limited-entry fishery and hence it is not possible for persons external to the fishery to operate pots although it is possible for non-licence-holders to own pots. There is a minimum pot-holding of 63 pots and a maximum of 150. The current average pot-holding, for all areas combined, is 116 pots per vessel. Once the minimum pot-holding is reached, there are no further restrictions on the issuing of a boat licence so that these pots can be operated. Owners of pots who do not have the minimum holding of 63 commonly lease their pots to operators in the fishery. These operators must, of course, comply with the minimum and maximum pot-holding requirements, restrictions on the number of pots per foot of boat length, the current temporary 18% reduction in pots able to be used, and other management measures.

At current pot prices of approximately $27 000, the annual return for leased pots is approximately 8-9% although lease prices are sometimes influenced by other considerations, such as ensuring through-put for processing factories. The fishery is also subject to a number of input-controls, which have been detailed in an earlier contribution to these case studies. These include measures such as minimum and maximum size limits, closed seasons *etc.*

3.   MEASUREMENT OF FLEET CAPACITY
3.1   Characterising fleet capacity

In all limited entry fisheries in Western Australia, the primary measure of fleet capacity has been the number of vessels licensed to fish in the fishery. In the western rock lobster fishery, the number of vessels has slowly decreased as amalgamation of pot-holdings has taken place. Changes in the number of vessels operating in the rock lobster fishery are shown in Figure 1 from the introduction of transferable property-rights in 1963 to 1999/2000, together with changes in the number of pot-lifts over the same period. The current (1999/2000) number of vessels in the fishery is 594, compared with 689 in 1990/91 and 836 in 1964. Apart from the primary measure of number of vessels, a number of other important measures of fleet capacity are also used in this fishery.

An important measure of fleet capacity in the rock lobster fishery has been the pot or trap. Various restrictions have been in place over the history of the fishery to limit vessel-numbers and pots per vessel, *etc.*, but the primary aim of these measures was to limit the total number of pots in the fishery. In 1965, the number of pots in the fishery was capped at 76 623 by a series of input-controls that restricted the number of vessels, the number of pots per unit length of the fishing vessel, and the size of vessels. These measures have largely remained in place although there has been some attrition of pots due to forfeiture, *etc.*, so that in 1999/2000 the total number of pots in the fishery was 69 288. Again, details of changes in the total number of pots in the fishery has been included in Morgan (2001a).

As vessel (license) numbers have declined in the fishery (see Figure 1) and the number of pots have declined at a much lesser rate, both the average size of vessels, and the average number of pots operated per vessel have increased. Table 1 shows these date from 1990/91 to the present with a comparison with 1962/63 and 1973/74. Earlier data on average vessel size is not available.

The frequency with which pots are utilized during the open season has been documented since the introduction of the limited-entry management-regime in 1963 through compulsory monthly fishing returns (which also record catches and areas fished) and a voluntary system of fishermen's log books. The log books are completed by approximately 30% of the fleet and provide daily information on catches and fishing activities in much greater detail than the compulsory monthly returns.

Figure 1

Changes in number of licenses and number of pot lifts in the western
rock lobster fishery, 1964/65-1999/00

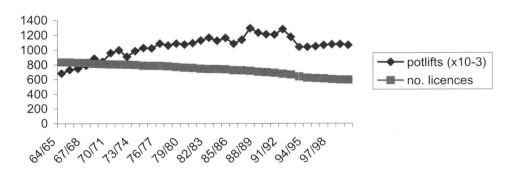

Table 1

Average length of vessels and average number of pots used per vessel in the western rock lobster fishery,
1990/91-1999/00 with a comparison with 1962/63 and 1973/74.
Data for 1962/63 and 1973/74 from Morgan (1977).

Year	Average length of vessel (m)	Average pots/vessel*
1962/63	8.71	90.7
1973/74	9.16	88.7
1990/91	10.93	102.7
1991/92	11.44	102.5
1992/93	11.59	104.2
1993/94	12.02	108.5
1994/95	12.34	111.6
1995/96	12.45	112.7
1996/97	12.52	113.4
1997/98	12.67	114.9
1998/99	12.82	116.3
1999/00	13.05	116.6

* Under temporary management arrangements introduced in 1993/94, only 82% of these pots can be utilized at any one time.

The pot-lift has therefore been the primary measure of fishing-capacity in the western rock lobster fishery since it incorporates both the number of pots being used and the frequency with which they are being used. This information is available daily on a spatial basis of an approximate 10x10 nautical mile grid that covers the entire fishing area.

The increased efficiency of pots in the fishery, brought about by technological advances in pot, vessel and equipment design, has been measured (Brown, Caputi and Barker 1993) as has the seasonal vulnerability to capture of rock lobsters (Morgan 1974). However, these changes in efficiency and vulnerability are not incorporated into the measure of fleet capacity as a matter of course, but are used in specific studies.

3.2 Changes in fleet capacity arising from the introduction of transferable property-rights

Since the number of pots in the western rock lobster fishery was effectively capped in 1965 through a series of input-controls, there has been no increase in the number of pots being used in the fishery since that time. In fact, through a process of slow attrition, the number of pots has declined from 76 623 in 1965, to 69 288 in 1999/2000.

However, the frequency with which these pots are used has increased dramatically, despite a six-week reduction in the length of the fishing season since 1978. As a result, the primary measure of fishing-capacity, the pot-lift, has changed since 1963, and is currently 56% higher than it was when tradable access-rights were introduced in 1963 (see Figure 1). Initially, the number of pot-lifts increased steadily as a result of more frequent usage of pots. In the late 1980s, there was a recognition that this increased fishing-capacity was having an adverse impact on the abundance of the stock, particularly the spawning stock. As a result, additional input-controls, detailed in Morgan (2001a), and including an 18% 'temporary' pot-reduction, were introduced which succeeded in stabilizing fishing-capacity. These measures, together with changes in size limits and additional direct protection for the spawning-stock have rebuilt the spawning-stock levels significantly. Table 2 presents data on changes in fishing-capacity (number of pot-lifts) since the introduction of tradable access-rights in 1963.

As a result of changes in fishing capacity, outlined in Table 2, catch rates have also varied since tradable access rights were introduced in 1963. Figure 2 shows these changes.

Table 2
Catches (t) and Fishing-Capacity (millions of pot-lifts) in the western rock lobster fishery of Western Australia, for the fishing seasons 1964/65 – 1999/2000. Seasons begin on 15 November each year and ended on 15 August the following year up to 1977/78, and 30 June the following year for 1978/79 onwards.

Year	Catch (t)	Pot-lifts (m)	Year	Catch (t)	Pot-lifts (m)
1964/65	7882	6.812	1982/83	12884	11.625
1965/66	8550	7.301	1983/84	11349	11.214
1966/67	9089	7.480	1984/85	9682	11.601
1967/68	10375	7.898	1985/86	8166	10.822
1968/69	8506	8.861	1986/87	8529	11.343
1969/70	7285	8.408	1987/88	12066	12.883
1970/71	8437	9.589	1988/89	12312	12.306
1971/72	8625	9.956	1989/90	10298	12.080
1972/73	7263	9.091	1990/91	9220	12.032
1973/74	7234	9.864	1991/92	12164	12.773
1974/75	8877	10.255	1992/93	12303	11.733
1975/76	9110	10.202	1993/94	11040	10.374
1976/77	9286	10.861	1994/95	10802	10.373
1977/78	10549	10.594	1995/96	9800	10.462
1978/79	12105	10.842	1996/97	9902	10.621
1979/80	11024	10.724	1997/98	10463	10.734
1980/81	10328	10.890	1998/99	13009	10.750
1981/82	11050	11.255	1999/2000	14437	10.635

Figure 2
Catch Rates in the western Rock Lobster Fishery, 1964/65-1999/00

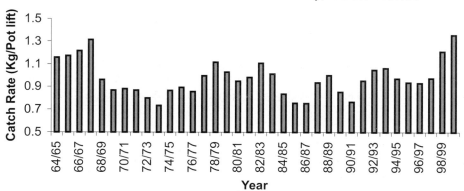

Of particular note is the decline in catch-rate through to about 1974/75. This was a time of increasing fishing-capacity as prices for rock lobsters increased. In recent years (1991/92 onwards), fishing-capacity has been reduced through a series of measures, including an 18% 'temporary' pot-reduction and this, combined with other measures designed to protect the breeding stock, has resulted in a trend of increasing catch-rates.

3.3 Consequences of changes in fleet capacity

The consequences of the changes in fleet capacity have been evident on a number of fronts. From a stock-abundance point of view, it was apparent as early as the mid-1970s that fishing-capacity was continuing to increase despite the range of input restrictions that were in place (Morgan 1979, 1980a, 1980b). This resulted in further restrictions, the most notable of which was the additional six-week closed-season, introduced in 1978.

More recently, concerns over the decline of the breeding-stock resulted in a range of measures, introduced in 1993/94, to increase its abundance. These measures included an 18% 'temporary' reduction in the number of pots allowed to be used as well as measures designed to provide direct protection to the breeding stock. The 18% pot-reduction had the effect of reducing the number of pot-lifts by 11.6% between 1992/93 and 1993/94 (Table 2), although the reduction in fishing-capacity between 1992/93 and 1999/00 has been a more modest 9.4%.

Clearly, compensatory mechanisms are operating so as to result in continuing upward pressure on fishing-capacity, despite the 'temporary' pot-reduction with approximately half of the potential benefits of an 18% pot-reduction being eroded over a seven year period. The additional measures that were put in place in 1993/94 to directly protect the breeding-stock (such as a maximum size for females that may be landed, adjustment to the minimum size permitted, and extending protection to those females which are mature rather than just carrying eggs) have obviously been highly successful. Breeding-stock abundance has increased by a factor of about three since 1992/93 (Anon. 1999) and has now returned to levels not seen since the late 1970s. This appears to be having a positive effect on subsequent recruitment to the fishery, as measured by the abundance of settling *puerulus larvae*.

However, continuing upward pressure on fishing-capacity may erode those increases in abundance in future years, although the potential for increased capacity (in terms of latent capacity available) is significantly less than it was in the early days of the fishery.

From an economic point of view, the most significant change in the fishery has been the rapid appreciation in the value of the access-right. This appreciation in asset-value seems to have been a direct result of the limited-entry nature of the fishery, aided in large part by the increase in prices received for the product in export markets.

Table 3 shows the approximate free-market transfer-value of pots since 1990/91 and demonstrates the rapid and consistent escalation of prices. The Table is separated into prices received for pots in Zones A and B combined (*i.e.* the northern part of the fishery) and Zone C (the southern part). This is a result of consistently higher prices being paid for Zone A and B licences because of generally higher catch-rates and the better ability, because of weather conditions, to work later into the season in northern areas.

Table 3

Approximate range of prices ($A) paid per pot on transfer for northern (Zones A and B) and southern (Zone C) areas of the Western Rock Lobster Fishery, 1990/91-1999/2000

Season	Price – Zones A and B	Price – Zone C
1990/91	$7000-$8500	$6200-$7000
1991/92	$9500-$12000	$8000-$9500
1992/93	$12000-$13000	$10000-$11500
1993/94	$20000-$25000	$13000-$20000
1994/95	$22000-$25000	$18000-$20000
1995/96	$30000	$21000-$23500
1996/97	$25000	$18000-$21500
1997/98	$29250-$30000	$21500-$25000
1998/99	$18000	$18000-$19000
1999/00	$24000-$25000	$22500-$28500

Using the data from Table 3, the total asset value of the access-rights to the western rock fishery can be calculated, and these are presented in Table 4. Average pot-values (for all zones) have been calculated as the means of the median pot-price shown in Table 3. These have then been multiplied by the number of pots in the fishery each season to arrive at a total asset-value of the access rights. Table 4 also presents data on the revenues generated each year (based on average prices and catches) and compares these revenues with the annual asset-value as a measure of earnings/asset-value. This latter measure should approximate earnings/capital employed since the value of equipment (boats, gear, *etc.*) is only a small, and relatively constant, proportion of total capital with the largest proportion being the asset-value of the access-right.

The total value of the access-rights to the western rock lobster fishery (*i.e.* excluding vessels, fishing gear, etc.) in 1999/2000 is therefore around $A1.42 billion. This value has been created since the initial allocation of access-rights in 1963.

Table 4
Total asset-value of access-rights to the western rock lobster fishery, together with
earnings/asset-value for the years 1990/91-1999/00.

Season	Average pot price ($A)	Total asset value of access rights ($A millions)	Revenue (*i.e.* earnings) $A millions	Earnings/ asset value (%)
1990/91	7275	514.78	189.98	36.9
1991/92	9750	675.58	249.36	36.9
1992/93	11625	805.53	223.55	27.8
1993/94*	19500	1108.61	280.75	25.3
1994/95*	21250	1207.61	297.07	24.6
1995/96*	26125	1484.81	232.27	15.6
1996/97*	22375	1271.25	241.33	19.0
1997/98*	26440	1502.15	211.87	14.1
1998/99*	18250	1037.30	263.43	25.4
1999/00*	25000	1419.84	389.80	27.5

*18% temporary pot reduction in effect which reduced available pots to 82% of licensed
pots. This has been taken into account in establishing total asset values.

Two things are worth noting about the transfer prices presented in Tables 3 and 4. First, the largest increase in pot-prices occurred in 1993/94 when the 18% reduction in pot-usage was introduced. It appears that increased confidence about the long-term management arrangements for the fishery contributed significantly to this increase in price, despite the loss of the use of 18% of fishing-capacity.

Second, the large drop in prices in 1998/99 was a result of a high catch (see Table 2) reducing market prices. This can be seen from Table 3 where, despite the high catch in 1998/99, total revenues only increased a small amount because of the reduced prices.

It is interesting to note the general trend of a decline in earnings/asset-value, although the last two years have seen an increase. The increase in 1999/00 was due to a large increase in earnings. This, in turn, was a result of the unusual co-incidence of a record catch of 14 437t and high prices paid to fishermen of around $A27/kg. In the longer term investors and operators in the fishery are apparently willing to accept a lower (although still substantial) return on their assets employed, and this may partly be attributed to the stable management environment of the fishery.

The high returns available in the fishery (see Table 4) have apparently resulted in a trend towards an increasing number of pots being owned by non-operators in the fishery, although precise data on this aspect is not readily available. A consensus of industry and Government opinion, however, indicates that, currently, 20-40% of all licensed pots are owned by non-operators with these pots being leased to fishermen. The non-operators include investors as well as entities such as processing companies which have taken a strategic stake in pot-ownership to ensure supply of product to their factories.

Costs of management of the fisheries have also been documented in recent years. Morgan (1997) showed that compliance costs had increased significantly in the fishery, apparently in response to the more complex management environment. In 1998/99, costs of management of the fishery totaled $A6.016 million with a breakdown of those costs being given in Table 5.

Table 5
Costs of management of the western rock lobster fishery for 1998/99

Component	Cost ($A)
Compliance and surveillance	2 140 000
Research	1 072 000
Policy and programme management	314 000
Corporate support (overheads)	700 000
Levy for fisheries development fund	1 790 000
Total	6 016 000
Cost per licensed pot	87

Source: Anon. 1998.

The total cost of management therefore represents approximately 2.3% of the gross earnings from the fishery or $A462 per tonne. This compares with a management cost for South Australian rock lobster fisheries of $A784 - 1137 per tonne (see Morgan 2001b). Since 1999/00, the total cost of management is now collected from pot-holders as a licence fee and is administered by the Government of Western Australia in order to carry out the various management functions outlined above. This recovery of costs from pot licence-holders was phased in, beginning 1995/96 when 85% of attributable costs were recovered.

4.　CONCENTRATION OF OWNERSHIP
4.1　Status prior to programme

When access-rights were restricted in 1963, there were 845 vessels operating in the fishery. The number of pots licensed to operate in the fishery was capped at that time at 76 623 through a series of measures which have been outlined in Morgan (2001a). This resulted in an average number of pots per vessel of 90.7 immediately prior to the restriction of access-rights. The fishery was, up to that time, an open access fishery with no restrictions on access-rights.

By 1974, the number of vessels had been reduced to 798, operating a total of 72 367 pots, an average pot holding of 88.7. Since that time, concentration of pot-holdings onto fewer, larger vessels has continued. There are currently 594 licence-holders in all zones of the rock lobster fishery with the numbers of licences having declined slowly over recent years as amalgamation of pot entitlements and structural adjustment of the industry has taken place. This aspect is further discussed below. Table 6 shows the number of licensees and the average number of pots per licence-holder for the period 1990/91-1999/00.

Table 6
Licence numbers and average pot-holdings in the WA rock lobster fishery

Year	Abrolhos Islands Zone A		North Coastal Zone B		South Coastal Zone C		Total	
	Licences	Pots/ licence*	Licences	Pots/ licence*	Licences	Pots/ licence*	Licences	Pots/ licence*
1990/91	180	100.4	166	96.8	343	106.7	689	102.7
1991/92	175	99.1	166	97.2	335	107.0	676	102.5
1992/93	168	100.4	7	99.4	330	108.6	665	104.2
1993/94	153	106.2	164	105.0	322	111.3	639	108.5
1994/95	149	108.8	158	109.7	314	113.9	621	111.6
1995/96	149	110.6	155	110.1	311	114.9	615	112.7
1996/97	149	111.4	154	110.1	308	116.1	611	113.4
1997/98	149	113.3	151	110.4	303	118.0	603	114.9
1998/99	148	114.3	150	110.8	298	119.9	596	116.3
1999/00	148	113.5	150	111.7	296	120.7	594	116.6

* Under temporary management arrangements introduced in 1993/94, only 82% of these pots can be utilized at any one time.

There is less information on the extent of concentration of the beneficial ownership of pots and how this has changed since 1963. When access-rights were first restricted in 1963, all pots were owned and operated by the fishermen. However, as indicated above, the high returns available in the fishery (see Table 4) has apparently resulted in a trend towards an increasing number of pots being owned by non-operators in the fishery. A consensus of industry and Government opinion, indicates that currently 20-40% of all licensed pots are owned by non-operators, with these pots being leased to fishermen.

In summary, since 1963, there appears to have been a significant concentration of ownership of pots and also a concentration of pots being operated from a diminishing number of larger vessels.

4.2　Restrictions on transfer of ownership

There are currently no restrictions on the ownership or the transfer of pots. However, there are a number of restrictions on the operation of these pots, including minimum and maximum pot-entitlements. These have been described above.

4.3　Prices received

Data have been collected on prices received for rock lobsters for a number of years and are shown in Table 7.

The price differentials between the northern zones (Zones A and B) and the southern zone (Zone C) are usually related to the average size of rock lobsters caught, with the generally more desirable smaller lobster being more common in northern areas.

Table 7
Prices received by operators in the western rock lobster fishery, 1990/91-1999/00. The data has been separated into those zones (Zone A and B) in the northern part of the fishery and those in the southern part (Zone C)

Season	Price Zones A & B ($A/kg)	Price Zone C ($A/kg)
1990/91	20.65	20.65
1991/92	21.00	20.00
1992/93	18.50	17.75
1993/94	28.00	26.50
1994/95	29.00	26.00
1995/96	25.50	22.00
1996/97	27.50	26.00
1997/98	20.50	20.00
1998/99	20.25	20.25
1999/00	27.00	27.00

4.4   Effectiveness of regulations governing ownership of rights

The effectiveness of regulations governing ownership of rights has been high from a compliance point of view, and this is reflected in the high cost of compliance activities (see Table 5). In addition, the high monetary value of the access-right (see Table 4) itself results in a high profile of the industry and the early detection and reporting of illegal activities.

5.   DISCUSSION
5.1   Reduction in fleet capacity

Since a reduction in fleet capacity was not originally a policy objective for the introduction of tradable access-rights, fleet capacity has not been reduced. Rather, the policy objective was to stabilize fleet capacity and, using the measures of fleet capacity available, this has been achieved (see Tables 2 and 3). However, because of the input-managed nature of the fishery there is constant pressure for fishing-capacity to increase. As a result, an increasing array of restrictions has needed to be applied in order to stabilize fishing-capacity. This has included a recent 'temporary' 18% reduction in allowable pots in the industry.

A major concern is that the measures of fleet capacity being used do not explicitly take into account efficiency increases brought about by technological innovation, and unfortunately no data-collection programmes are in place to collect such information on a regular basis. However, specific studies (*e.g.* Brown *et al.* 1993) suggest that this may become an important issue.

5.2   Concentration of ownership

The limited-entry nature of the fishery has not prevented a concentration of both ownership of pots and the utilization of those pots. This has been discussed above in Section 4.1 and data presented in Table 6 demonstrating the concentration of pot usage to a fewer number of larger vessels.

One of the most important issues surrounding the concentration of ownership has been the large profits which have been derived from the fishery. The value of the access-right to the fishery, together with earnings from the fishery, has continued to increase dramatically in recent years (see Tables 3 and 4). This follows a trend established when the fishery was made a limited-entry fishery in 1963.

Despite this increase in value of the access-right, Table 4 shows that consistently high returns on investment of over 25% per annum have been achieved in the fishery, with individual vessel earnings in 1999/00 averaging over $A650 000. Earnings-growth in the fishery has averaged some 18% per annum in recent years compared with Western Australian average earnings-growth of 1.6% per annum.

Rent generated from the fishery is therefore substantial and, using the total value of the access-rights and a discount rate of 3%, was approximately $A43 million in 1999/2000. Of this, only some $A6 million is collected as license- or access-fees.

The 'surplus' rent generated, of approximately $A37 million per annum, has been invested in ever more lavish vessels and equipment, and also in a range of other non-fisheries investments. The marginal returns on recent investment in vessels and equipment has reputedly been low or negative and such 'capital stuffing' has become a significant management and social issue.

The importance of social issues whereby the attitudes in the small coastal community can influence individual decisions to concentrate ownership has not been well recognized or studied in southern areas of Western Australian fisheries. Although these issues have emerged as major drivers of the effects of ITQ-management in South Australia, they are likely to be of lesser importance in Western Australian fisheries.

6. ACKNOWLEDGEMENTS

Special thanks are due to Chris Chubb, Eric Barker and staff of the Western Australia Fisheries Department, for making data available for the analyses presented in this paper.

7. LITERATURE CITED

Anon. 1998. The western rock lobster fishery – cost recovery and managed fishery fees. Information Paper, Fisheries WA; 6pp.

Anon. 1999. The effects of five years (1993/94-1997/98) of stable management in the western rock lobster fishery. Comm. Fish. Res. Bull. Fisheries WA; 8pp.

Brown, R.S., N. Caputi and E. Barker 1993. A preliminary assessment of increases in fishing power on stock assessment and fishing effort expended in the western rock lobster fishery. Proc. 4th Int. Workshop on lobster Biology and Management, Japan, 1993.

Morgan, G.R. 1974. Aspects of the population dynamics of the western rock lobster (*Panulirus cygnus*) II. Seasonal changes in the catchability coefficient. Aust. J. Mar. Freshw. Res., 25:249-259.

Morgan, G.R. 1977. Aspects of the population dynamics of the western rock lobster and their role in management. Ph.D. Thesis, University of Western Australia.

Morgan, G.R. 1979. Assessment of the stocks of the western rock lobster, *Panulirus cygnus*, using surplus yield models. Aust. J. Mar. Freshw. Res. 30:355-363.

Morgan, G.R. 1980a. Increase in fishing effort in a limited entry fishery - the western rock lobster fishery 1963-1976. J. Cons. Int. Explor. Mer 39(1):86-91.

Morgan, G.R. 1980b. Population dynamics and management of the western rock lobster fishery. Marine Policy 4(1):52-60.

Morgan, G.R. 1997. Individual quota management in fisheries – methodologies for determining quotas and initial quota allocation. FAO Fish. Tech. Pap. No.371; 41pp.

Morgan, G.R. 2001a. Initial allocation of harvesting in the Rock Lobster fishery of Western Australia. 136-143. *In:* Shotton, R. (Ed.) 2001. Case studies on the allocation of Transferable Quota Rights in fisheries. FAO Fish. Tech. Pap. No. 411. FAO, Rome.

Morgan, G.R. 2001b. Changes in fishing practice, fleet capacity and ownership of harvesting rights in the fisheries of South Australia. *In:* Shotton, R. (Ed.) 2001 Effects of transferable fishing rights on fleet capacity and concentration of quota rights. FAO Fish. Tech. Pap. No 412 pp 89-97. FAO, Rome.

Sheard, K. 1962. The Western Australian Crayfishery, 1944-1961. Paterson Brockensha, Perth, 145pp.

CHANGES IN FISHING PRACTICE, FLEET CAPACITY AND OWNERSHIP OF HARVESTING RIGHTS IN THE FISHERIES OF SOUTH AUSTRALIA

G. Morgan
106 Barton Tce. West, North Adelaide, SA 5006, Australia
<garymorg@ozemail.com.au>

1.   INTRODUCTION

For the past 25 years South Australia has managed its fisheries through a system of limited-entry with the first of these fisheries (the important rock lobster fishery) being made a limited-entry fishery in the 1970s. In addition to the limited-entry nature of the fisheries, there are three important quota-managed fisheries: those for southern rock lobster (*Jasus edwardsii*), for abalone (*Haliotis laevigata* and *H. rubra*), and for pilchards (*Sardinops neopilchardis*). These three fisheries, and the way in which the initial allocation of the quota was managed, have been described by Morgan (2001). Figure 1 shows the offshore area of South Australia.

Figure 1

The South Australian Rock Lobster Fishery began in the early 1870s as a hoop-net fishery. The first commercial lobster pots were used in 1889 and around the turn of the century small industries began to emerge in different parts of the state. By the late 1940s a thriving lobster-tail export market to America had been developed. Industry then developed rapidly with vessels becoming more sophisticated and catch increasing. The fishery is now a significant and expanding industry in South Australia, generating a business turnover of more than $A230 million/year and supporting of 2200 jobs (EconSearch 1999).

There are currently 254 licence-holders in the two zones of the rock lobster fishery with the numbers of licences having declined slowly over recent years as amalgamation of pot-entitlements and structural adjustment of the industry has taken place.

The abalone industry in South Australia began in the 1960s in response to the emerging market for this product in South East Asia. In 1971 the number of licences was restricted and the fishery was divided into three management zones that remain the basis of management of the fishery. The number of licences in 1971 was more than 100 but this was reduced over the years by a policy of non-transferability. There are currently 35 licences on issue. The 1999/00 total catch of abalone in South Australia was approximately 832t whole weight with the fishery being valued at approximately $A30 million/year based on producer prices. Table 3 gives the recent catch-history of the fishery.

The pilchard fishery in South Australia, like other *Sardinops* fisheries, exhibits large annual variations in abundance. This has been magnified in recent years by major fish-kills in 1995 and 1998 that have been linked to a herpes-like virus infection in pilchards. It has been estimated that these fish-kills resulted in the loss of up to 60% of the total adult population, although juvenile fish were not affected.

Despite these large fluctuations in abundance the pilchard fishery is managed using a total allowable catch (TAC) that is set each year in response to estimates of abundance based on annual egg- and larval-surveys. The 2001 quota of 9100t (compared with 3600t in 2000) is divided equally between the 14 licence-holders to produce an annual ITQ with daily monitoring of individual quotas taking place. There are further restrictions on areas permitted to be fished that are designed to separate the larger and smaller operators between offshore and inshore waters.

Annual TACs in the pilchard fishery have ranged from 3600t to over 11 000t during the period 1997-2001 with the vast majority of the fish being utilized locally as feed for the burgeoning tuna-aquaculture industry. The fishery, nevertheless, supplies only about 20% of the tuna-aquaculture industry's feed needs.

The history of the introduction of ITQs in these fisheries has previously been documented as part of this case study series (Morgan 2001). This paper examines the changes that have taken place in fishing practices, fleet-capacity and ownership of harvesting-right subsequent to the introduction of those transferable individual quotas in these three fisheries. The data for this examination rely heavily on annual economic performance reports of the fisheries, which, however, were only introduced as a reporting measure in 1997. No social impact data has been collected on the subsequent effects, and hence comments on social impacts are based on *ad hoc* observations. Catch and fishing-effort data are a mandatory reporting requirement in all South Australian fisheries and have been collected over the history of all fisheries since the introduction of ITQs and usually for some considerable period prior to their introduction. These data have also been used in this analysis as appropriate.

2.   THE NATURE OF THE HARVESTING-RIGHT

The nature of the harvesting-right varies with each of the three fisheries being considered. In the Southern Zone rock lobster fishery (the other part of the fishery, the Northern Zone, is managed by input-controls) the fishery is managed through a system of ITQs with the annual TAC being set each year by a joint industry/government management advisory body. This advisory committee (known as the Southern Zone Rock Lobster Fisheries Management Committee, or FMC) consists of rock lobster fishermen and processors, government managers, researchers, members of the Fishermen's Industry body and recreational fisher's representatives, and is chaired by an independent chairman. The FMC reports directly to the Minister for Primary Industry and its function (established in legislation) is to provide the minister with advice on management of the fishery, including annual quotas. Unfortunately, the government's fisheries department also provides separate advice to the minister and this often leads to conflicts between the two advisory bodies and a diminution of the effectiveness of the FMC.

The TAC has not changed in recent years in response to steady (or slightly increasing) indices of both overall stock-abundance and breeding-stock levels. Although the Southern Zone of the rock lobster fishery has been managed by such output-controls for a number of years, various elements of input-control still remain. The industry, in fact, has been fiercely protective of its management arrangements and cites stock sustainability and socio-economic issues as reasons for not moving to greater deregulation of the fishery.

Once the TAC is established, the number of registered pots (currently 11 900) is divided into the TAC to produce an allocation per pot. This allocation is currently 144kg/pot.

The tradable entity is therefore the pot (with its catch-allocation attached) and transfers occur freely in an open market. The market, however, is restricted to those who have a rock lobster licence to fish in this limited-entry fishery and hence it is not possible for people external to the fishery to purchase pots.

The fishery has been a limited-entry fishery for some 30 years, although licences to fish are freely transferable. The value of the access-right to participate in the fishery is related to the pot-entitlement that the licence has attached to it, although there is a maximum of 100 pots per licence. The value of the access-right has increased substantially as a result of this limited-entry, the profitability of the fishery and the certainty of the management structure.

In the Southern Zone a 15% reduction in pots (1984) and an industry-funded buy-back programme, which removed 41 licences (1987), have been implemented. The system of individual transferable quotas (ITQs) was introduced in October 1994. The current management arrangements include:

i.   total allowable catch of 1720t allocated at 144kgs/pot
ii.   pots limited to a total of 11 900
iii. limited-entry
iv.   legal minimum size of 98.5mm (carapace length)
v.   closed season from 1 May to 30 September
vi.   minimum mesh-diameter on pots of 50mm
vii. maximum of 100 pots per licence with 80 allowed to be worked and
viii. prohibition on taking berried (*i.e.* spawning) females.

The abalone fishery is also a limited-entry fishery and, like the rock lobster fishery, the transferable harvesting-right has acquired considerable value, with the most recent estimates being around $A4 million. However, no licences have changed hands for several years and hence the actual market value may be higher than this.

Quota-management has been in place since the late 1980s (the timing of the introduction of annual TAC system being slightly different for the three zones). A separate TAC is set for the two species of abalone   (*H. laevigata* and *H. rubra*) in the 3 management zones, ITQs established as equal shares of these TACs. The minor exception to this is that for one part of the Western Zone fishery where a combined quota (all species) of 600kg meat-weight per diver is allocated. As with the rock lobster fishery, the TAC is set annually by a joint industry/government management advisory body (the Abalone FMC), which provides advice to the Minister for Primary Industries. However, as noted above, the minister receives separate advice from the government fisheries department and the final decision on the TAC lies with the Minister for Primary Industries.

Some input-controls remain in the fishery, particularly a size-limit that is currently 130mm for blacklip abalone *(H. rubra)* in the Western and Central zones, and 125mm in the Southern zone. The size limit for greenlip abalone (*H. laevigata*) is 145mm in the Western Zone and 130mm in the Central and Southern Zones. There is also a restriction of 2 divers per licence with only one diver being able to operate on any one day.

In order to retain local ownership and control, no more than 15% of any licence may be held by a foreign citizen or Company.

Once the TAC is set, each licence is allocated an individual quota that is an equal proportion of the TAC for blacklip and greenlip abalone for that zone of the fishery. Quota is transferable between licence-holders only within each zone in a fishing period. The fishing period is from 1 September to the following 31 August in the Southern Zone, and 1 January to 31 December in the Western and Central Zones. Quota however, may not be permanently transferred.

The tradable entity in the abalone fishery is therefore the individual, divisible quota within a fishing period but the entire licence (with quota attached) can only be transferred on a permanent basis.

The harvesting-right in the pilchard fishery is, like the abalone and Southern Zone rock lobster fisheries, an equal share of the annually established TAC. This ITQ is allocated to vessels that are licensed to fish in the limited-entry pilchard fishery. Quota may be freely traded between the 14 licence holders only. Persons external to the fishery cannot purchase and use quota unless they have purchased a licence to operate in the fishery.

As indicated above, there are also some input-control measures in the pilchard fishery which principally relate to restrictions on areas fished.

3. MEASUREMENT OF FLEET-CAPACITY

3.1 Characterising fleet-capacity

In all the limited-entry fisheries in South Australia (including the three fisheries being considered here) the primary measure of fleet-capacity has been the number of vessels licensed to fish. In the rock lobster, abalone and pilchard fisheries, these numbers have remained unchanged since the introduction of limited-entry in those fisheries, which, in all cases, preceded the introduction of ITQs.

Apart from the primary measure of number of vessels, a number of other secondary measures of fleet-capacity are used in these fisheries.

In the rock lobster fishery, the main measure of fleet-capacity has been the pot or trap. Various restrictions have been in place over the history of the fishery to limit vessel-numbers and pots-per-vessel, *etc.*, but the primary aim of these measures was to limit the total number of pots in the fishery. The increased efficiency of pots in the fishery has, however, not been addressed although research in Western Australia (N. Hall, Fisheries Western Australia, pers. comm.) has indicated that technological improvements, such as improved echo-sounders have the capacity to increase pot-efficiency by up to 6% per annum. There are no ongoing programmes to address this issue in South Australia.

In the abalone fishery, the main measure of fleet-capacity has remained as the number of vessels or number of divers because the number of divers per vessel has remained fixed by regulation. Although there has been ongoing work on the use of the number of diver-hours as a measure of capacity, data are not collected on a regular basis and its use as a legitimate measure of capacity is still under some considerable discussion.

The pilchard fishery has used the measure of the number of vessels as a measure of fleet-capacity. There has been little change in the characteristics of the fleet over the past decade and hence this measure may be adequate. No ongoing programme for measuring changes in vessel characteristics or gear efficiency is in place.

3.2 Changes in fleet-capacity arising from the introduction of transferable property-rights

Boat numbers in the rock lobster fishery over time are an indicator of structural adjustment in the fishery (Table 1). It should be noted that for the first five years of quota management, transfers in the Southern Zone were only allowed within the fishery and this undoubtedly would have slowed the rate of adjustment.

Both Northern Zone (which is managed by input-controls) and Southern Zone also maintain upper pot-limits, which are an artificial impediment to free market adjustment in the respective fleets. The licence numbers over the last 10 years and average pots/licence are set out in Table 1 since 1992.

As Table 1 shows, licence numbers have declined by about 3.7% and 13.4% in the Southern and Northern Zones respectively over the past 10 years. Over the last five years the decline has been 2.1% and 9%, respectively.

By contrast, licence numbers in both the abalone fishery and the pilchard fishery have remained constant over the past 5 years although a recent reduction in pilchard licence numbers has occurred following a dispute over original entitlements to the fishery. This reduction was unrelated to any natural structural adjustment of the fishery.

With constant numbers of pots in each zone of the rock lobster fishery since 1992, the average pot-holding per licence has increased as licence numbers have declined.

Table 1
Licence numbers and average pot holdings in SA Rock Lobster Fisheries

Year	Southern Zone (ITQ managed)		Northern Zone (input managed)	
	Licence number	Av. pots/licence	Licence number	Av. pots/licence
1989	190		82	
1990	192		82	
1991	191		83	
1992	192	62.1	80	49.4
1993	189	63.1	79	50.0
1994*	187	63.8	78	50.6
1995	186	64.1	77	51.3
1996	186	64.1	75	52.7
1997	183	65.2	73	54.1
1998	183	65.2	71	55.6

* ITQs introduced in Southern Zone rock lobster fishery in October 1994.

The small reduction in the number of vessels in the ITQ-managed Southern Zone is a result of vessels leaving the industry, and their pot-entitlements being re-distributed among the remaining vessels in the fleet. Therefore as the slow reduction in vessel numbers has occurred, the average number of pots/vessel has increased. However, it should be noted that a much larger reduction in number of vessels has occurred in the Northern Zone of the fishery which is managed by input-controls.

The abalone fishery has not seen any change in fleet-capacity (measured as number of licences or vessels) since ITQs were introduced in the late 1980s. These remain at 23 for the Western Zone of the fishery, 6 in the Central Zone and 6 in the Southern Zone.

Similarly, the number of vessels operating in the pilchard fishery has not changed since the inception of the quota managed regime in the fishery, apart from an addition and then a recent reduction of 6 temporary licences which was unrelated to any structural adjustment in the fishery.

3.3 Consequences of changes in fleet-capacity

The consequences of the changes in fleet-capacity have been minimal in all fisheries because of the negligible changes in that fleet-capacity. However, the consequences of the move both to limited-entry and to ITQ-management have been profound. Often the effects of the two management measures cannot be separated.

The most significant change in all fisheries has been the rapid appreciation in the value of the access-right. This has been evident particularly in the rock lobster fishery and the abalone fishery, although it should be noted that the appreciation of licence-value in the limited-entry Northern Zone of the rock lobster fishery has matched, or even exceeded, that of the ITQ-managed Southern Zone. This seems to indicate that the primary cause of such escalation in asset-value has been the limited-entry nature of both fisheries rather than the ITQ-management of the Southern Zone part of the fishery.

Costs of management of the fisheries have also been documented in recent years. Over the period 1997/98-1999/2000 (the only years for which data are available), the total costs of managing the rock lobster fishery for research, compliance, management and industry development have fallen. In the Southern Zone they have declined from $A2 372 000 to $A1 955 000 in aggregate or from $A1408/t to $A1137/t. In the Northern Zone the costs have declined from $A988 000 to $A706 000 or $A1049/t to $A784/t. A summary of management costs for the rock lobster fishery is shown in Table 2.

From Table 2, it is apparent that costs of management of the ITQ-managed Southern Zone portion of the fishery are more expensive than the input-managed Northern Zone. Although the cost per tonne of management, compliance, research and development, does vary over the time, it is approximately $A350 per tonne higher in the ITQ-managed Southern Zone quota-fishery than the input-managed Northern Zone fishery for 1999-2000.

These higher management costs (which are paid by industry under a cost-recovery process) may explain the slightly higher asset value of the access right to the Northern Zone fishery than to the ITQ-managed Southern Zone fishery.

Compliance is the key component impacting on licence fees: in the Southern Zone, the cost of compliance has increased over the three-year period from $A914 000 to $A1 011 000 or from $A542/t up to $A588/t. This has happened in the face of declining total licence-fees.

In the Northern Zone, the compliance cost has fallen from $A278 000 to $A236 000 or from $A295/t to $262/t. According to the budgeted fees for 1999-2000, the difference in compliance costs accounts for $A326/t of the total difference of $A353/t between the two fisheries. It should be noted that the industry is currently considering restructured compliance costs in the Southern Zone.

In 1999/2000 the cost of compliance per tonne of lobster caught was approximately twice as high in the Southern Zone than in the Northern Zone (Table 2).

Another cost associated with the ITQ management system of Southern Zone is the need for greater monitoring and management-related costs by the Industry/Government Fisheries Management Committees (FMCs), which administer the management of both Northern and Southern Zone.

Since the time of the introduction and management of the quota system in 1994, the Southern Zone Rock Lobster FMC has met on 53 occasions. The Northern Zone FMC has met on 27 occasions. The budget for the Southern Zone for the financial year 1999-2000 was $A70 000, whereas the Northern Zone budget was $A45 000.

Costs of management of the abalone fishery for the period 1996/97-1999/2000 is shown in Table 3 from which it is apparent that management-costs are substantial although they have been reduced since 1997/98 through an active process of increased efficiency of service-delivery by government service-providers in research, compliance and management. Compliance costs remain the largest single cost of management of the fishery. Management costs represented 10.4% of the gross value of production in 1999/2000.

Table 2
Cost of management in the South Australian rock lobster fisheries, 1997/98 – 1999/2000

	1997-98	**1998-99**	**1999-2000**
Total licence fees			
SZ Management cost ($A)	2 372 000	2 105 000	1 955 565
SZ Catch (tonnes)	1 685	1 714	1 720
SZ Management costs/tonne ($A)	1 408	1 228	1 137
NZ Management costs ($A)	988 000	807 000	706 000
NZ Catch (tonnes)	942	1 016	900
NZ Management cost/tonne ($A)	1 049	794	784
Total management cost diff. SZ – NZ ($A/t)	359	434	353
Compliance costs			
SZ Compliance costs ($A)	914 000	974 000	1 011 000
SZ Compliance cost/tonne ($A)	542	568	588
NZ Compliance costs ($A)	278 000	280 000	236 000
NZ Compliance cost/tonne ($A)	295	276	262
Compliance cost difference: SZ – NZ ($A/t)	247	293	326

Table 3
Costs of management of the South Australian abalone fishery, 1996/97-1999/2000

Year	No. of licence-holders	Total cost of management ($A million)	Costs/licence-holder ($A)
1996/97	35	2.217	63 339
1997/98	35	2.608	74 519
1998/99	35	1.890	53 993
1999/2000	35	1.781	50 896

Costs of management of the pilchard fishery are not available.

All quota-managed fisheries in South Australia are highly profitable, with returns on capital ranging from 10.1% for the abalone fishery to 4.4% for the Southern Zone rock lobster fishery. This rate of return includes the return to the value of not only the vessel and gear, but also the licence value that is considerable, and is the major component of capital employed. For example, the imputed value of an abalone licence in 1998/99 was $A4.02 million (although none have changed hands for several years) compared with $A0.12 million for the value of the fishing vessel and equipment. The total capital employed in this fishery was therefore $A4.15 million.

There is little difference in return to capital invested between the two zones within the rock lobster fishery. A rate of return of 4.5% was estimated in the Northern Zone and 4.4% in Southern Zone in 1997/98.

Another economic indicator that may vary with differences in management of the fishery is the economic-impact that the fishery has on the regional economy in which it is located. The economic-indicator reports (EconSearch 1999a, b) for the Northern and Southern Zone rock lobster fisheries suggest, however, that there is very little difference between the two fisheries The impacts, measured in terms of employment, household-income, business-turnover and value-added, per tonne of lobster are generally greater in the Southern Zone but not significantly so (Table 4).

4. CONCENTRATION OF OWNERSHIP
4.1 Status prior to programme

Table 1 shows the number of vessels and pots in the Southern Zone rock lobster fishery both prior to, and after, the introduction of ITQs in 1994. While the overall number of vessels has declined somewhat (not necessarily and solely in response to the ITQ-management regime), the number of pots in the fishery has remained fixed in the fishery since the introduction of ITQs. This has resulted in a small increase in the number of pots per vessel.

Table 4
Economic impacts of South Australian commercial fisheries, 1997/98

	Southern Zone (Output managed)	Northern Zone (Input managed)
Turnover		
Fishing (direct) ($Am)	50.9	27.7
All other sectors (indirect) ($Am)	99.5	53.0
Total ($m)	**150.4**	**80.7**
Total/direct	3.0	2.9
Total/tonne ($A)	90 000	86 000
Value Added		
Fishing (direct) ($Am)	34.7	19.2
All other sectors (indirect) ($Am)	50.1	26.6
Total ($Am)	**84.8**	**45.7**
Total/direct	2.4	2.4
Total/tonne ($)	50 000	49 000
Employment		
Fishing (direct) (jobs)	710	312
All other sectors (indirect) (jobs)	780	418
Total (jobs)	**1 490**	**730**
Total/direct	2.1	2.3
Total/tonne (jobs)	0.89	0.77
Household Income		
Fishing (direct) ($Am)	20.0	9.6
All other sectors (indirect) ($Am)	21.8	11.5
Total ($Am)	**41.8**	**21.1**
Total/direct	2.1	2.2
Total/tonne ($A)	25 000	22 000

Source: EconSearch 1999a, b.

With the high profitability of the fishery, a number of licence-holders have acquired additional licences, often through company or other structures where the eventual beneficial owner is sometimes obscure. Some processors have also been active in seeking to acquire a portfolio of vessels to ensure supply. However, anecdotal evidence suggests that this is not currently a major issue and the maximum number of licences with beneficial ownership to a single person is probably no more than four or five.

More importantly, this trend is also evident in the Northern Zone of the fishery, which is managed by input-controls, and hence the concentration of ownership is probably unrelated to the introduction of ITQs, but relates more to the limited-entry nature of the fishery and its growth in profitability over a number of years.

A policy of owner-operator remains in place in South Australia: this has discouraged concentration of ownership in all fisheries. However, this policy has been applied in a haphazard way over several decades and, together with the more complex corporate structures being used for ownership, has not been a total impediment to the concentration of ownership of harvesting-rights.

Prior to the introduction of ITQs in the abalone fishery in the late 1980s, ownership of the licence to participate in the fishery was to a single operator as a result of the owner-operator policy. Again, like the rock lobster fishery, anecdotal evidence suggests that has been some small concentration of ownership but the prohibition on the permanent transfer of quota has significantly hindered this process.

Perhaps the most important impediment in the concentration of ownership has been the social issue of a small number of licensees living in coastal villages, and the community attitudes against such concentration, particularly from investors or fishermen from outside the region. These social restrictions have also been identified (Econosearch, 1999) as the most probable reason for the lack of re-structuring of the Southern Zone rock lobster fishery to benefit from the theoretical advantages of ITQ management.

4.2   Restrictions on transfer of ownership

A number of restrictions on transfer of ownership are currently in place in the ITQ-managed fisheries. No more than 15% of any licence may be held by a foreign citizen or company and, in the abalone fishery, permanent transfers are prohibited.

The most important restriction on transfer for all the fisheries is that transfers can only occur between licence-holders that have an access-right to these limited-entry fisheries. Because most of the fisheries are small, this has resulted in a small market for tradable access-rights unless the full licence for access to the fishery is also purchased.

4.3 Prices received

Data has been collected on prices received for abalone and rock lobsters for a number of years. The data for rock lobsters are shown in Table 5 separated in to Northern Zone and the ITQ-managed Southern Zone.

Table 5
SA rock lobster catch and value of catch, 1990/91 - 1998/99

Year	S. Zone Catch (t)	S. Zone value ($A million)	S. Zone average price ($A/kg)	N. Zone catch (t)	N. Zone value ($A million)	N. Zone average price ($A/kg)
1990/91	1 562	26.7	$17.09	1 104	18.2	$16.48
1991/92	1 940	36.3	$18.71	1 222	21.4	$17.51
1992/93	1 754	34.8	$19.84	1 064	20.5	$19.27
1993/94	1 669	43.2	$25.88	930	23.4	$25.16
1994/95*	1 720	48.6	$28.26	891	25.5	$28.62
1995/96	1 684	44.6	$27.32	903	23.8	$26.36
1996/97	1 635	47.0	$28.75	893	24.4	$27.32
1997/98	1 680	50.9	$30.30	942	27.7	$29.40
1998/99p	1 713	47.2	$27.55	1 016	26.7	$26.28

* ITQ management introduced to the Southern Zone fishery

From the Table, there is no discernible change in prices received as a result of the introduction of ITQs in the Southern Zone of the fishery in 1994. Rather, both zones have experienced a steady increase in price for the past decade.

For abalone, average prices received since 1990/91 have also increased from $A16.22/kg in 1990/91 to $A33.90 in 1994/95 to $A29.15 in 1998/99. Prices received have been influenced to some extent by the development of the abalone aquaculture industry though, because the aquaculture industry produces a different size product, this competition for markets has been minimal.

Both the abalone industry and the rock lobster fishery are export-orientated fisheries and hence the prices received are influenced by currency exchange-rates and international demand. The introduction of ITQs therefore seems to have little impact on prices. The quality of the product produced has been maintained at a high level (to meet export standards) and hence price increases as a result of quality improvements have not been noticeable.

No price data is available for the pilchard fishery.

4.4 Effectiveness of regulations governing ownership of rights

A significant cost of management in all ITQ-managed fisheries in South Australia is the cost of compliance. Hence the effectiveness of regulations has been high from a compliance point of view. However, as noted above, the policy of owner-operator has been haphazardly applied and new corporate-ownership structures have allowed concentration of ownership without significant impediment. The most important impediment to such concentration seems to be both the limited entry nature of the fisheries (hence resulting in a very small market for trading of access-rights) and the social issues whereby attitudes in small coastal communities influence individual decisions to concentrate ownership.

5. DISCUSSION
5.1 Reduction in fleet-capacity

In no case in the ITQ-managed fisheries of South Australia was fleet-capacity reduced. Rather, the policy objective was to stabilize fleet-capacity and, using the measures of fleet-capacity available, this has been achieved (see Tables 1 and 2). However, a major concern is that the measures of fleet-capacity being used do not take into account efficiency-increases brought about by technological innovation, and no data collection programmes are in place to collect such information. Studies from elsewhere suggest that this may become an important issue. It is therefore likely that real capacity has increased by an unknown amount in all the ITQ fisheries of South Australia.

All of the ITQ-managed fisheries in South Australia retain significant input-controls and these are often more effective in managing capacity and fishing-effort than the ITQ-regime alone. However, the retention of these input-controls has also meant that the beneficial effects of industry restructuring, which should be available under an ITQ-management system, have not occurred. As an example, significant improvements in return on capital could be had in the rock lobster fishery, by taking quota at different times of the year than is currently the case. However, the input-control of closed seasons (promoted as much by industry as by Government) remain as an impenetrable impediment to such benefits.

A comparison of ITQ-managed and input-control managed segments of the rock lobster fishery have failed to demonstrate any convincing differences in fleet-capacity or economic benefits as a result of the move to ITQs. However, as noted above, this is almost certainly a result of the significant input-controls that remain and the impediment that these have in realizing the benefits of ITQ management.

5.2   Concentration of ownership

The limited-entry nature of the fisheries and the restrictions applying to change of ownership of access-rights has meant that concentration of ownership has been minimal in all fisheries. In fact, the opposite has occurred in the abalone fishery where change of ownership of access-rights (licences) are almost unknown, the last (half share) changing hands in 1996.

The importance of social issues, whereby small coastal community attitudes influence individual decisions to concentrate ownership, has not been well recognized or studied in South Australian fisheries. These issues have emerged as major influences of the effects of ITQ-management in South Australia.

6.   LITERATURE CITED

EconSearch 1999a. Economic Indicators for the SA Southern Zone Rock Lobster Fishery 1997/98. Report prepared for Primary Industries and Resources South Australia, February 1999.

EconSearch 1999b. Economic Indicators for the SA Northern Zone Rock Lobster Fishery 1997/98. Report prepared for Primary Industries and Resources South Australia, February 1999.

Morgan, G.R. 2001. Initial allocation of harvesting in the Rock Lobster fishery of Western Australia. 152-158. *In:* Shotton, R. (Ed.) 2001. Case studies on the allocation of Transferable Quota Rights in fisheries. FAO Fish. Tech. Pap. No. 411. FAO, Rome.

CHANGES IN FLEET CAPACITY AND OWNERSHIP OF HARVESTING RIGHTS IN AUSTRALIA'S NORTHERN PRAWN FISHERY

A.E. Jarrett
Pro-Fish Pty. Ltd.
8 Harwood Close, Cairns QLD 4870, Australia
<annie.jarrett@internetnorth.com.au>

1. INTRODUCTION

Australia's Northern Prawn Fishery (NPF) covers approximately 800 000km^2 of ocean with waters extending to the edge of the Australian Fishing Zone (AFZ)[1]. The NPF was established by the late 1960s when the Federal Government called for expressions of interest for fishing companies to establish shore-based processing factories at strategic points between Cape York and Darwin in Northern Territory, and to operate fleets in the fishery as part of its policy to develop Northern Australia.

Four joint Australian/Japanese ventures and three Australian-owned companies were granted permits for the fishery. Joint venture companies were allowed initially to operate foreign trawlers on the condition that they were to be phased out and replaced with Australian-built trawlers. By 1970, 228 Australian-owned and 24 foreign-owned trawlers were operating in the NPF and processing plants had been established at Darwin, Groote Eylandt, Katherine and Karumba. At this time the majority of the catch was made up of banana prawns (*Penaeus merguiensis* and *Penaeus indicus*). The annual gross value of production of the NPF ranges between $A100 million and $A150 million.

The Australian trawlers were 'wet' boats, which held product on-board on ice or in refrigerated tanks, and returned to port once or twice a week to unload at shore-based processing establishments. Joint venture companies employed foreign refrigerated mother ships to handle catches from trawlers. Some were carrier ships and transported product to shore bases for processing; others processed prawns on board for export.

The fishery expanded beyond the Southern Gulf of Carpentaria during the 1960s with the first banana prawns caught in Joseph Bonaparte Gulf off the western coast of

A typical trawler in the Northern Prawn Fishery

northern Australia in 1967. While there is no formal documentation of the number of boats fishing in the NPF in the mid to late 1960s, records exist which indicate that at least 65 boats landed product in 1968. Between 1968 and 1970 receivals of prawns from the NPF by Australian processing establishments doubled to 3500t.

During the 1960s, fishing gear was limited to a single otter-trawl net towed off the stern, but by the early 1970s most vessels were towing double rigs, known as twin gear, with one net towed from a boom on each side of the vessel, which dramatically improved prawn catches.

The remote location of the fishery meant that larger, purpose-built trawlers with considerable freezer capacity were needed to replace the existing wet boats (those that held prawns in ice or chilled brine only) if the fishery were to expand. The first Australian freezer vessels entered the fishery in 1970 and by the end of the 1970s more than 120 new freezer vessels had been built. These vessels, ranged in length from about 20 to 30m and were the most technologically advanced trawlers of the time, having the latest electronic fish-finding

---

[1] The 200 mile Exclusive Economic Zone around Australia and its external Territories.

equipment, lazy-line winches[2], snap freezers, prawn-sorting conveyors, Kort nozzles and controlled-pitch propellers. By the mid 1980s no wet boats remained in the fishery.

The arrival of the freezer boats caused the fishery to expand and tiger prawns (*Penaeus esculentus* and *Penaeus semisulcatus*) became an increasingly important part of the catch. The freezer boats spent most of the year committed to the Northern Prawn Fishery. Prior to the introduction of freezer boats in the fishery, vessels spent only 11% of their available fishing year in the NPF.

Interest in the fishery escalated in 1974 when, as a result of huge monsoonal rains, in excess of 12 500t of banana prawns were caught. The exceptional season, coupled with the 'open door' policy (no limit on the number of licences) of the government of the day, plus increasing industry interest, resulted in vessel numbers and fishing effort rising rapidly over the next couple of years. In the 1980s all Australian/Japanese joint ventures ceased operations in the NPF and many Australian-owned companies were operating fleets of up to 14 trawlers. At this time the fishery was an open-access fishery with no input- or output-controls in place and large catches of both banana and tiger prawns were being recorded.

Limited-entry was introduced into the NPF in 1977 through a three-year moratorium on licences and new trawlers entering the fishery. This was part of an interim management plan when the first formal advisory committee NORPAC (Northern Prawn Advisory Council, made up of industry, scientists and managers) was formed in 1977 to provide advice to government on management of the Northern Prawn Fishery. Up until 1988, when jurisdiction for management of the NPF passed to the Federal Government, the fishery was jointly managed by the Federal, Queensland, Western Australian and the Northern Territory governments.

The 1977 moratorium was introduced as a result of the concerns of industry members of NORPAC that the lack of restrictions on trawler numbers and the resultant increasing fishing effort were having a negative impact on prawn stocks. The effect of the limited-entry policy was to provide the participants in the fishery with exclusive commercial access to the NPF. This was the first step in the establishment of property-rights in the NPF. Licences issued under the limited-entry policy remained the only access-rights in the fishery until 1984.

2.   THE NATURE OF THE HARVESTING-RIGHT

Despite the agreed moratorium on new vessels accessing the fishery, liberal entry-criteria resulted in a total of 292 licences being issued, compared with the 145 trawlers which had fished the year before. The limited-entry policy also failed to control fishing-effort as licences were transferable and a liberal boat-replacement policy allowed licences attached to small trawlers to be transferred to larger purpose-built trawlers committed to the NPF, thus resulting in considerable expansion of fishing-capacity.

The expanding effort and capacity in the fishery was further exacerbated by a decision taken in 1980 to allow boats below the size that attracted a ship-building subsidy to be replaced with 'subsidy size' trawlers[3] with no penalty. Trawlers over the 'subsidy size' could be replaced on a one-for-one basis. This decision was taken as a concession to small-boat owners whose trawlers had not qualified for the ship-building subsidy. This proved to be retrograde step as it resulted in wide-scale boat-replacement and substantial increases in vessel size and capacity to the detriment of the biological and economic sustainability of the fishery. Between July 1977 and June 1982, 117 trawlers were built under the ship-building subsidy scheme.

As a result of the failure of the limited-entry policy to control the expansion of fishing capacity and effort, an industry/government working group was established in 1981 to review management arrangements in the NPF. The working group considered a number of options for addressing the over-exploitation of prawn stocks and over-capacity in the fishery. These included, but were not limited to: the proposal to measure and control the fishing-capacity of each trawler by a unitisation scheme based on hull size and main engine horsepower; the possible introduction of a buy-back scheme; a more stringent boat-replacement policy; and area-closures to protect prawn stocks.

Following two years of discussion and negotiation on the proposals, the NPF 'unitisation' scheme was introduced in 1984 with each trawler being allocated a B-class unit (the right to fish in the NPF) and a number of A-class units (a rating based on the underdeck tonnage and main engine horsepower of each trawler). It was through this scheme that individual transferable harvesting-rights were allocated in the fishery, however the units were not incorporated into a management plan until the first NPF Management Plan was legislated in 1985. Additional units, known as C-class units were granted in 1986. While these units also represented the right to

---

[2] A lazy line is the rope that is wrapped around the top of the cod-end and then wrapped around the capstan-head of either an electric- or hydraulic-winch, as a means of bringing the net on board when the shot is finished.

[3] A direct government subsidy of approximately 20% of the construction cost of the trawler paid to the ship-builder, which lowered the cost to the buyer and allowed vessels to be upgraded to 150 gross construction tons and approximately 23m in length.

fish in the NPF, vessels to which C-class units were allocated were restricted to fishing only in the Joseph Bonaparte Gulf (see Figure 1).

Figure 1
Australia's Northern Prawn Fishery

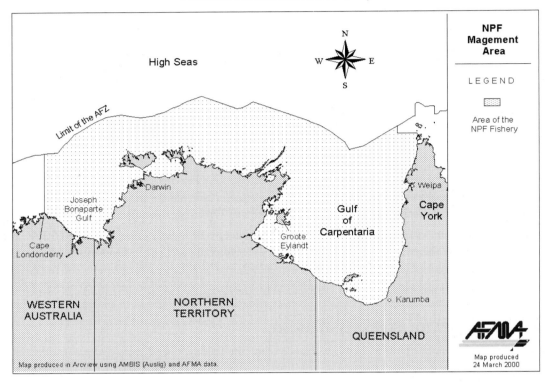

As a result of agreement by industry at the time that the minimum size of trawler to qualify for the ship-building subsidy was 375 A-class units, all trawlers that measured less than 375 A-class units were allocated a minimum of 375 A-class units. This concession was provided on equity grounds because of the number of operators who had already taken advantage of the ship-building subsidy to significantly upgrade to boats of 'subsidy size'. The concessional units (the difference between the number of A-class units determined by the hull size and engine horsepower rating of each small trawler, and the minimum allocation of 375 units) were called "suspense" A-class units. The total number of B-class units issued was determined by the number of licences which had been issued as at 1984.

Transferability of A-class units allowed fishers to use any size boat they chose, provided that each boat had a B-class or C-class unit attached to it. The requisite number of A-class units was attached to each boat corresponding to its size and the existing total pool of A-class units did not increase. NPF units quickly achieved a market value and financial institutions began accepting the units as collateral against loans. This resulted in units being accepted as the 'currency' and the property-rights in the fishery. While all units could be transferred by sale there was no provision for leasing of units under the *Fisheries Act 1952* in place at the time.

The Australian Fisheries Service (AFS), the federal government department charged with managing the NPF, was responsible for issuing licences and maintaining ownership records. This was confined to maintaining records of licence- and vessel-ownership prior to the introduction of the unitisation system. From 1985 to 1994 the provisions of the NPF Management Plans in effect at the time required the AFS to keep a formal register known as the Central Boat Unit Register (CBUR).

The Central Boat Unit Register recorded:

i.   the name and address of every unit-holder;

ii.   the number and type of A-class units available to be allocated, or that had been allocated, to each unit-holder

iii. the serial number of each B-class or C-class unit allocated to each unit-holder in relation to each boat to which units were assigned:
      (a) the name and distinguishing number (registration) of the boat
      (b) the dimensions specified for the purpose of measuring the vessel for unitisation

    (c) the number of hull units,

iv.   the type of engine installed in the boat

v.   the number of engine-power units

vi.   the applicable number of A-class units

vii.   the serial numbers of the B-class and C-class units and

viii.   any other fishing entitlements.

Following the establishment in 1991 of the Australian Fisheries Management Authority (AFMA) and its enabling legislation (the *Fisheries Management Act 1991)*, a new Management Plan for the NPF came into effect in 1995 under to replace the previous Plan which had been in place under the *Fisheries Act 1952*. The fishing rights in effect under the *Fisheries Act 1952* through the NPF Management Plan (being the A- and B-class units) were automatically rolled over into the new Plan, and became A- and B-class Statutory Fishing Rights (SFRs). This meant each operator was given the equivalent number of A- and B-class rights under the NPF Management Plan 1995 as they had held under the previous Plan. All C-class units had been cancelled by 1995.

The Statutory Fishing Rights established under the *Fisheries Management Act 1991* are considered to have more security than the rights which existed under the *Fisheries Act 1952*. The rights established under the *1991 Act* are a form of property-rights established by statute, have a life of at least ten years, are automatically 'rolled over' if the NPF management plan is revoked and a new plan is implemented in the fishery, and compensation may be payable by the federal government if the rights are acquired on unjust terms. It is worth noting that the rights issued under the *1952 Act* were issued for a period of 12 months, were renewable at the discretion of the government/Minister, had no on-going tenure and there was no provision for compensation.

The grant of Statutory Fishing Rights based on A- and B-class units under the *Fisheries Management Act 1991* has firmly entrenched the fishing-rights allocated under the NPF unitisation scheme as the 'currency' and property-rights in the NPF. This is accepted and recognised by industry, AFMA, other stakeholders and most importantly, by financial institutions, which accept NPF units as security against collateral for loans.

The *Fisheries Management Act 1991* requires the AFMA to maintain a formal register of Statutory Fishing Rights on which is recorded:

i.   the name of the person to whom the fishing right is granted

ii.   a description of the fishing right

iii.   the period (if any) for which the fishing right is granted

iv.   the managed fishery in relation to which the fishing right is granted

v.   the conditions (if any) of the fishing right and

vi.   such other particulars (if any) as are prescribed.

The current Federal government 'user pays' policy requires industry to pay 100% of the attributable management costs in Commonwealth fisheries, which includes the cost of maintaining the Statutory Fishing Rights register. Management costs are recouped though industry levies based on the number of A-class fishing rights held.

The *Fisheries Management Act 1991* also provides for additional interests to be registered. This may be relevant where a financial institution or a loan guarantor requires their interest in the SFRs to be registered. As the *1991 Act* also makes provision for the leasing of SFRs, the interests of both lessors and lessees can be included in the register. Where the AFMA has established that an instrument (document) exists verifying the existence of the interest, the AFMA must, on application, register any claim of interest by entering in the register the names of the persons and the particulars of the interest involved.

A- and B-class Statutory Fishing Rights are easily transferable, by either sale or lease upon application to AFMA. Provided there is no impediment the transfers are effected on application and are recorded on the Statutory Fishing Rights Register. Transfers of NPF Statutory Fishing Rights may be subject to other State or Federal government imposts such as capital gains tax or stamp duty, however these imposts are not regulated by the AFMA.

3.   MEASUREMENT OF FLEET-CAPACITY
3.1   Characterizing fleet-capacity

The primary purpose of the unitisation system was to measure and control fishing capacity in the fishery. The unitisation system resulted in each trawler licensed for the NPF being allocated a B- or C-class unit representing the trawler's right to access the fishery, and a quantum of A-class units based on a combination of hull size and engine power determined through an agreed formula which was later incorporated in the NPF Management Plan 1985 and subsequent Plans:

Number of A-class units = Length (m) x Breadth (m) x Depth (m) x 0.2120141 HP

where: the linear dimensions are measured in metres.

In total 133 269 A-class units, 292 B-class units and 10 C-class units were granted under the unitisation scheme. All A-class and B-class units were, and remain the "currency" in the fishery today.

Provided that the total number of A-class units in the fishery was not exceeded, individual trawlers were not restricted in size and vessel replacement was encouraged as a result of the introduction of the Government's ship-building subsidy[4] which was designed to maintain a viable Australian ship-building industry and to assist Australian ship-builders to compete internationally. The average age of the fishing vessels when unitisation was introduced was approximately seven years.

The majority of operators at the time were towing four otter-trawl nets, commonly referred to as 'quad gear', made up of two nets towed on either side of a 'sled' or 'skid' from a boom on each side of the boat. However some operators were still towing two or three nets. Many operators were also towing a 'try' net, which was a small net (having a headline length of about 2-3 fathoms) used as a sampling device to help them find, and then remain on, locations of high prawn-density without losing valuable fishing-time by shooting and hauling-up their main gear.

3.2   Changes in fleet-capacity arising from the introduction of transferable property-rights

The NPF fleet has undergone major decreases in fishing-capacity since unitisation was introduced. This has been reflected through a significant reduction in A-, B- and C-class units. As the total number of units has been restricted through legislation since the first NPF Management Plan was introduced in 1985, no increase has, or could, occur in the total numbers of these rights. This reduction in fleet-capacity can be contributed to a number of factors including:

i.   stringent restrictions requiring licence- and unit-forfeiture upon boat-replacement
ii.   the existence of the NPF Voluntary Adjustment Scheme (*i.e.* a buy-back scheme) and
iii.   the NPF restructuring scheme in place from 1990 which culminated in the compulsory surrender of 30.76% of all A-class units in April 1993.

A 'two-for-one' boat-replacement policy was implemented in 1987 and remained in place until the end of 1992. This policy required any owner who voluntarily replaced a trawler of any size, to surrender one extra B-class or C-class unit. For trawlers rated at over-375 units, an equal number of A-class units to the total A-class unit-holdings of the replacement, had to be surrendered in addition to the surrender of one B-class or C-class unit. The two-for-one boat replacement policy resulted in 11 B-class units and 7 C-class units being forfeited. There is no identifiable record of the number of A-class units which were forfeited under the two-for-one boat replacement policy, but it is believed to be several thousand.

A buy-back scheme known as the Voluntary Adjustment Scheme (VAS) was implemented in 1985 to remove the excess capacity/units in response to concerns about the viability of the resource. Under this scheme, units were voluntarily sold to the VAS, and then cancelled (so they could not re-enter the fishery). The operation of the VAS was the responsibility of the Northern Prawn Fishery Adjustment Scheme Committee (VAS Committee).

The scheme was initially funded by a $A3 million Government grant and levies paid by unit-holders in the fishery. At this time, a 'user pays' policy was imposed by the Commonwealth government, requiring industry to pay levies for management of the fishery. These NPF levies were calculated based on the number of A-class units held, although initially vessels rated under-375 units paid levy at half the rate of those vessels over-375 units. The implementation of the buy-back scheme required industry levies to be increased substantially to repay the VAS loan.

From the commencement of the VAS scheme in 1985 to September 1986, a total of 12 380 A-class units were purchased by the VAS. A total of 25 B-class units were also surrendered during this time but they had no value and did not attract a premium price under the scheme. The price paid (under VAS) for A-class units was initially $A100 per unit, but by the end of 1986 this had risen to approximately $A400 per unit. The impact of the VAS on unit-prices is further explored in Section 4.3.

---

[4] From 1980, boats below the size that attracted a ship-building subsidy could be replaced with minimum 'subsidy size' vessels (150 gross construction tons and about 23m length) with no penalty. Vessels over the 'subsidy size' were replaced on a one-for-one basis.

By 1986 a serious decline in brown tiger prawn stocks in the western Gulf of Carpentaria was evident, indicating recruitment over-fishing. Scientists from CSIRO recommended a reduction in fishing effort of 25% to overcome this over-fishing. Falling export prices for prawns further exacerbated the economic problems resulting from poor catches. To address these issues it was agreed in 1987 to reduce the number of A-class units in the fishery to 70 000 by 1990 through the buy-back scheme.

Operators were encouraged to remove trawlers from the fishery: a premium was paid by the VAS if B-class or C-class units were forfeited and cancelled as part of the unit package sold to the buy-back scheme. To facilitate the redeployment of trawlers leaving the NPF, an industry-registered company was established (the NPF Trading Corporation) whose role was to assist the VAS Committee in buying units, and assist with the re-sale of trawlers withdrawn from the fishery by identifying potential markets for those trawlers. By the end of 1989, 45 B-class units, 2 C-class units and 20 810 A-class units had been sold to the buy-back scheme.

The reduction-target of 70 000 A-class units set by NORMAC in 1987 was not reached by 1990, and stock-depletion and low vessel-profitability were ongoing problems. In 1990, NORMAC (the management advisory committee which replaced NORPAC) agreed to a further reduction in A-class units with a target of 50 000 A-class units by the beginning of the 1993 prawn season. This was to be achieved by an accelerated buy-back scheme funded through by a $A5.0 million grant and a $A40.0 million government-guaranteed loan to be repaid by industry. In return for providing the loan guarantee the government required a guaranteed outcome that the accelerated restructuring would achieve its target. It was agreed that if the target of 50 000 A-class units was not reached through the buy-back scheme, a compulsory surrender and cancellation of A-class units would take place in April 1993 to reduce the fishery to 50 000 A-class units. The target of 50 000 A-class units was subsequently revised to 53 844 units following agreement by industry that a concession be provided for vessels under-375 A-class units which exempted 'suspense units' from compulsory surrender.

In the accelerated restructuring programme, A-class units were purchased by the VAS at $A450 per unit. However, to encourage operators to withdraw trawlers from the fishery, an additional $A500 per A-class unit was paid, provided that the B-class unit to which the A-class units were attached was surrendered to the VAS and subsequently cancelled. A total of 27 863 A-class units and 44 B-class units were purchased through the buy-back scheme between 1990 and 1992.

At the end of 1992, the target for reduction to 53 844 A-class units was not reached, and on 1 April 1993, 30.76 % of the remaining Active A-class units were compulsorily surrendered and cancelled in accordance with the surrender provisions contained in the NPF Management Plan. The compulsory surrender of units resulted in a reduction in the number of active fishing vessels from 171 in 1992, to 124 in 1993, although 137 B-class units remained in the fishery in 1993.

By regulation, a trawler is not able to fish unless the requisite number of A-class units are attached to the vessel at all times, thus concession-holders who did not have enough units to meet their surrender obligations withdrew from the fishery, selling or leasing their residual units to others remaining in the fishery. Most concession-holders who remained in the fishery either bought or leased units to cover their surrender provisions from those who were leaving the fishery. However some concession-holders replaced their engines with lower horsepower-rated ones in order to free up A-class units to cover their surrender requirements and remain in the fishery. Some company operators reduced their fleet size and used the units previously allocated to vessels which were removed from the fishery, to cover their surrender provisions. Table 1 indicates the reductions in the size of the NPF fleet since unitisation was introduced.

Since the conclusion of the NPF restructuring programme in April 1993 there have been minimal changes in the size and overall capacity of the NPF fleet up to 1999. There are currently 53 844 A-class fishing-rights and 132 B-class fishing-rights in the

Table 1

Changes in the NPF fleet size (numbers of A-, B- and C-class units) (1985 – 2000)

Relevant information before 1985 is unavailable.
The data presented are only indicative averages.

Year	A-class units	B-class/C-class
1985	128 000	292
1986	120 000	267
1987	117 000	264
1988	110 000	240
1989	104 550	218
1990	96 000	212
1991	77 300	174
1992	68 800	174
1993	53 800	137
1994	53 800	137
1995	53 800	136
1996	53 800	134
1997	53 800	132
1998	53 800	132
1999	53 800	132
2000	53 800	132

NPF. However, there has been considerable redistribution of fishing-capacity within the fleet. The existence of individual harvesting rights has allowed operators to buy, sell and lease units from within the existing pool on an 'as needs' basis so as to maximise their individual catching-capacity. By transferring the rights between themselves, operators have been able to upgrade their vessels and engines, or vary their fleet structures to optimise their individual fishing operations. The changes within the fleet structure are further explored under Section 4.

The lifting of the boat-replacement policy restrictions (no forfeiture of additional units or licences) at the end of 1992 allowed operators to build new trawlers without penalty, provided that they held the requisite number of A-class fishing rights required by the formula in the NPF Management Plan. This resulted in the replacement of a number of older trawlers with new trawlers. With innovations in vessel- and engine-designs, technological changes and clever manipulation of the unitisation rules, trawlers built in recent years have more fishing power than most of the older designed and constructed trawlers in the NPF, and yet are rated as having considerably less A-class units (fewer hull- and engine-units).

While the overall fishing-capacity of the fleet (numbers of A- and B-class fishing-rights) has not increased, the effective fishing effort (catching capacity) of the fleet has increased substantially. This is a common occurrence in input-controlled fisheries where new technology, more efficient fishing gear and increased knowledge of fishers contributes to the effective fishing-effort actually exerted, a factor known as 'effort creep'. Increases in effective fishing-effort, and the flaws in the existing unitisation system, have resulted in agreement to move from the existing unitisation system to a gear-based system of management in order to provide a better measure of fishing-effort and a more flexible system when further reductions are required to balance effort and sustainability. The new system was due for implementation in the NPF in July 2000 and is expected to result in further reductions in fleet-capacity as vessels are withdrawn from the fishery to enable operators to transfer gear to the remaining vessels[5].

3.3 Consequences of changes in fleet-capacity

The reduction in the size of the NPF fleet since 1985 has resulted in both social and economic changes. There has been a significant reduction in the numbers of large company operations since unitisation was introduced in the NPF. The low catches and the poor economic climate which prevailed during the 1980s caused many large companies to withdraw from the NPF. Many of them sold their licences to the VAS. The reduction in the numbers of private operators in the NPF is disproportionately fewer than the reduction in company operations (see Section 4).

Many of the large companies operating in the NPF were also responsible for establishing the shore-side support base for the fishery, including unloading facilities and processing plant built in the late 1960s and early 1970s in Darwin, Groote Eyelandt, Karumba, Normanton, Cairns and Townsville (Figure 1). The withdrawal of these company operations, coupled with the overall reduction in trawlers operating in the NPF, and the move to processing- and value-adding activities on-board trawlers (rather than on-shore) has brought about

A Northern Prawn trawler goes bananas *(Fenneropenaeus spp.)*

the financial demise and subsequent closure of the majority of the shore-based facilities. This had significant impacts in terms of loss of employment and revenue to the regional economies.

Over time the reduction in the size of the NPF fleet was accompanied by a significant reduction in the number of crew engaged in the fishery. The majority of trawlers carry between 4 and 7 crew. There has been no

---

[5] See also Section 4.3.

economic analysis undertaken of the 'flow on' effects of the reduction in fleet size since 1985, however it is estimated that there are approximately 800 - 1000 fewer positions available for crew in the NPF today, compared to 1985.

In contrast to the negative impacts of fleet-reduction, the positive economic benefits to operators remaining in the fishery, and to local economies, which benefit from increased business generated from a profitable fishing industry, have been significant. With fewer operators sharing the total catch as a result of fleet-reduction in the NPF, those remaining in the fishery have become more profitable. This has been particularly evident in relation to the restructuring programme which culminated in the compulsory surrender of units in 1993. The catch per unit effort (CPUE) per trawler, and the profitability of those trawlers remaining in the fishery has increased substantially, although total catches in fishery have not increased.

Since the mid-1990s the profitability of the NPF fleet has also been influenced beneficially by external economic factors such as high prawn-prices, low interest-rates, favourable foreign currency exchange-rates and lower fuel-prices in recent years. However the same levels of individual profits per trawler would not exist with a larger fleet size. Trends in the profitability of the NPF fleet between 1985 and 1998 are indicated in Table 2.

4. CONCENTRATION OF OWNERSHIP

4.1 Status prior to programme

As indicated in Section 2, the only rights which existed in the NPF prior to the introduction of the unitisation scheme were fishing vessel licences issued under the limited-entry scheme. With the introduction of unitisation, each trawler which was licensed to fish in the NPF was allocated a B-class[6] unit and a number of A-class units (as described in Section 3.2). In the cases where there was no particular vessel attached to a licence, the allocation was based on the licence details (boat length and engine horsepower) applicable to be previous boat to which the licence was attached. As a result, it is assumed the ownership structure of the fleet immediately after the allocation of individual transferable rights through the unitisation system was the same as the structure of the fleet prior to the allocation[7].

4.2 Restrictions of transfer of ownership

From the time A-, B- and C-class units were allocated they were, and remain, fully transferable, either between existing operators who wish to upgrade their vessels or engines, or increase their fleet size; or to new entrants in the fishery either as a total package, (with or without a vessel) or in smaller packages of A-class units. The only restriction on transferability was that 100 A-class units had to be maintained on a licence package in order to retain the B-class unit to which the A-class units were allocated. If the A-class units was reduced below 100 A-class units, the B-class unit attached to the package was cancelled. This restriction was introduced so as to deter operators from stripping licence-packages and creating unattached (or 'limbo') B-class units. This restriction remains in place today.

When the *Fisheries Management Act 1991* was implemented, the leasing of units was approved under provisions of the *NPF Management Plan 1995*. Leasing can occur between existing operators to upgrade vessels or engines, or to bring vessels into the fishery, provided that the requisite A- and B-class fishing-rights are leased in accordance with the provisions in the Plan.

4.3 Prices received

All A-, B- and C-class units issued under the unitisation scheme were granted free of charge, but as the units were tradable from the date of issue, they quickly developed a market value. There is no formal register of prices paid in the commercial market place due to the confidential nature of this information, however, anecdotal evidence of average prices paid in the commercial market-place has been provided by boat/licence brokers and is shown in Table 3. A price as low as $A50 per A-class unit was paid on the open market in 1985, which included the price of the B-class unit. The highest price paid for B-class units ($A100 000) was reached on the open market during the years (1987-92) that the two-for-one boat-replacement policy was in effect (as described in Section 3.2). This was due to the demand for B-class units to meet the forfeiture provisions of that policy when trawlers were replaced and upgraded.

Price structures are on record for the years that the NPF Voluntary Adjustment Scheme was in place. The VAS set the floor-price for units sold on the open market and in some cases, units which could have been sold to the VAS were sold in the commercial market place for higher than the VAS paid. However the poor catches and

---

[6] C-class units were not granted until 1986

[7] The first Central Boat Unit Register was printed in 1985. No formal information is available on the ownership of units as allocated in 1984.

Table 2
Estimated financial performance of the Northern Prawn Fishery between 1985 and 1998
(financial performance per trawler averaged over the total fleet, expressed in $A)

	1985-86 [b]	1986-87 [b]	1989-90	1991-92	1993-94	1995-96	1997-98
Receipts [a]	746 400	640 600	448 050	592 940	964 350	947 890	1 120 350
Costs [a]							
Administration	na	na	42 150	30 940	31 580	50 430	9 720
Crew costs	na	na	109 000	149 870	255 970	239 040	299 250
Freight and marketing	na	na		17 220	24 030	17 130	15 980
Food	na	na	1 790				
Fuel	na	na	94 370	113 220	140 250	122 560	134 150
Insurance	na	na	20 320	21 670	30 650	31 660	37 790
Interest paid	26 500	25 100	41 910	28 590	11 450	17 080	27 310
Licence fee and levies [b]	na	na	30 950	41 500	45 320	52 270	56 830
Packaging	na	na	17 690	10 400	13 900	15 780	14 370
Repairs and maintenance	na	na	79 180	112 640	185 260	206 290	174 950
Other costs	na	na	23 180	19 790	31 850	29 680	38 440
Total cash costs	594 900	484 300	460 540	545 830	770 260	781 920	808 790
Boat cash income [a]	na	na	-12 490	47 110	194 090	165 970	311 560
Less depreciation [c]	53 300	58 600	44 920	28 400	55 320	56 470	55 250
Total costs	648 200	542 900	505 460	574 230	825 580	838 390	864 040
Financial performance [a]							
Boat profit			-57 410	18 710	138 770	109 500	256 310
Plus interest, leasing, rent	26 500	25 100		43 070	13 800	29 240	33 580
Profit at full equity	71 700	72 600	-15 500	61 780	152 570	138 740	289 890
Capital (excluding licences)	589 700	671 300	755 860	574 940	898 980	985 950	1 064 300
Capital (including licences)	na	na	na	na	2 153 180	3 492 320	3 729 380
Rate of return to boat capital [d]	15.0 %	12.0 %	-2.1 %	10.7 %	17.0 %	14.1 %	27.0 %
Rate of return to full equity [e]					7.1 %	4.0 %	8.0 %

a All figures are adjusted for relative standard errors
b Results are for vessels rated at 375-or-more A-class units
c Depreciation adjusted for profit and loss on capital item sold

d Excluding value of licence
e Including value of licence
na not available

low prices, which prevailed in the mid to late 1980s, meant that commercial markets for units were limited, and the VAS was successful in removing a large number of units.

Table 3
Indicative prices ($A) paid by the VAS and the commercial market-place for A-, B- and C- class units from 1985 to 2000

Year	VAS A -class Units	B-class purchased	VAS B -class Units	Market A-class Units	B-class purchased	Market B-class Units
1985	100-200	yes	0	50-100	yes	not available
1986	250-425	yes	0	250-300	yes	0-50 000
1987	280-450	yes	0-20 000	535-600	yes	0-100 000
1988	450	yes	0-20 000	600	yes	0-100 000
1989	450	yes	0-20 000	600 –1265	yes	0-100 000
1990	450-950	yes	0	1200-1300	yes	0
1991	450-950	yes	0	1350	yes	0
1992	450-950	yes	0	2000-2300	yes	0
1993	VAS Closed	VAS Closed	VAS Closed	2300-2500	yes	0
1994				3250-3500	yes	0
1995				4000	no/yes	0
1996				4500-5500	yes	0
1997				6000	yes	0
1998				6000	no/yes	0
1999				6250	yes	0
2000				6500	yes	0

While the specific price paid by the VAS Committee for any unit purchased under the VAS was considered confidential, typical prices paid from the beginning of the VAS in 1985 to its conclusion in 1989, ranged from $A100 to $A450 per A-class unit. Prices paid for B-class and C-class units from 1985 to 1989 ranged from nothing upto $A20 000. In the accelerated restructuring programme from 1990 to 1992, an additional $A500 per A-class unit was paid, provided that the B-class unit to which the A-class units were attached was surrendered to the VAS, and thus subsequently cancelled.

The effectiveness of the VAS (as a mechanism for removing units from the fishery) was reduced when it was agreed that if the target of 50 000 A-class units was not achieved by the end of 1992 there would be a compulsory surrender of units to meet any shortfall. The pending compulsory surrender of units in 1993 stimulated the commercial market as operators traded units between each other to accumulate additional A-class units to allow them to meet their compulsory surrender requirements and maintain a certain fleet size. The demand for units drove the commercial market price well above the price paid by the VAS, and few units were sold to the scheme in 1991-92 and so it was stopped in 1992.

The agreement to implement a new system of management in the NPF, based on gear, further stimulated the commercial market for A-class units during the late 1990s. The gear-based system will replace the existing unitisation system, which has become ineffective as advances in technology, including innovations in trawler design and engine configurations, have resulted in the unitisation being manipulated in many instances. The inability of legislators to enforce the rules on boat-size and engine-horsepower, practically and cost-effectively (particularly at sea), has resulted in considerable uncontrolled 'effort creep'[8], and over-fishing of both species of tiger prawns in the NPF.

B-class fishing-rights will continue to exist under the new system, and the transition to gear-units will be based on the existing A-class fishing-rights. However, in order to reduce the effective effort in the fishery, the total amount of gear which will be allocated is to be reduced by approximately 15% compared to that which was being towed in 1997. This has already caused a renewed demand for A-class units as operators have now purchased units to offset the impact of the new gear-unit and effort-reduction package. This package was due to be implemented in July 2000, and the demand for units has kept A-class unit prices at a premium. B-class units have not generally attracted a specific value (see Table 3) in recent years as the focus of management measures, particularly restructuring programmes, have been targeted at A-class fishing rights. It is anticipated B-class rights will continue to have little value under the gear-units system.

There is a clear trend in that the various management arrangements introduced into the NPF over time have influenced both the demand for, and the price of, fishing-rights units. This has been particularly so during periods

---

[8] Increases in effective fishing effort resulting from technological advances, skipper ability, new hull and engine designs etc.

of restructuring, when considerable trading occurred as operators responded to such programmes by either buying or selling units. Table 2 indicates how the introduction of various management measures such as the VAS, the compulsory surrender of units in 1993, and the pending gear-unit system have influenced the price of NPF fishing-rights.

4.4   Effectiveness of regulations governing ownership of rights

Because NPF fishing-rights have generally been freely transferable and tradable since their initial allocation there have been no regulatory constraints on ownership of these rights. The rights can be owned or leased by individuals, companies, corporations and trusts. They can be transferred and traded at any time under the provisions of the NPF Management Plan 1999.

A temporary restriction on the formal transfer of NPF fishing-rights occurred from December 1992 to end March 1993 when a period of non-transferability was imposed to allow the AFS to consolidate the unit register as a result of the compulsory reduction. During this period it is understood that certain private contractual agreements were entered into, whereby operators transferred units but with the transfers to take effect only following the implementation of the compulsory reduction on 1 April 1993, at which time the non-transferability period was lifted.

A second temporary restriction on the formal transfer of NPF fishing rights was imposed on 15 December 1999 to allow the Australian Fisheries Management Authority to consolidate the SFR register in preparation for the transition from A-class SFRs to gear SFRs due to occur in July 2000. It is understood that private contracts to transfer the ownership of units have been entered into during this period and that those transactions will be formalised when the formal period of non-transferability is lifted in July 2000.

4.5   Affects of programme

Considerable changes have occurred in ownership of NPF fishing-rights in the 15 years (1984-99) since their initial allocation. There are a number of factors which have contributed to these changes including (See also Sections 3.2 and 4.2):

i.   commercial decisions taken by operators to upgrade their vessels in size or engine capacity, increase or reduce the number of vessels owned, leave the fishery and sell their units to existing operators or new entrants

ii.   boat-replacement restrictions

iii.   the existence of the NPF Voluntary Adjustment Scheme (VAS buy-back scheme)

iv.   the NPF restructuring scheme in place from 1990 to 1993, including the compulsory surrender of 30.76% of all A-class units in April 1993 and

v.   the pending transition to a new system of management based on gear, including a reduction in the gear allocated compared to that towed in 1997.

While it is not possible to quantify to what degree the above factors have been responsible for changes in ownership within the fleet at any given time, the existence of individual harvesting-rights has allowed operators to respond to management changes through the market. As NPF rights are fully transferable and tradable rights, the operators have been able to buy, sell, lease or forfeit fishing-rights as and when required on an individual basis in order to optimise their individual fishing-capacity or to respond to management measures such as restructuring programmes.

However there is no evidence that suggests that there has been any long-term transfer of ownership from any one sector of the fleet to another, *e.g.* from private operators (*i.e.* owners of one or two vessels in the NPF) to companies[9]. There are no definable ongoing trends to indicate a concentration of ownership of either A-, B- or C-class fishing-rights in any one sector of the fleet, and the distribution of the fishing-rights between the various sectors of the fleet has remained relatively balanced over time. Neither is there any evidence of an ongoing trend of concentration of rights on either a geographical or regional basis, although the fleet is spread over a wide geographical area with vessels based in various parts of Western Australia, Northern Territory and Queensland.

The only definable trend has been a significant reduction in the number and size of large fishing companies over time (see also Section 4.3.). Poor catches and economic down-turns in the 1980s resulted in large companies withdrawing their equity from the fishery. Between 1985 and the end of 1990 a large number of companies withdrew from the fishery, and either sold their licences to private operators or to the buy-back scheme.

Alternatively, companies have reduced the size of their fleets to offset the compulsory-surrender provisions in 1993, by withdrawing vessels from the fishery and using their excess units to meet the forfeiture requirements

---

[9] For the purposes of this document, part or total common registered ownership of 3 or more B- or C-class units; or 3 or more unallocated packages of A-class units, is deemed to constitute a company

on the remaining vessels. As a result, many of the fishing-rights held by those companies which have either withdrawn from the fishery or reduced their fleet sizes, have not been transferred to another sector of the industry. Table 4 shows indicative variances in ownership of NPF fishing-rights between private operators and companies since 1985.

Table 4

Numbers of licence-units owned by companies or private operators within the NPF fleet from 1985 to 2000
These data are indicative only. Averaged to account for daily variations in the Central Boat Unit Register.

Year	Company B/C-class	Company A-class	Private B/C-class	Private A-class	Total B/C-class	Total A-class
1985	154	67 000	138	61 000	292	128 000
1986*	116	53 000	151	66 000	267	120 000
1987	na	na	na	na	264	117 000
1988	106	na	134	na	240	110 000
1989	84	48 000	134	56 500	218	104 550
1990	74	39 000	138	57 000	212	96 000
1991	57	30 500	117	46 800	174	77 300
1992	59	24 800	115	44 000	174	68 800
1993**	57	24 800	80	29 000	137	53 800
1994	57	24 800	80	29 000	137	53 800
1995	59	25 000	77	28 800	136	53 800
1996	53	23 000	81	30 800	134	53 800
1997	57	21 800	75	32 000	132	53 800
1998	58	22 000	74	31 800	132	53 800
1999	48	20 000	83	33 800	132	53 800
2000	58	24 800	74	29 000	132	53 800

* A total of 133 269 A-class units and 302 B- or C-class units were granted by 1986 though this figure was subsequently reduced by sales to the VAS in the same year.
** The compulsory surrender of units in 1993 resulted in a total of 53 844 A-class units and 137 B-class units.
na = not available

5. DISCUSSION

The allocation of individual transferable fishing-rights has been a highly successful management strategy in the NPF. While there is some argument as to the extent that the unitisation scheme itself has been successful in controlling effective fishing-effort (See Jarrett 2001), the scheme has been a useful and successful tool for controlling and reducing fishing-capacity. A number of management measures have been implemented in the NPF which have been based on the unitisation scheme and which have successfully reduced fishing-capacity to various degrees (as described in Section 3.2).

Between 1984 - 2000, almost 80 000 A-class rights and 170 B- or C-class units were taken out of the fishery. This represents a reduction in A-class units of approximately 60%, and a reduction in B- and C-class units (vessel licences) of approximately 56%. This fleet-reduction exercise in the NPF is considered to be one of the largest and most successful examples in the world.

The allocation of fishing-rights through the unitisation system has also been successful in establishing an accepted and legally defensible form of fishing property-rights which previously did not exist in the NPF. The A- and B-class units allocated under the *Fisheries Act 1952* were the subject of a legal challenge in 1993. In this case, the Full Bench of the Federal Court of Australia accepted that the NPF fishing-rights (A- and B-class units) were property-rights. These rights were automatically rolled over when the *Fisheries Management Act 1991* replaced the *1952 Act*, at which time the rights became A- and B-class Statutory Fishing Rights (SFRs) - a form of property-rights established by statute. The NPF Statutory Fishing Rights have a life of at least ten years; are automatically 'rolled over' if the NPF management plan is revoked and a new plan is implemented in the fishery, and compensation may be payable by the federal government if the rights are acquired on unjust terms.

The security of the NPF property-rights has allowed managers to implement management measures to reduce fleet-capacity from time to time in a legally defensible manner without bringing about a diminution of the relative value (share of the fishery) of the existing fishing-rights. Without these rights it is doubtful that measures in the NPF to reduce fleet-capacity could be as successful as they have been.

6.   ACKNOWLEDGEMENTS

P. Pownall, SPC Consultants, Canberra
F. Meany, Consultant, Canberra
Graeme Stewart & Associates, Boat-brokers, Perth
Efrem Gamba, Director/Shareholder, Austfish Pty. Ltd., Australia
David Carter, CEO, Newfishing Australia, Fremantle, Western Australia
Ian Hopkins, Markwell Marine, Cairns, Australia
Map of the Northern Prawn Fishery by courtesy of the Australian Fisheries Management Authority

7.   LITERATURE CITED

ABARE – (Australian Bureau of Agricultural and Resource Economics) Fisheries Surveys Reports "Estimated financial performance of boats in the northern prawn fishery" 1985-86, 1986-87, 1989-1990, 1991-1992, 1993-1994, 1995-1996 and 1997-1998.

Australian Fisheries Magazine, November 1993. 'NPF Restructuring - lessons for the future' AFS pp3.

Jarrett, A.E., 2001. Initial Allocation of Unitisation (Boat/Engine Units) as Harvesting Rights in Australia's Northern Prawn Fishery. 202-211. *In:* Shotton, R. (Ed.) Case studies on the allocation of transferable quota rights in fisheries. FAO Fish. Tech. Pap. No. 411. FAO, Rome. 373 pp.

Northern Prawn Fishery Central Boat Unit Registers 1985, 1986, 1988, 1990, 1992, 1993, 1994, 1996, 1997, 1998, 1999 and 2000

Pownall, P.C., (Ed.) 1994. Australia's Northern Prawn Fishery: the first 25 years. NPF 25, 179pp.

Taylor, B., 1992. "Northern Prawn Fishery Information Notes" Special collated issue No.16 (February 192) pp3-12.

Appendix I
Current Management Arrangements in Australia's Northern Prawn Fishery (1 January 2000)

Australia's Northern Prawn Fishery (NPF) covers approximately 800 000km^2 of ocean with waters extending to the edge of the Australian Fishing Zone (AFZ). The Northern Prawn Fishery is Australia's most valuable Commonwealth fishery managed by the Australian Fisheries Management Authority (AFMA), a statutory authority appointed by the Federal Government to manage Federal fisheries on its behalf.

The target commercial prawns catch includes: white banana (*Penaeus merguiensis*), Indian banana (*Penaeus indicus), brown tiger (Penaeus esculentus), grooved tiger (Penaeus semisulcatus)*, giant tiger (*Penaeus monodon*), blue endeavour (*Metapenaeus endeavour*), red endeavour (*Metapenaeus ensis*), western king (*Penaeus latisulcatus*) and red spot king (*Penaeus longistylus*).

The NPF is managed under the *Northern Prawn Fishery Management Plan* (1995, with subsequenty updates). The fishery is a limited-entry, input-controlled fishery. The current inputs regulated under the Plan are: the number of boats that may fish in the fishery, the size of each boat used for fishing and its main engine horsepower.

Entry to the fishery is through the holding of Statutory Fishing Rights (SFRs) which under the *Fisheries Management Act 1991* are recognised as being a form of property-right created by statute. The *Northern Prawn Fishery Management Plan 1999* provides for the granting of two Classes of SFRs being: A-class and B-class SFRs. The former are based on the vessel size and engine power of the participating boat, and may be further divided into the categories of "active", "surplus" and "suspense" on the basis of their historical and current status. B-class SFRs limit the number of boats in the NPF.

Under the Plan, a boat used by a concession-holder to operate in the fishery must be nominated against one B-class SFR, and against a threshold number of A-class SFRs (which is known as the "applicable number of Class A SFRs") which is calculated from the sum of the underdeck hull volume and the maximum continuous rated main engine power expressed in kilowatts. The Plan limits the total number of B-class SFRs to 133, and the total number of A-class SFRs to 53 844. All SFRs in the fishery are fully transferable and may be bought, sold or leased.

The Plan provides the objectives to be pursued in managing the fishery, the measures by which the objectives will be pursued, and the performance criteria by which management of the fishery can be assessed. The objectives in the Plan parallel the AFMA's legislative objectives as set out in Section 3 of the *Fisheries Management Act 1991* and include (but are not limited to) objectives relating to ecologically sustainable development of fish resources, the precautionary principle, and the economic efficiency of the fishery.

The Plan also provides for the making of fishery 'Directions' that prohibit holders or operators from specified activities. Directions are a flexible management tool currently used in the NPF to establish:

i.   permanent closures of trawling around nursery areas and sensitive habitats
ii.   seasonal-closures that are used to prevent over-fishing of small-sized prawns, protect spawning prawns, and more recently to reduce fishing-effort,
iii. bans on daylight trawling, to protect tiger prawns in the tiger prawn season
iv.   bans on vessel movements in sensitive areas and times, to prevent disturbance to banana prawn aggregations
v.   requirements to have fishing-gear sealed for compliance purposes
vi.   specific areas and times that are exempt to closures, for the purpose of permitting gear-trials
vii. limits on bycatch, determined under Memorandums of Understanding between the Commonwealth and States and
viii. gear-restrictions limiting operators to single or dual net-rigs and specifying the dimensions for try-nets (small sampling nets).

THE EFFECTS OF THE INTRODUCTION OF INDIVIDUAL TRANSFERABLE QUOTAS IN THE TASMANIAN ROCK LOBSTER FISHERY

W. Ford
Department of Primary Industries, Water and Environment
GPO Box 44A, Hobart, Tasmania, Australia 7001
<Wes.Ford@dpiwe.tas.gov.au>

1. INTRODUCTION

In Tasmania four fisheries operate under a management regime where the access-rights are provided as a share of a total allowable catch (TAC). Individual transferable quotas (ITQs) were introduced into the abalone fishery in 1985, the rock lobster fishery in 1998, and the giant crab fishery in 1999. A combination of a competitive TAC and individual quota are used to manage the jack mackerel fishery. This paper provides a summary of the rock lobster fishery prior to quota, some rationale for quota-management, describes the effects of moving to ITQs, and discusses the success or otherwise of the process[1].

The Tasmanian rock lobster fishery targets the southern rock lobster (*Jasus edwardsii*) in the waters adjacent to Tasmania. Since 1986 Tasmania has had jurisdiction for the fishery in waters generally south of 39° 12'S, and out to 200 nautical miles from the coastline. This jurisdiction was conferred on Tasmania by way of an Offshore Constitutional Settlement agreement with the Commonwealth Government. However the fishery has been operating within the coastal waters (inside 3 nautical miles) off Tasmania since around 1830.

Prior to 1966 access to commercial rock lobster licences was not limited, but there were limits on the number of rock lobster pots that could be used by each vessel. During the period from 1956 to 1966 the number of pots in the fishery increased form 4000 to 8900 (Harrison 1986). In 1966 the number of commercial licences was limited to 442 (Harrison 1986). During the following 30 years there were a number of effort- and licence-reduction strategies that resulted in the number of licences falling to 321 in 1997 with a limit of 10 507 rock lobster pots.

During the early 1980s the fishery was still expanding as fishers were moving their operations further offshore. Catches peaked at 2217 tonnes in 1985, and then began to decline. This decline in catch occurred despite an increase in effort, which peaked at nearly 2.1 million pot-lifts, or about 50 000 vessel-days, in 1992. The fishing effort increased by about 30% during the period from 1985 to 1992, and as the catch was declining (Figure 1) the catch-rate began to fall dramatically.

Following the peak catch in 1985 (the fishing year being March 1985 to February 1986)[2], and the subsequent decline coinciding with a substantial increase in effort, the industry and scientists became increasingly concerned about the sustainability of the fishery. Given that the fishery had been exploited commercially since around 1830, and there had been heavy fishing pressure during the 1960s, there was general concern that the stocks were declining.

During the period November 1993 to February 1997 the Government used a number of additional seasonal closures to try to limit catch. This had some success but Figure 1 shows that the fishing effort (in terms of vessel-days fished) was not significantly reduced. This was because only 65% of the available vessel-days were actually used in 1993 and fishers could shift their effort into the rest of the open season.

During the early 1990s various management options were explored and debated within the industry and with government. The Government established a working group to investigate the management options and their suitability for the rock lobster fishery. It was apparent that there were only two real options: either a substantial reduction (about 30%) in the fishing effort, or the introduction of a total allowable catch (Anon. 1993). The industry was divided on the issue, with the majority recognising that there needed to be reductions in the catch and effort, but no agreement about how to do it. Finally, in August 1996 the Government decided that the fishery would be managed by output-controls and that individual transferable quotas would be introduced.

---

[1] The views expressed in this paper are those of the author and may not be the views of the current or previous Tasmanian Governments.

[2] For the purpose of this paper the fishing-year analysis has been done on a March-to-February year so that past catches and fishing effort can be compared with the two years under quota-management, which has a March-to-February year.

Figure 1
Rock lobster catch, and fishing effort (1970/71 to 1999/00)

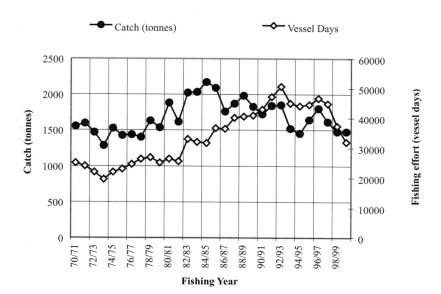

The Government had two objectives in mind, the first was to reduce the catch to a level which would be sustainable and allow the biomass to rebuild over time. The second was to provide a mechanism whereby the industry could restructure and allow those who wished to leave the fishery to achieve a reasonable return for their previous access. It was recognised that whatever management option was adopted, reducing the catch would inevitably lead to fewer fishers participating in the fishery, or a less viable industry. The Government, supported by a majority of the industry (finally about 67%), believed that quota was the better option for achieving its two objectives.

During the year prior to the introduction of quota (in March 1998) there were a maximum of 305 active fishing vessels working in the fishery out of a possible 321 licences. These 305 vessels were operated by fishers who owned and operated a licence, by family operations and lease-holders. It was estimated that about 850 people were employed directly on fishing vessels at that stage.

2.   THE NATURE OF THE HARVESTING RIGHT

Prior to the introduction of quota-management the fishers had a fishing licence that allowed them to use a number of pots to take rock lobster. It was an annual licence to which was attached a common law expectation that it would be renewed. The Minister had the power to cancel licences and could exercise that power if a person was convicted of serious offences.

The licences were transferable and additional rock lobster pots could be purchased from any fisher who was leaving the industry, subject to certain constraints related to the total number of rock lobster pots that could be held. In the final year before quota was introduced, fishers were able to own up to 25% more rock lobster pots that they could use on their vessel; this was to allow the restructuring process to begin.

The nature of the fishing-right was such that fishers did not have access to any given share of the resource, they essentially used their licence to fish as hard and fast as they could in order to catch their perceived "share" before someone else caught it. There was little incentive for individuals to fish sustainably, because someone else would catch the fish if they did not. However, some fishers only fished as long as they needed to make a reasonable living. It was this problem which lead to the increased fishing-effort and hence the decline in the stock. Changes in the management-regime were required to ensure that the resource was fished sustainably and in an economically viable way.

The unit of access-right to the fishery was the number of pots held on the licence. Licences were traded on the basis of the number of pots attached to the licence, regardless of the historic catch associated with the licence. In 1993 licences were traded for about $A4000 per pot; this increased to around $A10 000 in late 1997.

The introduction of ITQs in March 1998 gave licence-holders a defined share of the resource, which provided incentive for fishers to move away from the "race-to-catch" mentality. The resulting rock lobster fishing licence and attached rock lobster quota-units are described in the following paragraphs.

The rock lobster quota-units were allocated in perpetuity to the people who held commercial rock lobster licences by means of legislation (the *Living Marine Resources Management Act 1995*). The initial allocations included partial recognition of past catch-history with 9% of the total allowable commercial catch (TACC) being allocated on that basis in 1998, phasing out to no recognition in 2001. Then the quota-units will provide exclusive access to take 1/10 507 of the TACC.

The licence is an annual licence which must be renewed by the Minister as long as (a) the licence fees are paid, and (b) the person has not been convicted of a relevant offence under the law of another State, Territory or the Commonwealth. If the licence renewal is refused under (b) the licence-holder has a right of appeal to an independent appeal tribunal. Case history suggests that the refusal to renew a rock lobster licence on these grounds would require a serious offence in a rock lobster fishery. Therefore, this provision cannot be used for minor breaches of other fisheries laws.

The Act provides that only the fishing licence is forfeited if a total of 200 demerit points[3] ($A20 000 in fines) are reached in any five-year period, the rock lobster quota-units are not forfeited. As the greatest asset-value lies with the quota-units, the licence-holder's investment is protected from forfeiture. While any licence-holder who is convicted of the offences that resulted in the demerit points would be excluded from holding a licence for 5 years, they would be able to sell or transfer their quota-units, thereby retaining the asset-value of the quota. Any licence-holder who leased the licence to another fisher would not be prevented from buying another licence and transfering the quota units to it.

Tasmanian lobster fishing vessels showing traps on foredecks
Credits: Tasmanian Aquaculture and Fisheries Institute

The Act requires the holders of existing commercial fishing licences to be allocated the rock lobster quota, regardless of the instrument that creates the licence. This means that a new type of licence cannot be created to transfer the ownership of the quota-units to a new group of licence-holders.

The above three characteristics effectively mean that the rock lobster quota-units and rock lobster licences are issued to the licence-holder in perpetuity, and they cannot forfeit the asset-value if convicted under State law, or other relevant law. However, as with any criminal proceedings, profits from crime can be seized by the Crown.

Rock lobster quota-units can be freely transferred between the 316 licence-holders on a permanent or seasonal basis. Only licence-holders may own quota-units, this means that the maximum number of unit-holders recognised by the Government is 316. This will be one of the issues for the future, in that there are already a number of licence-holders who believe that the ownership of quota should be separate from the fishing entitlement. Currently the majority of licence-holders want ownership of quota to remain with the fishing licence.

This issue is further compounded by industry's desire to have legislation introduced to allow it to register financial interests in licences and quota-units. Already it is apparent that there are a number of licence-holders

---

[3] Demerit points are a system of assigning one point for each $A100 of a penalty imposed for a conviction under the *Living Marine Resources Management Act 1995* or its subordinate legislation. A licence ceases to have effect·if 200 demerit points are assigned in a five year period. Assigning the points is not at the discretion of the courts or Minister but is mandatory.

with investor partners who have funded the purchase of additional quota-units (Ford 2000). This trend may continue, and there will be increasing pressure to allow these investor partners to be formally recognised (Ford 2000). Such legislation may provide a legal register of lending interests and may require the Government to seek the approval of any interested parties before any quota-units can be transferred.

3.   MEASUREMENT OF FLEET-CAPACITY
3.1   Characterizing fleet-capacity

Prior to the introduction of ITQs the capacity of the rock lobster fleet was measured in terms of vessel numbers, vessel size, and the number of rock lobster pots that could be used. During the period from early 1993, when restructuring was seriously proposed, and leading up to ITQs, the number of licences declined from about 336 to 316, but the capacity in terms of rock lobster pots did not decline, with the number of pots remaining stable at 10 507.

In 1997 the rock lobster fishing fleet was made up of 315 vessels ranging in length from 6-26 metres. The majority of vessels were used primarily for rock lobster fishing but had the capacity to diversify into other fisheries on a seasonal basis. The vessels were a mixture of wooden and steel hulls plus a few fibreglass hulls. The majority of the fleet was of the displacement-hull style with a small number of planing-hull vessels. The average age of the fleet exceeded 15 years with few new vessels operating. The majority of vessels were owner-operated, but there was a trend toward the leasing of vessels and licences. The market value of vessels participating in the fishery at this stage varied from a low of $A15 000 to more than $A750 000 each

Each vessel had a rock lobster pot-allocation based on either the length or gross tonnage of the vessel. Prior to 1997 the pot-allocation varied between a minimum of 15 and a maximum of 40 pots, however this was increased to a maximum of 50 pots in 1997 when licence-holders could increase their pot-allocation by up to 25% by purchasing pots from other licence-holders (Figure 2).

Figure 2
Frequency distribution of rock fishing vessels in terms of number of rock lobster
pots holdings, January 1997

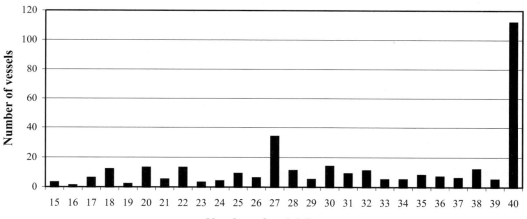

Fleet-capacity in terms of the fishing-effort applied was one of the key concerns that the move to ITQs sought to address. The total fleet-capacity has been considered in terms of potential effort compared with actual effort, this became important during the period from 1993 to 1997 when there were extended seasonal closures designed to reduce fishing-effort, and therefore catch. In 1992/93 prior to the reduced fishing seasons, assuming that each vessel could have worked for a maximum of 80% of the available days of the season, then the total fleet-capacity for 332 vessels would have been in the order of 80 000 vessel-days.

In 1992/93 just over 50 000 vessel-days (or 63% of available vessel-days) were fished. During the following four years the limits on season and fleet size reduced the available vessel-days to about 74 600 in 1993/94, and to 63 500 in 1997/98, while the number of vessel-days actually fished was fairly stable at between 44 000 and 46 000 (Figure 1). The number of available vessel-days fished increased from 63% in 1992/93 to 72% in 1997/98. This increase demonstrated that the fleet had more than enough capacity to respond to reductions in the fishing season by fishing at other times when the season was open. This can be seen in

Figure 3, which shows that the reductions in the length of the fishing season imposed during the Tasmanian summers over 1993-96, reduced the fishing-effort in December, whereas the fishing-effort in August (winter) increased.

Figure 3
Fishing effort in the months of August, or December, each year

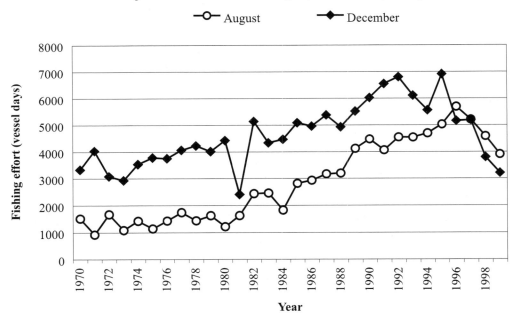

Fleet-capacity in terms of total vessel tonnage only reflects the size of the fleet, with the average size of vessel remaining fairly stable over a number of years at about 29-30t. This would have put the fleet-capacity at 9990t in 1992/93, 9150t in 1997/98 and 7590t in 1999/2000.

For the purpose of this paper measures of fleet-capacity are analysed in terms of changes in fishing-effort and changes in the number of vessels. More detailed work will be undertaken over the next two years to apply standardisation techniques to measures of fishing-effort and fleet-capacity. While the following analyses have not been standardised, they demonstrate clear trends showing that fleet-capacity and fishing-effort have been much reduced as a result of the introduction of ITQs.

3.2   Changes in fleet-capacity arising from the introduction of transferable property-rights

The introduction (in March 1998) of ITQs into the rock lobster fishery has reduced the number of fishing vessels and fishing-effort (vessel-days in Figure 1) by about 17% and 28% respectively in two years. Figure 4 shows frequency of vessels by gross tonnage for the March-to-February fishing-year (1997/98) before ITQs, and for the second year under ITQs (1999/00). It can be seen that the total numbers of vessels had been reduced by about 17% but that their size-frequency had not changed substantially.

The shift to ITQs has resulted in a reduction in fishing-effort. Figure 3 illustrates the reduction in effort in 1998 and 1999 for December or for August. This has been brought about because the catch has been reduced, therefore the effort required to catch the rock lobster has fallen and there is no need to compete for the catch at times of low prices. The smaller catch has meant fewer vessels are needed to take the rock lobster, which has seen the number of vessel fishing-days fall by 28% over 1998 and 1999 (Figure 1), where it can also be seen that ITQs have reduced fishing-effort back to the level that was employed to take the peak catches of the early 1980s. Under ITQs the catch is about 9% lower than it was for the fishing year prior to quota. The larger reduction in effort means there have been substantial improvements in catch-rates.

Figure 3 shows the reduction in fishing-effort in 1998-99: the effort had fallen in December of both years below that for August, because there was less need to race-to-catch the rock lobster when the beach price is low. However, the reduction in effort in both months mainly reflects the reduction in the total number of vessels.

Figure 5 shows two indexes of catch per unit effort (CPUE) comparing the eight years prior to ITQs (90/91 to 97/98) with the first two years under ITQs (98/99 to 99/00). This figure shows that both indexes show improving catch-rates under ITQs, and while these data have not been standardised for changes in the dynamics of the fleet, there is a clear trend emerging. This fits well with the anecdotal information provided by fishers who generally agree that they are seeing an improvement in catch-rates. Changes in annual catch-rate are heavily influenced by recruitment into the legal fishery: during 1995/96 and 1996/97 there was a good recruitment pulse of rock lobster entering the fishery.

Figure 4
The size-frequencies for vessels used in the rock lobster fishery, before ITQs, and currently

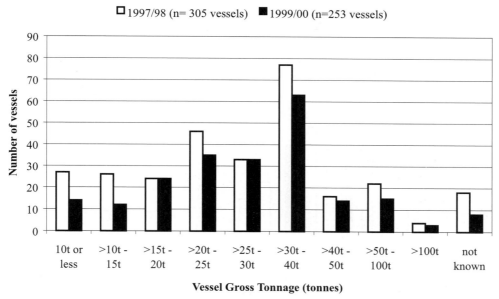

☐ 1997/98 (n= 305 vessels)   ■ 1999/00 (n=253 vessels)

Note: There are a number of vessels for which there is no information on the gross tonnage, only their overall length, but as additional information is obtained it is entered into the licensing information system.

Figure 5
A comparison of rock lobster catch rates for the prior to ITQs and the first two years under ITQs

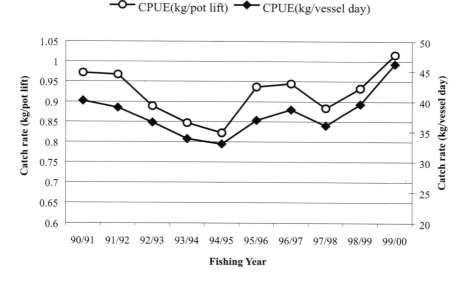

—○— CPUE(kg/pot lift)   —◆— CPUE(kg/vessel day)

The problem of trying to standardise catch and effort information to allow statistically valid comparisons to be made between pre- and post-quota effort-data is of such importance that a two-year research project has just commenced to look at the impact of ITQs on the fishery-modelling process, which is heavily reliant on past catch and effort data.

One of the claims made by those who opposed ITQs, in particular the equal allocation proposal, was that it would remove the opportunity for high-catching fishers to take as much as they had in the past. They claimed they could not afford to buy or lease additional quota and would be limited to about 7 tonnes if they had held the maximum of 50 rock lobster pots. Figure 6 shows that this has not occurred and that there is little difference between the number of high-catching fishers (say 8t and above) in the two years before ITQs and in the first two years under ITQs. This is also supported by Figure 8, which shows there are now about 15 more licence-holders who hold more than 50 quota-units, than those who held 50 pots prior to quota.

Figure 6
Numbers of vessels arranged by magnitude of their catches, for the two fishing years prior to ITQs (1996-98) and the first two years under ITQs (1998-2000)

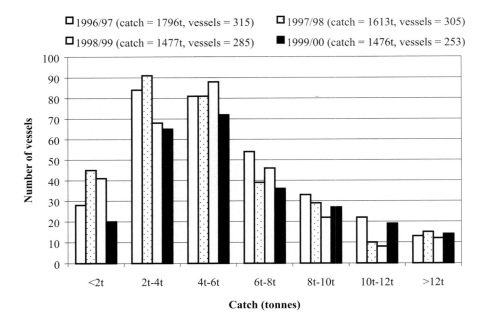

Figure 6 shows that there has been a reduction in the number of vessels catching small quantities (less than 2 tonnes) of rock lobster, and it is these fishers who have left the industry under ITQs. It is likely that such a pattern has emerged for a number of reasons. One may be that some fishers (aged in their 50s and 60s) may have been debt free and only caught enough to live comfortably, and have since decided to retire and sell or lease their quota. At this stage little work has been done to identify what has motivated some fishers to leave the fishery, whether it be retirement, or loss of access to leased licences. Some work has been done to look at individual fishers to see how their fishing has changed as a result of ITQs. Williamson *et al.* (1999) have done some preliminary survey work in order to look at how fishers have changed their operations as a result of ITQs, being a follow-up to their survey work (Williamson *et al.* 1998) done prior to the introduction of quota.

Quota-management has also seen the fleet change the way it fishes across seasons. Prior to the ITQs the traditional season would commence in November and run through to August, with September and October being closed. Many of the rock lobster moult during this September period and so the seasonal opening in November was known as the "new shellers". The catches and catch-rates peaked in November and fell away through to July when there was a slight rebound. On the other hand the beach price did the opposite, starting low and increasing during winter (June-August). In the mid-1990s the beach price would start around $A19/kg for live rock lobster in November and increase to around $A45-$A50/kg in September. Figure 3 also shows the changes which have occured since 1996 in the magnitude of fishing effort in August (winter) as opposed December.

One of the benefits seen with a move to ITQs was that fishers could choose to fish at a time of higher beach prices and thus improve their returns. Figure 7 shows that there has been a shift of effort (an increased percentage of total effort) into the March-September period, resulting in an increased proportion of catches during that period. This shift was anticipated under the quota, so the system was designed with the fishing year starting in March. This means that fishers are able to fish as hard as they wish during winter (March through August) to try to maximise the value of their catch, knowing that come November/December their remaining uncaught quota should be relatively easy to catch.

3.3 Consequences of changes in fleet-capacity

A number of consequences have arisen from the restructuring of the fleet, resulting from the move to ITQs. While there have been rapid improvements in the stock, with both increases in biomass and egg production (Gardner 1999), there have been a number of social costs associated with the reduction in the fleet size. The fishers most directly affected by the fleet restructuring were those who lease licences, of which there were 112 in January 1997, and 85 in November 1999 (Gardner 1999, Williamson *et al.* 1999).

Figure 7
A comparison of rock lobster catch and effort for the winter period for the three fishing-years prior to ITQs and the first two under ITQs

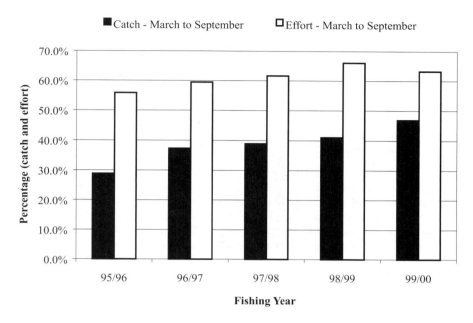

One of the objectives of ITQs was to rationalise the fleet to reduce the fishing-capacity. This inevitably meant that some fishers would have to leave the fishery and there would be people who would become unemployed as a result. Unfortunately many have been unable to move into other Tasmanian fisheries as they too were being rationalised. The loss of 52 vessels from the fishery in two years has meant the loss of about 120 direct jobs for fishermen, mainly lease-skippers and deck-hands.

As at November 1999, 84.8% of the 315 active licences were held by Tasmanian owners. This proportion has remained the same since January 1997. An analysis of ownership of the licences is summarised in Table 1.

Table 1
Ownership of rock lobster licences by Tasmanian residents and Interstate residents and by company or individual, January 1997 and November 1999

Licence holder group	January 1997		November 1999	
	No. of licences	%	No. of licences	%
Tasmanian individuals	230	71.6	230	73.1
Tasmanian companies	39	12.1	37	11.7
Interstate individuals	32	10.0	32	10.1
Interstate companies	20	6.2	16	5.1

The industry is made up of: fishers who own and operate a licence, family operations, investors who lease quota, and fishers who lease licences and quota. However the distribution of ownership has changed under ITQs: in January 1997 309 licences were attached to fishing vessels, of these 176 were operated by the owner, or by the nominated person if the holder was a company or partnership. Twenty-one licences were operated by a family-member of the owner (usually a son, brother, or husband). At that time 112 licences were leased or operated by someone other than the owner or the owner's family, and 12 were not being used. Since quota have been introduced there has been a decrease in the number of owner-operators and of leased licences (because of the number of licence-holders who only lease out their quota). The change in the way licences are used is shown in Table 2.

Another consequence of reducing the fleet and limiting catch is likely to be greater fishing-effort in the input-controlled scalefish fishery. To date there has been no work done to substantiate this, but one of the focuses of the two-year project will be to assess the impact of introducing ITQs.

Table 2
An analysis of how rock lobster licences are operated
comparing January 1997 and November 1999

Licence-holder group	January 1997		November 1999	
	No. of licences	%	No. of licences	%
Tasmanian owner operators	160	49.8	140	44.4
Tasmanian family operators	20	6.2	8	2.5
Tasmanian owned, leased licences	83	25.9	64	20.3
Interstate owner operators	16	5.0	8	2.5
Interstate family operators	1	0.3	2	0.6
Interstate owned, leased licences	29	9.0	21	6.7
Licences not used	12	3.7		
Tasmanian owned and quota only leased			55	17.5
Interst ate owned and quota only leased			17	5.4

The rapid decrease in the number of fishing vessels has meant that there is a higher-than-normal number of vessels on the market, and there appears to be little interest in buying old rock lobster vessels. While those who have sold out of the fishery have obtained a good return for the quota-units, they may not have been able to sell their vessel, or have sold it for less than it was previously worth. There is little real data available at this stage to better describe such 'flow-on' consequences of rationalising the rock lobster fleet.

4.   CONCENTRATION OF OWNERSHIP
4.1   Status prior to individual transferable quotas

Prior to ITQs, licence-holders could hold a maximum of 40 rock lobster pots on a licence, this was increased to 50 in the year before the ITQs commenced, in order to facilitate some preliminary restructuring. There were no limits on the number of licences held. While most licence-holders only held one licence, some held two and one licence-holder even held seven. One year prior to ITQs there were 316 licences amongst 294 licence-holders. However some licence-holders were the same entity, in that a person may have held two licences in different company names, so in reality there were about 286 separate licence-holders. This concentration in ownership began in about 1991 as fishers came to see that the industry needed to be rationalised, and some took the view that they should prepare for it, and bought additional licences with the view of protecting their stake in the fishery.

Figure 8 shows the distribution of the numbers of rock lobster pots on licences in October 1994 and in October 1997. Comparison of these figures for 1994 and 1997 shows the changes that resulted from allowing fishers to hold up to 25% more pots than they could use. This demonstrated that restructuring began at the time that ITQs were announced and again after quota were issued.

4.2   Restrictions of transfer of ownership

The *Living Marine Resources Management Act 1995* requires all rock lobster quota-units to be held on a licence, of which there are 315. This means that licence-holders are the only people who can hold quota. Quota-units can be transferred freely between licence-holders within the minimum and maximum limits on holding quota. The management plan restricts a licence-holder, or any entity having a beneficial interest, to holding a maximum of 200 quota-units, with no more than 100 quota-units held on any one licence. Licences can be held by individuals, partnerships, companies or trusts.

4.3   Prices paid for licences and quota

The market price for access to the rock lobster fishery has increased dramatically since 1993 when the price was about $A4000 per rock lobster pot. In late 1997 just prior to the introduction of ITQs the price had increased to about $A10 000. Currently quota-units are trading for between $A18 000 and $A20 000 for access to a unit of 143kg of rock lobster with a return from leasing of about $A1200 per quota-unit (6%-6.5%) after licence fees are deducted. In addition to the market for quota there is also a market for the fishing licences, which are traded for at least $A20 000, with anecdotal information to suggest that the price may be as high as $A50 000.

Market-price information is available to the Government - all such purchases are subject to stamp duty and so must be assessed before transfers can be processed. However, this information is not released publicly.

Figure 8
A comparison of the ownership of the access-rights prior to ITQs and two years later

☐ Oct 1994 (Access unit = number of pots, n= 336 licences)
☐ Oct 1997 (Access unit = number of pots, n= 316 licences and 294 licence holders)
■ Oct 1999 (Access unit = number of quota units, n= 316 licences and 277 licence holders)

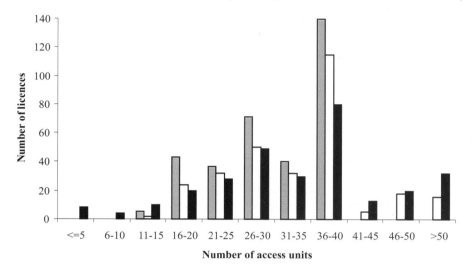

4.4 Effectiveness of the regulations governing ownership of rights

The rock lobster management plan, which is a set of statutory rules, provides the minimum and maximum quota- holding limits, and the *Living Marine Resources Management Act 1995* contains provisions that prohibit the acquisition, or the receiving of any benefit from holding more quota-units than specified in the management plan.

While this is yet to be tested in court there has been one case where a licence-holder exceeded the limits and was required to divest himself of some units, which he did without the need for prosecution. Aside from that case the minimum- and maximum-holding rule has not been breached, as far as can be determined. However, as there are no records kept of real beneficial ownership, it is possible that someone has, or will, exceed the limit but not be detected.

Anecdotal information exists that there has been an increase in the number of quota-investors who are silent partners, in the sense that the Government does not recognise that they

A modern Tasmanian lobster vessel
Credits: Tasmanian Aquaculture and Fisheries Institute

may have a share in a licence or quota. There are apparently a number of skippers who have bought extra quota to be held on the licence that they lease, and have it secured with various contractual arrangements with the licence-holder. This also occurred in the abalone fishery and led to the separation of the diving licence from the quota-units in 1991. At this stage the industry representatives are likely to seek to use the proposed legislation to register financial-lending interests to prevent a similar separation occurring in the rock lobster fishery.

4.5   Affects of individual transferable quota

Just prior to ITQs there were 316 licences amongst 294 licence-holders (some fishers held two or more licences), however some licence-holders were the same entity in that a person may have held two licences in different company names. In reality there were about 286 separate licence-holders (Anon. 1997). The number of licence-holders and licences have fallen since long-term management changes were first discussed back in 1991, as some licence-holders had been preparing for change through buying additional licences or buying additional pots. The number of licence-holders has dropped in the last few years: in 1987 there were 345 licences held by about 339 licence-holders, while as at January 1997 there were 321 licences held by 301 licence-holders. In March 2000 the 315 licences were held by 277 licence-holders, which in reality is more likely to be 270 separate entities.

Under ITQs the minimum-holding for quota is 5 units and the maximum is 200. Figure 8 compares the frequency of ownership of numbers of access-rights prior to ITQs (October 1994 and October 1997), and two years later (October 1999). It clearly shows that the number of licence-holders with a small number of units (5 through 15) has increased. This is mainly due to some fishing skippers buying into the fishery for the first time, as well as some investors buying small holdings of quota. At the other end of the spectrum the number of licence-holders with large numbers of units (above 41) has increased as some have bought up quota-units to maintain previously high catch-levels.

5.   DISCUSSION
5.1   Reduction in fleet-capacity

The introduction of ITQs had two main objectives: (a) reducing the catch to a sustainable level and (b) restructuring the fleet to remove excess fishing-capacity. Both objectives have progressively been achieved within the first two years of the quota system (Williamson *et al.* 1999). Such achievements at this early stage are greater than the Government had anticipated. After only one year the biomass of legal-sized rock lobster increased by 11% (Gardner 1999) and the number of vessels fell by 10%, with a further 11% reduction in the number of vessels after the second year, or a total reduction of 17% over two years.

The introduction of ITQs has driven the restructuring of the fishery and has resulted in fewer vessels catching the rock lobster with less effort. This has been at the cost of a reduction in the number of leased licences, the number of crew employed (Williamson *et al.* 1999), and the operators who were the higher-catchers having to reinvest in the fishery, as well as older vessels becoming hard to sell. These impacts were not unexpected and were identified by Williamson *et al.* (1998) in their study that looked at the socio-economic profile of the fishery and the possible impacts of quota-management, predicting that many of the above changes would be able to be detected within three years of the introduction of ITQs. They also concluded that similar changes would occur regardless of how the fishery was restructured to take a lower catch.

Williamson *et al.* (1999) reported similar findings as a result of surveys conducted since the introduction of ITQs. This work forms part of an ongoing doctoral study looking at the socio-economic impacts of ITQs in the rock lobster fishery.

5.2   Concentration of ownership

The rock lobster management plan and supporting provisions in the *Living Marine Resources Management Act 1995* establish mechanisms to limit the concentration of ownership of rock lobster quota-units. These provisions should provide sufficient deterrent against trying to exceed the limits, because the penalty for a conviction could include the cancellation of the licence and a maximum fine of $A100 000, a prison term of up to one year, or both.

The onus will be on the Government to conduct periodic, detailed, checks of quota-holders to detect any breaches of the limits on holdings, or benefits received from, more than 200 quota-units each.

6.   LITERATURE CITED

Anon. 1993. Restructuring of the Tasmanian Rock Lobster Fishery. Unpublished report, Rock Lobster Working Group, Department of Primary Industry and Fisheries, Tasmania.

Anon. 1997. Draft Rock Lobster Management Plan and Policy Document. Unpublished. Department of Primary Industry and Fisheries, Tasmania.

Ford, W. 2000. Will improving access rights lead to better management - Quota management in the Tasmanian rock lobster fishery. *In:* Shotton, R. (Ed.) Use of property rights in fisheries management. Proceedings of the FishRights99 Conference, Fremantle, Western Australia, 11-19 November 1999. Workshop presentations. FAO Fish. Tech. Pap. 404/2, pp289-295.

Gardner, C. 1999. Tasmania rock lobster fishery 1998/99 – Fishery Assessment Report. Unpublished. Tasmanian Aquaculture and Fisheries Institute, University of Tasmania.

Hansard 1997. Legislative Council Hansard, Tasmanian Parliament, 19 November 1997, parts 1-2, pp 1-107.

Harrison, A.J. 1986. The Development of Existing Rules in the Tasmanian Rock Lobster Fishery. *In:* Bear, S. (Ed.) Tasmanian Rock Lobster Seminar 1986, Department of Sea Fisheries Technical Report 25. pp7-10.

Williamson, S., L. Wood and M. Bradshaw. 1998. Socio-Economic Profile of the Rock Lobster Industry in Tasmania and the Effects of a Shift to a Quota Management System on Four Port Communities. Unpublished report, Department of Geography and Environmental Studies, University of Tasmania.

Williamson, S., L. Wood and M. Bradshaw. 1999. Restructuring the Tasmanian Rock Lobster Industry: Some Socioeconomic Consequences of the First 12 Months of an ITQ System. FishRights99 Conference, Perth, Western Australia, 11-19 November 1999 – unpublished workshop paper.

CHANGES IN FISHING CAPACITY AND OWNERSHIP OF HARVESTING RIGHTS IN THE NEW SOUTH WALES ABALONE FISHERY

A. McIlgorm and A. Goulstone[1]
Dominion Consulting Pty Ltd,
Suite 7&8, 822 Old Princess Highway, Sutherland NSW 2232, Australia
<mcilgorm@tradesrv.com.au>
and
NSW Fisheries, Cronulla Fisheries Centre, PO Box 21, Cronulla NSW 2230, Australia
<goulstoa@fisheries.nsw.gov.au>

1.   INTRODUCTION

The abalone (*Haliotis rubra*) fishery is a single-species high-value fishery fished by divers exploiting a gastropod mollusc, which lives on the seabed among seaweed adjacent to shore, generally in water depths of less than 20 metres. The divers prefer "Hookah gear" which enables them to breathe using airlines, from small boats whose length is generally not in excess of 8m. A diver would have a crew member assisting him, though there are strict rules over who is entitled to dive in the fishery.

Abalone *(Haliotis rubra)*
Photo credit: Dianna Watkins, NSW Fisheries

The fishery is driven by overseas demand for abalone with Australia supplying approximately 60-70% of the world market for this species. The abalone sector across Australia has approximately 300 licensed divers and had a value of $A181 million (price at first sale) in 1997-98, being approximately 12% of the total value of Australian fisheries (McIlgorm and Tsamenyi 2000). The NSW abalone fishery is small, producing 333 tonnes in the 1997-98 fishing season and having a value at first sale of $11m for the 1997-98 season. International demand plays a crucial part in keeping the NSW abalone fishery profitable.

The fishery commenced in the 1960s. The largest catches were in the early 1970s, though prices were low until the development of exports to Japan in the late 1970s. Regulations changed causing the shucking of shellfish at sea to be moved: first to wharfs, and then to within processing premises. A minimum shell-size limit of 100mm was introduced in 1977 to curtail over-exploitation, and at that time many transient fishers moved to dive in the neighbouring state of Victoria. By the end of the 1970s the impact of good markets in Japan was being felt by the fishery, and many enthusiasts and amateur dive-club members decided to become specialised abalone divers. Many divers bought fast "sharkcat" vessels giving the industry a glamorous image.

The abalone fishery is the oldest managed fishery in NSW, and access was first restricted in 1980. Since then the fishery has seen a reduction in numbers of diver/operators due to a combination of regulations, restructuring, and further development of fishing-rights regimes. This case study examines the changes in fishering-capacity as the fishery has moved from being managed through a general fishing licence, to species-specific limited-entry licensing, going through several adjustment-schemes and the eventual introduction of Individual Transferable Quotas (ITQs). In NSW limited-entry and ITQ-management has been followed by the share-management system, which augments quota-holdings with recognised and compensatable fishing-rights under the NSW *Fisheries Management Act 1994* (Watkins 2000).

---

[1] The paper should not be taken as representing the policy of NSW Fisheries.

As a fishery with low numbers of divers and significant biological and management data, the NSW abalone fishery provides an opportunity for a study of capacity over a 20-year period. The change in the fishery has been in adjusting the human-capacity, as opposed to vessel-capacity, and in this way may be in contrast to other studies. It may provide useful information for fishery managers wishing to augment rights-based fishery management, and in understanding the people-element in fishing-capacity issues.

2.   THE NATURE OF THE HARVESTING RIGHT

In the 1970s the fishery was administered by the requirement of a general fishing licence that enabled abalone to be taken by any commercial fisher in NSW, *i.e.* a general access-right to the fisheries of NSW. At that time abalone prices were low and the fishery undeveloped. Market prices increased in the late 1970s and the potential to export into the Japanese market led to greater interest in abalone harvesting. By the late 1970s some concerned fishers were calling for control of the fishery due to its rapid development and diminished catch-rates.

In 1980 a more restricted licence was introduced, enabling fishers who had evidence of previous abalone fishing involvement to hold a fishing endorsement specifying access to the abalone; holders of general licences could no longer harvest abalone. This abalone licence was not transferable except under exceptional circumstances such as sickness, and at the discretion of the Director of Fisheries.

In 1985 saleability was brought in with the "2 for 1" buy-back programme. Under this system an entrant had to buy two licences to get one new consolidated licence in the fishery. This forced divers to discuss the price of the right (which was $A60 000 each), representing an entry price to the fishery of $A120 000 in 1985. For this the divers had an expectation of catches of 15-20 tonnes at $A3-4/kg (Smythe, pers comm.). The floating of the Australian dollar in 1985 led to prices of $8-$9/kg by the end of 1985. This affected the levels of effort by new divers who had to work harder to repay debts incurred in entering the fishery, and it gave incentives for existing divers too, to work harder. This led to new concerns about levels of effort and over-exploitation of the resource.

In this period the licence was technically an annual permit and was renewed each year by the Director of Fisheries. Banks did not recognise licences as collateral and loans were made against divers' assets (*e.g.* houses, *etc.*). Several interstate divorce settlement cases (such as Kelly v Kelly 1990), led to licences being regarded as an asset having "a proprietary interest". Subsequently the attitude of banks in recognising licences as collateral was mixed "depending on the bank and the different personalities involved" (Smythe, pers. comm.)[2].

NSW abalone diver's boat
Photo credit: Dianna Watkins, NSW Fisheries

In 1988 as a result of over-exploitation concerns, an Individual Quota (IQ) was allocated equally at 10 tonnes of abalone per diver, an initial equal allocation at which fishers were all financially viable. In 1990 a minimal transfer provision of being able to receive 4 tonnes of abalone from other divers, or to sell 2 tonnes, was implemented. For example a diver could trade-up to holding 14 tonnes of abalone, or trade-down to holding only 8 tonnes. Further quota-transfer was implemented to stem "year end arrangements" between fishers who had caught less quota than envisaged and thus arranged figures to keep within the minimal transfer provisions.

The IQs were subsequently reduced to 9 tonnes per diver in 1992 and were made into fully transferable ITQs in 1994. By this time the banks were more comfortable with lending against ITQs, though lending practices varied between banks and with the financial circumstances of the client. The ITQ per diver was reduced to 8.24 tonnes per diver in 2000.

---

2 NSW Abalone Management Advisory Committee.

In the mid-1990s, as a result of the property-rights working-groups and subsequent developments (see Young 1995, Goulstone 2000 and Waktins 2000) the NSW share-management regime was introduced. Provisional share allocations were made in 1996 with a fixed number of shares in proportion to the ITQ being allocated. This was equivalent to 100 provisional shares per diver.

The shares are recognised under the *Fisheries Management Act 1994*, are transferable, are perpetual in duration, with guaranteed renewal every 10-year period. The management plan is current for a 5-year period. The share-right entitles fishers to compensation under the *Fisheries Management Act 1994* if the government decides to close the fishery.

The NSW Fisheries Department has been responsible for maintaining the ownership records under each of the schemes of management *e.g.* general fishing-licence, restricted abalone-entry license, ITQs and share ownership registries. However there is no obligation on the Department to record financial arrangements such as liens against entitlements. Sale of rights was limited under the original pre-1980 fishing licence, but possible under the restricted abalone-licence after 1985, although requiring the approval of the Director of Fisheries.

In moving to share-management from a restricted fishery, there was a provisional share-management period to confirm initial allocations, settle appeals and develop a management plan before full share-management commenced. Under provisional share-management, transfers required the Director's approval and only complete bundles of 100 shares could be sold. On introduction of full share-management and the management plan in the year 2000, shares could be sold in lots of 10.

Under full share-management, 100 shares were issued to each of the 37 divers, making 3700 shares available in the fishery. If the Total Allowable Catch (TAC) was to increase there are still only 3700 shares, thus this is the incentive to steward the resource. The permission of the Director is not required to transfer shares, though the management plan specifies the conditions for holding shares. A new entrant diver must have 70 shares to dive and no one entity can hold over 210 shares. This was a design feature of the rights-regime to prevent concentration of ownership (Young 1995).

Under the share-management scheme shares can be sold to anyone who is an Australian citizen, foreign ownership being limited to 20% of a body corporate. The abalone must be taken from the water by a nominated diver holding 70 shares. Under the share-scheme greater third-party ownership of shares is possible than under previous management arrangements.

In 1999 when the management plan was being developed the divers discussed transferability in the consultative draft plan. Greater third-party investment through de-regulated ownership was seen as potentially increasing share value, but it was feared that this would lead to more divers being in the fishery. At a mimimum share-holding of 50 shares per diver, up to 74 divers could be in the fishery. Industry representatives feared that should prices fall then harvests may illegally exceed the quota, in proportion to the number of divers, and threaten the viability of the resource. Divers on the Management Advisory Committee recommended that the minimum provision should be 70 shares per diver. Conversely if the trade led to maximum holdings of 210 shares, the current TAC could be taken by 18 divers. The Management Plan addresses capacity-concerns through having triggers to initiate a review if diver numbers increase or decrease substantially.

In summary, tradeability has on occasions been restricted at the request of industry in order to prevent potential resource depletion. Under share-management trading happens within guidelines that protect the resource and yet release the economic benefits of rights-management.

3.   MEASUREMENT OF FLEET-CAPACITY
3.1   Characterizing fleet-capacity

The measurement of capacity in fisheries is proposed as having three elements (Kirkley and Squires 1999):

i.   Capacity - potential output
ii.   Capacity utilization - observed output to capacity output and
iii. Capital utilization - desired stock of capital and the observed capital stock.

As such, these measures give "...cursory consideration of the role of labour or crew...." (Kirkley and Squires 1999). The abalone fishery's "fleet-capacity" is not best measured in boat numbers or capital equipment. This makes it representative of many small boat fisheries in Australia where diving and netting from small-boats in estuaries represents approximately 20% of all licensed commercial fishing activity by number of licence-holders (McIlgorm and Tsamenyi 2000). Management experience has shown that dealing with the fishers in such fisheries involves fisher/diver-based capacity-measures, and will also involve catch, effort, regulations and capital values of fishing-rights.

The biology and catch trends with management controls have been plotted over time by Worthington *et al.* (1998) and are shown in Figure 1. The largest catches were at over 1,200t in the early 1970s prior to regulations on size-limits. The impacts of restricting licences, and of several size-limit regulations are also noted. In 1988 a quota of 10 tonnes per diver was introduced and this was reduced in 1992 to 9 tonnes per diver.

Figure 1
Impacts of regulation and rights changes on annual catch in the NSW abalone fishery
(Source: NSW Fisheries records, and after Worthington *et al.*1998)

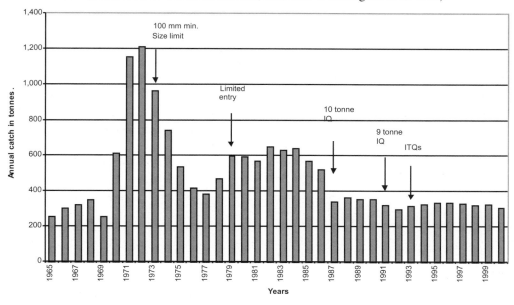

Capacity in terms of catch taken, can be seen to have been adjusted by biological size limits, limited entry, restructuring, and subsequent rights developments.

3.2   Changes in fleet-capacity arising from the introduction of transferable property-rights

The number of divers is proposed as representing a measure of fishing-capacity in this fishery. The historical trend in number of divers (1970- 2000) is shown in Figure 2.

Figure 2
The historical trend of numbers of divers in the NSW abalone fishery.
Timing of management measures are indicated (Source: NSW Fisheries records)

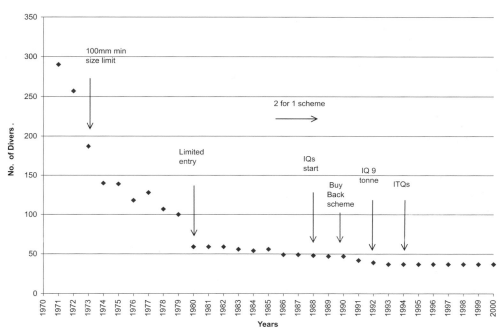

In the period 1970 to 1975, the year of peak activity under open-access was 1971 with 290 divers, dropping to 257, 187, and 140 in 1972, 1973 and 1974 respectively due to size-limit restrictions and depleted catch-rates. Fishers were all part-time participants. Groups of divers would arrive in winter, fish and then moved on to other states when previous catch-rates were not achieved (Smythe, pers. comm.).

Figure 2 illustrates the drop in divers from 100 to 59 at the onset of limited-entry to the abalone fishery in 1980. In 1990 a buy-back scheme cost $A1.32 million to remove 5 entitlements. The number of divers stabilised in the period 1993-2000 at 37.

While the number of fishers is the most significant measure of capacity in this fishery, other indices of capacity are effort (Figure 3) and catch-per-diver (Figure 4). Effort trends across the fishery are reported in Figure 3 and show the reduction in effort with advances in management from 29 000 hours in 1979, prior to limited-entry, down to 16 000 hours in the late 1990s. Figure 4 reports the catch-per-diver and indicates how greatest total catch was taken by many divers at low rates of catch-per-diver in the early 1970s. Catch-per-diver increased with limited-entry and reduced with ITQs, a measure to contain capacity.

Figure 3
Effort in hours fished per year in the NSW abalone fishery (Source: NSW Fisheries records)

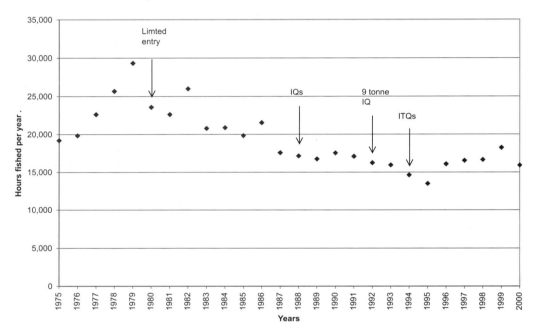

In 1992 there was discussion on the basis for ITQ allocations. Those fishers who had bought into the fishery under the "2 for 1" scheme, gaining a consolidated licence, reasoned that each original licence should have an initial quota-allocation. This eventually led to a legal challenge which did not uphold this view, and was only resolved in 1998. This caused much uncertainty in ITQ quota-trading.

A measure of potential capacity may be the current capacity (37 divers each harvesting 9 tonnes for a total of 333 tonnes of abalone), divided by the highest annual average catch-per-diver (in Figure 4, approximately 12 tonnes). This would indicate that the catch in the year 2000 could be taken by 27 divers. Under share-holding rules the industry indicated that as few as 18 divers could potentially take the current total allowable catch.

Capital measures of capacity in the abalone fishery are more difficult to measure because the boats have altered since the 1970s when they were larger as the fishers traversed the coastline by boat (Waugh 1977). the numbers of boats have in fact fallen in proportion to the reduction in numbers of divers: although many of the longer-term divers have kept larger boats, through time, smaller boats that can be towed by trailer have become more common place. This has been the case with the nominated divers who have tended to minimise catching costs and have preferred to tow smaller vessels, travelling along the improved roads rather than travelling long distances at sea (Smythe, pers. comm.).

Boat-capacity has been altered subtly by the development of the 'nominated diver'. As the original divers from the 1970s became older, they often found the physical nature of diving led to ill health. Nominated divers are now permitted under the *Fisheries Management Act 1994* and so enable sons, relatives and other persons to be employed by the original licencee.

Figure 4
Catch-per-diver (tonnes per year) in the NSW abalone fishery (Source: NSW Fisheries records)

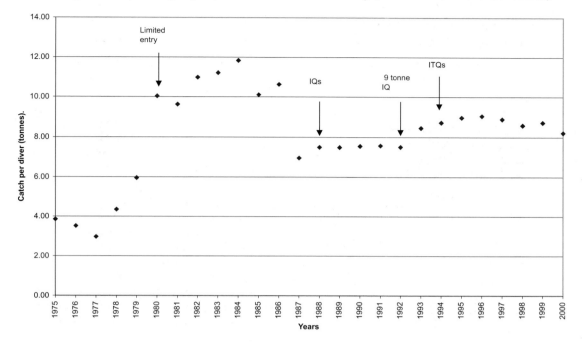

Over time this trend has developed with the nominated diver being a worker contracted on a percentage of the landed value or more usually a fixed price per kilo of product landed. While the owner settles management fees and community contribution payments, the nominated diver usually provides boat and diving services on contract. This has led to use of smaller boats (such as 6-metre mono-hulls instead of twin-hull catamarans of the 1980s) in order to reduce costs and lead to greater operating flexibility, as these boats can be easily be towed by 4-wheel drive vehicles. The advent of the live-export abalone market in the mid-1990s has resulted in the smaller newer boats being equiped with "live tanks", and with the product being cared for at all stages of production and distribution.

The number of nominated divers has increased from 9 of the 37 divers in 1996, to 27 of the 37 divers in 2001 (NSW Fisheries records). This implies that only 10 owners now dive for their own quota. Given the aging of the original licence-holders, this change was essential to prevent under-the-table arrangements in the face of ill-health. Nominated divers also mean that the fishery is now in the current year providing returns to 37 owners, contract payments/wages to 27 divers, wages to 10 owner-divers, plus wages to 37 deckhands. This division bewteen ownership and nominated divers is a significant development arising from the ability within the rights-regime to nominate another diver.

3.3   Consequences of changes in fleet-capacity

The reduction in numbers of divers through limiting entry, and the introduction of ITQ- and share-management have changed the nature of industry, which has become more professional and responsive to market needs and opportunities. The 'flow on' from ITQs has been greater industry viability for the 37 quota-holders and thus industry members are more willing to pay for extra services such as compliance, research and enhancement, as well as for the cost-recovery of management services.

Social impacts of the rights-regime have been the creation of diving jobs (nominated divers) for young divers who wish to enter the industry. Many entrant divers are sons, and sons-in-law of share-holders. This means that an old licence to one diver with deckhand (*circa* 1987), has now been replaced with a share which is supporting an owner, one diver and a deckhand. Many of the young divers will be able to run their diving actvities as a viable business, invest in shares themselves through time, and eventually hold full-shares (Smythe, pers. comm.).

The rights-system has given greater security for the families of share-owners. On the death of a share-holder the family can employ a nominated diver and still receive an income from the abalone shares. On contemplating retirement, the owners can sell-down their share-holding to 70 shares, and still work the smaller quota or use a nominated diver. The time operated by each diver depends on a range of factors such as: skill, experience, and the area being fished. However the quota-system has given more flexibility to the choice of when fishers fish.

4.   CONCENTRATION OF OWNERSHIP
4.1   Status prior to programme

In the late-1970s there were many divers operating under a licence open for any fishery in NSW. Figure 2 reports the large number of fishers prior to the introduction of limited-entry in 1980, and Figure 4 confirms the average catch of 3-6 tonnes per diver.

With the advent of limited entry from 1980 to 1988 there was a downward movement in numbers of divers, but with no aggregation of licence-ownership. On the introduction of ITQs, several transfers have had the potential for businesses to hold two permits, but this is rare. The cost of entitlements may be a barrier to entry and aggregation. The share-management system has led to one owner operating three divers within the limit of 210 shares. This is not regarded as a significant concentration of ownership across the 37 blocks of shares.

4.2   Restriction in the transfer of ownership

The original licence system restricted ownership of allocations to those with evidence of previous involvement. On commencement of limited-entry, licences were not transferable - unless under exceptional circumstances and subject to permission of the Director of Fisheries. Licences were made transferable as of 1985. With the introduction of IQs in 1988 the transferability of quota was not possible, until in 1992 limited transferability, and then in 1994 full-transferability, were authorized.

Under the new share-rights regime, in 1996 the provisional shares were transferable in total packages of 100 and with the permission of the Director. On full share-management in the year 2000, the shares could be transferred in lots of 10, though within the holding limits of the management plan and eligibility criteria (a diver must have at least 70 shares, and a business entity can have a maximum holding of only 210 shares.).

4.3   Prices received

Beach-prices (the price at first-sale received by the diver from the processors), are recorded by the Abalone Divers Association. Beach-prices are also monitored by government as they are used in estimating the community contribution due from share-holders under the *Fisheries Management Act 1994*. Figure 5 shows the average beach-prices in Australian dollars each year over over the last twenty.

Figure 5
Average beach-prices for the 1974-2000 period in the NSW abalone fishery
(Source: NSW Abalone Divers Association)

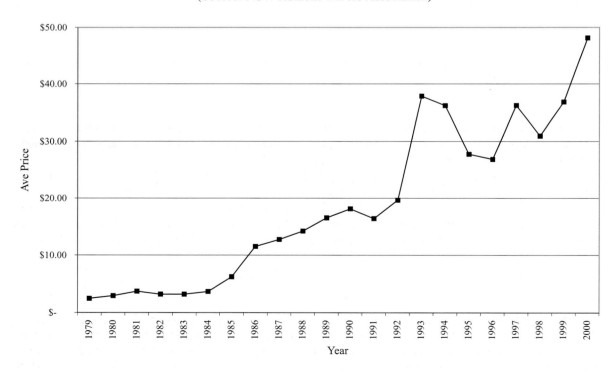

The annual rise in price has been at 14% over the 20-year period 1979-2000 and at 10% per annum since 1985. The reasons were: strong demand in Japan and the floating of the Australian currency as of 1985 and its depreciation against the Yen. With this demand and sound management, beach-price have resulted in higher values for licences, ITQs and now shares. Prices for licences, quotas and shares are not generally public knowledge, other than prices appearing in sales adverts *etc.*, but are recorded by industry, and monitored by government who must approve the transfer ensuring stamp duties are paid to Treasury. Transaction records of licences prior to ITQs, of ITQ and of share transfers are available from industry. As of 1999, a package of 100 provisional shares, based on an ITQ of 9 tonnes, was selling for approximately $A1.45 million.

Licence- and ITQ-price information was investigated by the Dominion Consulting Pty. Ltd. (McIlgorm and Campbell 1999). From financial theory the capital value per unit of quota is linearly related to beach price (assuming unit catching-costs, interest-rates, tax and finance arrangements are considered constant over time. Figure 6 shows changes in Capital Value Factor (capital values per unit of quota divided by beach price). These are from confidential transactions as recorded by industry and are graphed as annual averages of transactions. Capital value is the value of entitlements only and does not include boats and fishing gear.

Figure 6

Changes in Capital Value Factor (capital value per tonne of quota divided by beach price) for quota and share transfers in the period 1985-2000 in the NSW abalone fishery (Data source: NSW Abalone Divers Assn.)

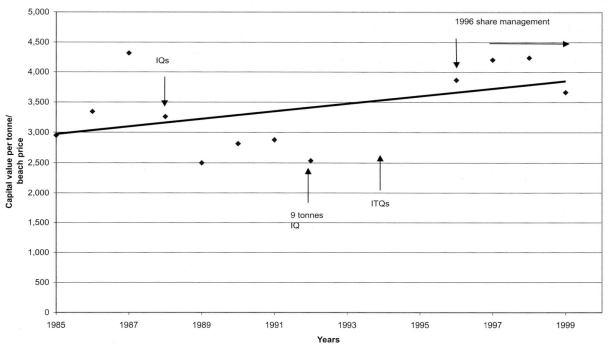

There is a definite relationship between the capital value per unit of quota and the beach-price. Figure 6 reports that over and above the beach-price there is a rise in the capital value per tonne of quota, of approximately 1.8% per year (1985-1999 period).

The impact of IQs and ITQs on the capital value of quota can be seen from the graph. In the years after the introduction of IQs the capital values adjusted by beach-price declined relative to pre-IQ levels. A reduction in catch limitation from 10 to 9 tonnes in 1992, and limited transferability did not increase IQ values (but there was only one trade in 1992). No trades happened in 1993-1995, inspite of the advent of full ITQs, but 12 trades occurred in 1996. This pattern of trades may reflect uncertainty due to the legal challenge by the "2 for 1" consolidated divers.

A linear regression analysis, using the following model, was done on the available data:

$$V/q = a_1 + a_2 P + a_3 D + a_4 D 1$$

Where capital value is (V), quota is (q), P is beach price, D is a step dummy variable to test whether the market price of licences/tonne of quota was affected by implementation of IQs in 1998, D1 is a step dummy variable to test whether the market price of licences/tonne of quota was affected by implementation of ITQs in 1994. The data are 34 confidential observations on capital transactions held by the Abalone Divers Association. The estimates produced under normal least squares assumptions are reported below:

Variable	Constant	Price	D	D$_1$
Estimate	−5,620.2	4,253.9	-20,799	16,221
t ratio	(-1.56)	(13.43)*	(-4.1)*	(3.1)*

n= 34, $\surd^2$ = 0.96,   * significant at a 1% level.

The results confirm a strong relationship between the capital values of quota and beach price. The change to IQ-management caused a significant decline in capital value per tonne of quota, whereas the change to full ITQs after 1994 raised capital values per tonne significantly.

The analysis is preliminary, subject to the assumptions of the model, and is made more complicated by the introduction of a community contribution charge to industry in 1996. As a form of tax this would be expected to reduce the capital value per tonne of the beach price (McIlgorm and Campbell 1999). It is also unclear to what extent the post-1994 results reflect the changes to share-management in 2000, announced in 1995. We would expect the new right to have more innate value than previous licences, IQs, or ITQs, as the share would reflect benefits from a recovering stock.

The clearest message is in the relationship between capital values of quota and beach price under a range of management regimes. The fishery has been kept profitable by rising beach-prices, and rights-management initiatives have contributed to this being capitalised in share values.

4.4   Effectiveness of regulations governing ownership of rights

Several issues have arisen as the rights-regime has evolved. The ownership rules were first devised in 1980 with divers as owner-operators only. In the course of time ill-health or old age meant that under original rules the licences could not be operated.

For consolidated licences, from the "2 for 1" scheme, there was an incentive to make legal arrangements, usually through a solicitor, to have ownership changed in practice, but without altering the records held by the Fisheries Department. This happened in several cases, but in the event of a fishing offence by the diver this arrangement could lead to cancellation of a licence.

However, the Fisheries Department recognised that the motivation for such arrangements was to let sons, other family members and contract divers into the fishery, and the advent of the 'nominated diver' provisions this informal practice. The message is that as fishers get older, the transfer of their rights to relatives is an important social consideration for the fishers.

4.5   Affects of programme

Through the move from licences to ITQs and shares there has been little concentration of ownership. Under share-management, entities are limited to 210 shares and only one investor has three divers. The family-nature of diving, the cost of entitlements, and the transfer provisions have limited the concentration of ownership to date. The rise in beach prices has also prevented restructuring caused by poor financial viability.

5.   DISCUSSION
5.1   Reduction in fishing-capacity

The analysis of fishing-capacity shows that limited-entry and quotas have reduced catch and effort, and hence capacity in this fishery towards more sustainable levels. However the development of ITQs and the share-management rights-system has enabled significant rises in the price of abalone to benefit the share-holders. The licence- and ITQ-systems have enabled 37 divers to remain in the fishery, but limiting fishing-effort. However original owner-operators have been able to use nominated divers, with 9 doing so in 1996 and 27 in 2001. Ten years ago there were 37 owner-operators with deckhands. There are now 10 owner-operators, 27 owners, 27 nominated divers and 37 deckhands. The division of owners and nominated divers does not increase fishing-capacity, and illustrates the diversity in outcomes that may follow rights-management – the same catch now supports more people.

There is currently no lack of profitability that might cause further restructuring. When in 2001 the ITQs were reduced from 9 to 8.25 tonnes per diver, the price rises associated with the fall in the Australian dollar more than compensated operators in terms of their overall revenue. Share-holders see this as an opportunity to rebuild the stock levels of abalone following scientific advice and potential economic returns. The situation would be different if prices were falling and it indicates that commercial viability and capacity in a "post-ITQ" fishery are essential for future management.

5.2   Concentration of ownership

In the post-ITQ fishery there have been concerns about the potential for concentration of ownership. To date these have not been realised due to the design limits of the share-management scheme and the position of industry in wanting to limit the number of divers in order to maximise the legal catch. Industry owners consider that they pay divers at rates which encourage the divers to comply with the harvesting regulation and not undertake illegal harvesting and marketing practices.

6.   DISCUSSION

The price rises in this fishery have led to illegal fishing outside the quota- or share-system. This has been estimated as high as 50% of the current legal TAC, and is a major capacity-issue (Worthington *et al.* 1998). This is the major risk in the management of capacity in this fishery where sustained price rises and the markets for abalone also give strong incentives to illegal fishers.

Rights-holders are aware of this and have funded additional compliance staff within government to protect their fishing interests. It remains to be seen whether this will be sufficient to reduce the illegal fishery.

7.   ACKNOWLEDGEMENTS

We acknowledge the information provided by the staff of the NSW Fisheries department, and by Mr John Smythe of the Abalone Management Advisory Committee and the NSW Abalone Divers Association.

8.   LITERATURE CITED

Goulstone, A.R. 2000. Rights-based Fisheries Management in New South Wales, Australia. *In:* Shotton, R. (Ed.) Use of property rights in fisheries management. Proceedings of the FishRights99 Conference. Fremantle, Western Australia, 11-19 November 1999. Workshop presentations. FAO Fisheries Technical Paper. No.404/2. Rome, FAO. pp78-83.

Kelly *v* Kelly 1990. 64 ALJR 234.

Kirkley, J.E. and D. Squires 1988. A limited information apporach for determining Capital Stock and Investment in a fishery. Fishery Bulletin, 2, No. 2, pp 339-349.

Kirkley, J.E. and D.E. Squires 1999. Measuring Capacity and Capacity Utilization in Fisheries. *In:* Gréboval, D. (Ed.) Managing Fishing Capacity: selected papers on underlying concepts and issues. FAO Fisheries Technical Paper No. 386, Rome., FAO, pp 75-116.

McIlgorm, A. and H.F. Campbell 1999. An Independent Appraisal of the Report by Hassall and Associates, and Reappraisal of the Community Contribution in NSW. A Report to NSW Abalone Management Committee and NSW Fisheries. Un-published Document by Dominion Consulting Pty. Ltd.

McIlgorm, A. and M. Tsamenyi 2000. Rights-based fisheries development in Australia; has it stalled? *In:* Shotton, R. (Ed.) Use of property rights in fisheries management. Proceedings of the FishRights99 Conference. Fremantle, Western Australia, 11-19 November 1999. FAO Fisheries Technical Paper. No.404/2. Rome, FAO. pp148-154.

Watkins, D. 2000. Abalone and the implementation of a share-based property rights in New South Wales, Australia. *In:* Shotton, R. (Ed.) Use of property rights in fisheries management. Proceedings of the FishRights99 Conference. Fremantle, Western Australia, 11-19 November 1999. FAO Fisheries Technical Paper. No.404/2. Rome, FAO. pp234-238.

Waugh, G. 1977. Costs and Incomes in the NSW Abalone Fishery: The First Year of Licence Limitation. Unisearch Ltd.

Worthington D.G., R.C. Chick, C. Blount, P.A. Brett and P.T. Gibson 1998. A Final Assessment of the NSW Abalone Fishery in 1997. Fisheries Research Institute, NSW Fisheries.

Young, M.D. 1995. The Design of Fishing-right Systems – the New South Wales Experience. Ocean and Coastal Management, Elsevier Science Ltd. Vol. 28, Nos 1-3, pp45-61.

CHANGE IN FLEET CAPACITY AND OWNERSHIP OF HARVESTING RIGHTS IN THE AUSTRALIAN SOUTHERN BLUEFIN TUNA FISHERY

D. Campbell
David Campbell and Associates
PO Box 228 Kippax Act 2615 Australia
<dcampbell.fish@bigpond.com>

1. INTRODUCTION

1.1 Resource biology

The Australian southern bluefin tuna fishery is part of a larger international fishery targeting southern bluefin tuna (SBT, *Thunnus maccoyii*). A single highly migratory 'straddling' stock exists in the Indian and Southern Ocean and individuals may live up to 40 years in age and weigh over 200kg. The juvenile phase begins in the spawning grounds south of Java with most of the two- to eight-year-old fish forming large surface schools that migrate anti-clockwise over the southern Australian continental shelf. At eight years of age, mature fish leave these Australian continental waters for the Southern and Indian Oceans (Figure 1).

Figure 1
Distribution of bluefin tuna catch off the Australian coast

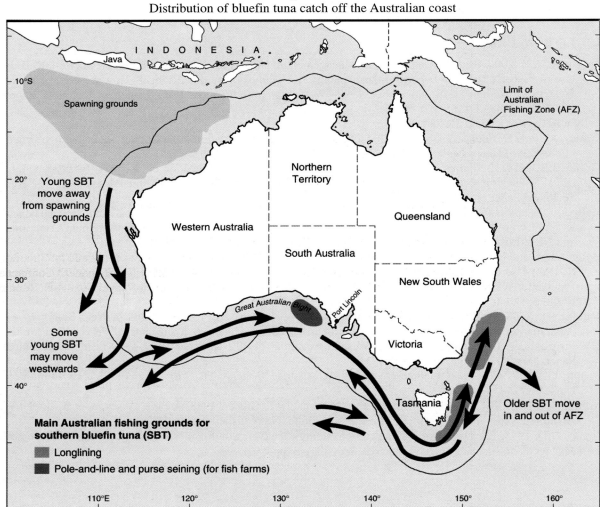

1.2 The global fishery

The highest reported annual global landings of SBT (more than 81 200t) occurred in 1961. The Japanese fishery landed 7800t or approximately 95% of global landings and the Australian fishery landed 3700t, or approximately 5% of the total. In 1983, the year before individual transferable catch quotas (ITCQ) were introduced into the Australian fishery, the reported global catch was 42 800t, or about half the record catch taken 22 years earlier. Of this, Australian landings were 17 700t (41%) and Japanese landings were 24 900t (58%). The reported landings by New Zealand and 'others' made up the remaining 1% of the reported total global landings. Since the 1989-90 fishing season, Australia's national quota has been 5262t; Japan, 6065t; and New Zealand, 420t, or 11 747t in total. From 1989-90 to 1998, the total landings of these three fisheries remained at around 11 300t. As a result of increased catch by 'others' and 1464t taken by Japan outside of its quota, global landings in 1998 were 1924t (Figure 2).

Figure 2
SBT global landings: 1966 to 1998

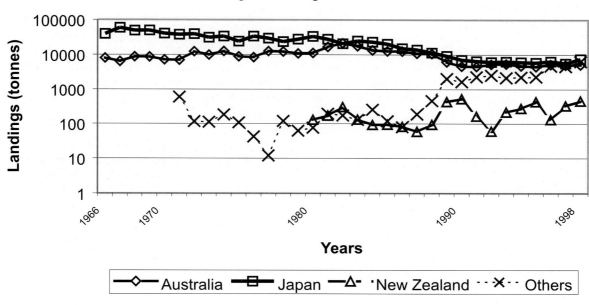

Note: 'Others' is primarily the catch of: the Republic of Korea, Indonesia and Taiwan (Province of China). There are a number of other fishing nations that catch and retain SBT: for instance, figures from the Japanese Department of Finance show Japanese imports of SBT from 19 national sources for the year 1997, and from 15 national sources in 1998.

Prior to the introduction of ITCQ, the Australian fishery relied on pole-and-line and purse-seine methods to catch immature surface-swimming fish within the Australian fishing zone. Much of this catch went into the low-valued Australian canning market, although an increasing proportion was exported to the higher-valued Italian and Japanese markets.[1] The Japanese fishery used long-lines to target deep-swimming mature fish for the Japanese sashimi market, while a large part of the relatively small New Zealand hand-line fishery went to the high-value Japanese sashimi market (Wesney, Scott and Franklin 1985).

1.3 A collapsing stock

Global stocks of SBT began to fall in the 1960s as a result of over-fishing, and continued to do so through the 1970s and the 1980s. In Australia, increasing fishing-effort, cooperative fishing between purse-seine and pole-vessels, increased use of aerial searching and expanded Western Australian fishing effort, culminated in a record Australian SBT catch of 2100t in 1982-83 (Majkowski and Caton 1984). This increase in Australian catch and increasing Japanese fishing-effort led to continuing falls in estimated levels of the parent-stock (Hampton and Majkowski 1986).

In the early 1980s, concern about a declining parent-stock triggered a series of trilateral consultations involving Australia, Japan and New Zealand. It was recognised that while parent-stocks were then depressed to about one-third of their pre-fishing levels, throughout 1975-80, the stock-levels had been reasonably stable. The participants agreed that stocks had been further reduced more recently as a result of high exploitation-rates. In response, they agreed on the need to implement fishing constraints to protect stock-biomass, and set as a conservation objective the return by 2020 of the biomasses of the parent-stocks to their 1975-1980 levels.

---

[1] In 1982-83 Australian exports to Italy and Japan were, respectively, 8900 and 2300t, valued at $A9.4 and $A3 million. In 1983-84 exports to were respectively: 2000 and 3000t, valued at $A3.2 and $A3.4 million (Smith 1986).

Sampling purse-seine-caught SBT from a tow-cage to determine average weight; this is used along with a count by under-water video (when the fish are transferred to a 'ranching' cage) to determine gross weight to be debited against quota.
Photo: Albert Caton, Bureau of Rural Sciences, Australia

This led to Australia and New Zealand implementing separate national quotas in 1984 and Japan implementing a national quota in 1985 (Wesney, Scott and Franklin 1985). These consultative agreements on a global quota were carried through to 1997.[2] The Australian share of the global quota was allocated among those with a recent history in the SBT fishery under an ITCQ fishery management programme.

This paper examines the effect that the introduction of ITCQ has had on the Australian SBT fishery in relation to changes in the distribution of quota and the capacity of the fishing fleet. To understand the changes that have occurred in the Australian fishery, it is important to include the Japanese fishery as well as others targeting the SBT stock. Both the Australian and Japanese fisheries take a large proportion of the global catch and changes in catch by one affect the operations of the other. In addition, the operation of Australia-Japan joint ventures within the Australian fishing zone, and the harvesting of Australian SBT quota by Japanese-operated vessels have affected the Australian fleet-capacity and the distribution of ownership and fleet operations. In addition, Japanese involvment in the development of SBT grow-out farms in South Australia has affected expected adjustments to the fishery following the introduction of ITCQ. Also, the arrangements between the Australian and Japanese fishing fleets appear to have been facilitated by the introduction of ITCQ into the Australian SBT fishery (Campbell, Battaglene and Brown 1996).

2.   THE NATURE OF THE HARVESTING RIGHT

Prior to 1976 the SBT fishery was an open-access fishery. Concerns over the long-term economic viability of the fishery in the mid-1970s (Wesney, Scott and Franklin 1985) led in 1976 to a freeze on entry to New South Wales and South Australia fisheries, while the Western Australia fishery remained open. The upgrading of boats, increased effort per boat and an expansion of fishing effort in Western Australia led to an overall increase in

---

[2] In May 1994 the trilateral arrangement between Australia, Japan and New Zealand was given greater international standing through the establishment of the Convention for the Conservation of Southern Bluefin Tuna. The Convention was implemented under the United Nations Conventions on the Law of the Sea, which obliges nations to cooperate, to ensure conservation and to promote optimal use of highly migratory species. The Convention provides the basis for the establishment of the Commission for the Conservation of Southern Bluefin Tuna (CCSBT). The Commission provides the framework for determining an annual global total allowable catch and national quota allocations among Australia, Japan and New Zealand. The CCSBT, however, failed to include 'other' countries (Figure 2), which had operated in the fishery since the early 1970s. While the quota, and total landings of the CCSBT members has fallen since 1980, the landings of 'other' countries, including those boats operating under flags of convenience, have continued to increase.

fishing effort and decreasing operator-returns. As a result, the freeze was terminated in April 1981 and the fishery returned to an open-access fishery.

Prior to the introduction of ITCQ in 1984, three fishing methods were used in the Australian southern bluefin fishery. The most widely used method was pole-and-line, which was carried out on boats of 10 to 30m in length. Larger boats were used to purse-seine, while a small amount of catch was taken using small (mostly inshore) boats trolling for surface-swimming fish. Some characteristics of the state-based fleets as they were in 1981-82 are presented in Table 1.

Table 1
Characteristics of the Australian southern bluefin tuna fleet 1981-82

	New South Wales - South Australia	Western Australia	Australian fleet (total)
Number: Pole boats	49	68	117
Number: Purse-seiners	5	-	5
Length (m): Pole boats	21.8 (4)	11.5 (3)	16.4 (2)
Length (m): Purse-seiners	31.0	-	31.0
Age: Pole boats	8 (18)	16 (12)-	
Age: Purse-seiners	11	-	12 (9)
Refrigerated	49	16	65

Note: Figures in parentheses are relative standard errors.
Source: Bureau of Agricultural Economics (1986).

In October 1983 a twelve-month interim management-programme involving a national total allowable catch (TAC) of 1900t, was put in place. Under the interim management-programme, the western sector (Western Australia) received an allocation of 400t (a reduction of 200t on the previous year's landings) and the eastern sector (New South Wales and South Australia) was allocated 1500t (100t below the sector's previous record landings). To ensure access to the resource by pole-and-line vessels, a 500t catch-limit was placed on purse-seiner catch with the additional requirement of no transhipment of catch to carrier boats. A minimum size for landed tuna was imposed to reduce the number of small fish caught.[3] Aside from these limitations, there were no constraints on boats' operations.

Following a review and recommendations by the Industries Assistance Commission (1984), a long-term management programme based on ITCQ, was introduced into the fishery in October 1984. The essential element of the programme was the allocation on a boat-by-boat basis of a fixed proportional entitlement of a national total allowable catch, where the national total allowable catch could be varied. Other than a restriction on the taking of SBT off the Western Australia west coast,[4] entitlements could be freely sold, or leased in whole or in part, between existing or new operators and between different regions of the Australian fishery, irrespective of fishing method used.

The joint industry-government SBT Management Advisory Committee used two criteria as a basis to allocate quota. These were: the best annual catch over the three seasons 1980-81 to 1982-83, and the assessed market value of the SBT boat and fishing gear in August to September 1984. In allocating quota-units, a weighting of three-quarters was given to the best catch, and a weighting of one-quarter was given to the assessed market value. The quota-units provided the quota-holder a proportion of the national annual quota thus allowing a smooth transition in the allocation of individual allowable catch with any change in national quota. The total allowable catch for the first year of operation was set at 14 500t, with 14 050t allocated; the difference being held back pending appeals. This was a substantial reduction on the average of 18 900t taken over the period 1981-82 to 1983-84 (ABARE 1989).

Initially, 136 boats were allocated quota-units equivalent to more than 5t of quota. Seventy, or slightly over half of these boats operated in the Western Australia fishery, which received 27 500t (20%) of the allocated quota. South Australia's 40 boats, which made up 29% of the fleet, accounted for 9300t (64%) of the initial quota allocation and the largest allocation of quota on a per boat basis. The 26 boats from New South Wales made up 19% of the national fleet, while their quota allocation made up 14% of the total allocation (Table 2).

---

[3] 225 t of undersized southern bluefin tuna were dumped in 1982-83 during the interim management-plan.

[4] This constraint is unlikely to have had a substantial affect on the distribution of quota holdings or in the distribution of fishing effort as the value of holding quota can be increased by targeting on larger SBT, which have a larger value per unit of weight.

Table 2

'Leavers' and 'stayers' in the fishery 1983-84 to 1985-86[a]

	New South Wales		South Australia		Western Australia		Total	
	No.	Average quota (tonnes)	No.	Average quota (tonnes)	No.	Average quota (tonnes)	No.	Average quota (tonnes)
Leavers								
- fished 1983-84	7	107.0	7	101.7	41	39.0	55	55.7
- did not fish 1983-84	16	55.3	3	145.0	8	17.7	27	54.1
- total	23	71.3	10	116.1	49	35.5	82	55.2
Stayers	3	128.1	30	270.4	21	48.2	*54*	176.3
Stayers & leavers	26	77.7	40	231.8	70	39.3	136	103.2

[a] Boats allocated 5t or less of quota in 1984 are not included. The Department of Primary Industry, of which the Australian Fisheries Service was part, retained 455t to accommodate the need for additional quota allocation. Leavers are defined as those boats for which quota was reduced to 5t or less as at 30 June 1986.
[b] Because only three owner-operators left the South Australian fishery, no personal data were collected for that state, for reasons of confidentiality.
Source: Australian Fisheries Service.

3.   MEASUREMENT OF FLEET-CAPACITY
3.1 Characterising fleet-capacity and their operational environment

By the early 1980s, the biological and economic status of the fishery was undergoing increasing pressure. The Bureau of Agricultural Economics[5] (1983, 1986) published survey data showing the average per boat return to capital and management had fallen from $A49 000 in 1980-81 to $A8285 in 1981-82 and an average per boat debt level of $A57 733 in 1981-82. In the same paper, the Bureau of Agricultural Economics (1983) presented CSIRO findings that the estimated global sustainable catch of 3300t had been exceeded by approximately 700t in 1982.

In setting a global quota, the objective was to return the parent biomass of SBT to the 1980 level by 2020. The recommendation by the Bureau of Agricultural Economics (1983) and the Industries Assistance Commission (1984),[6] for the use of ITCQ to manage the fishery within the national quota, was to achieve an economic objective.

At the time of the introduction of the ITCQ-based management plan into the Australian SBT fishery, it consisted of three fleets located in southern New South Wales, Port Lincoln in South Australia, and the Western Australian south coast. Not accounted for in this was a small amount of bycatch taken off Tasmania by local fishers with some Tasmanian boats receiving quota of less than 5t. The fleets differed in the nature of their operations and in their access to SBT stocks. Differences also existed in the relative opportunities to participate in other fisheries, in the non-fishing skills of owner-operators, and in opportunities to participate in non-fishing employment. These differences could be expected to have resulted in relative regional differences in the opportunity-cost for owner-operators to remain in or leave the SBT fishery.

In most part, the New South Wales fleet consisted of multi-purpose vessels that combined poling of 3-8 year old tuna from October to January with trawling in the southeast trawl fishery during the SBT off-season. By 1983-84, as a result of increases in catches in South Australia and Western Australia, and a decrease in SBT breeding stocks, the New South Wales catch had fallen to less than 1000t. While alternative employment skills and opportunities were limited, fleet operations were readily redirected on a full-time basis to the southeast trawl fishery, which had a developing catch of orange roughy (*Hoplostethus atlanticus*).

The South Australia fleet targeting 3- to 5-year-old fish in January through May, consisted of larger specialist vessels including five purse-seine vessels. In 1968-69 landings were just over 3000t, but had increased 10 600t, or almost 2/3 of the national landings in 1983-84, the season prior the introduction of ITCQ-based management. Unlike New South Wales, there were few, if any, alternative fisheries to which the boats or operators could move, while alternative employment skills and opportunities were limited.

The newly-developed Western Australian fishery was dominated by small multi-purpose boats that used pole-and-line to target 2-to 3-year-old fish during November through May. When not poling for SBT, these vessels were used in a multi-species wet fish fishery. Just over 4000t of SBT was landed in 1983-84. Alternative

---

[5] Superseded by the Australian Bureau of Agricultural and Resource Economics (ABARE).
[6] Quantitative work by Kennedy and Watkins (1984, 1985) was important to this. Campbell (1984) and Wesney, Scott and Franklin (1985) reviewed some of the issues involved in considering the application of individual transferable catch quota to the southern bluefin tuna Fishery.

employment opportunities were no better than those for southern New South Wales or for Port Lincoln. However, many of those who had entered the recently established Western Australia SBT fishery had working skills aside from fishing, and as a result, had more employment options available to them.

Larger SBT, whether for canning or the sashimi market, normally fetch a higher price per kilogram. As a result, SBT in New South Wales could be expected to be more valuable than the same volume of catch taken in South Australia or Western Australia. However, because of the regional differences discussed above, it could be expected that at the time of the introduction of ITCQ there would be region-by-region differences in the opportunity-cost of remaining in the SBT fishery consistent with such that: those in New South Wales > Western Australia > South Australia. These differences in opportunity-cost appear to have been important in the consequent restructuring of the fishery and the concentration of quota-ownership and fishing-effort in South Australia (Campbell, Brown and Battaglene 2000).

3.2 Changes in fleet-capacity: initial adjustment 1983-84 to 1985-86
3.2.1 Fleet-reduction

By introducing ITCQ, it was expected that, because of higher unit-value, profit maximising quota-holders would maximise the value of their rights by targeting larger fish, which receive a higher price on the Sashimi market. As a result, it was expected that the fishery would move towards the use of long-lines and the operational focus would move eastwards as stocks return to the New South Wales coast.

A rapid and substantial restructuring of the tuna industry fleet followed the allocation of quota in 1984: over half of the boats that were allocated quota left[7] the fishery in the first twelve months of the scheme, with quota from two-thirds of the boats being sold by the end of the 1985-86 financial year. This two-thirds-reduction in the number of boats accounted for one-third of the quota, as the boats taken out of the fishery had, on average, less quota and were likely to be older and smaller than those vessels retained in the fishing fleet (compare Tables 1 and 3).

More than half of the boats taken out of the fishery were from the Western Australia fleet, while the largest proportion of within-state adjustment was in New South Wales, where nearly all of the vessels allocated quota had left the fishery by the end of June 1986 (Table 2).

Table 3
Characteristics of 'leaver' boats from the southern bluefin tuna fishery

Details	New South Wales	South Australia	Western Australia	Australia
Number of boats [a]	23	10	49	82
Average quota sold [b] (t)	71.8 (8.9)	107.1 (16.91)	36.9 (6.35)	55.5 (5.76)
Average value of quota sold [b] ($)	114 200 (8.90)	143 800 (18.13)	63 200 (7.97)	87 600 (6.11)
Average price received [b] ($/t)	1 590 (4.48)	1 342 (10.54)	1 712 (3.50)	1 579 (2.74)
Average assessed boat value [b] ($)	373 926	494 160	75 615	211 993
Average age, October 1984 [a]	11	12	13	12
Average length [b] (m)	21 (1.38)	19.3 (1.88)	12.8 (5.22)	15.92 (1.89)

Note: Numbers in parentheses are relative standard errors expressed as percentages of the estimates.
Source: [a]Australian Fisheries Service. [b]Campbell and Wilkes 1988.

3.2.2 The nature of boats leaving the fishery

The main characteristics of the SBT 'leaver' boats are summarized in Table 3. All boats were pole-boats and, on average, were built in 1972 with a hull length of about 16m. Vessels taken out of the New South Wales fishery were constructed more recently and were larger than the Western Australia boats. As most were non-specialist vessels, only minor modifications were required to make them serviceable in other fisheries.

Many of the boats that had not operated in the SBT fishery in 1983-84 had already left the fishery and were unlikely to have returned to the fishery in the near future. Therefore, the following discussion is confined to those 55 boats which had operated in the fishery 1983-84, had received an initial allocation of quota, and for which quota was sold (or at least reduced to below the threshold 5t by the end of June 1986).

---

[7] Boats allocated more than 5 t of quota were assumed to have left the fishery once the quota allocated to the boat had been reduced through sale or lease to 5 t or less.

Overall, there was little difference between the boats exiting the fishery, in terms of the average amount of quota, according to whether they had or had not operated in the SBT fishery in 1983-84 (Table 2). On a state basis, however, there were substantial differences: in South Australia, the small number of 'leaver' boats that had not fished for SBT in 1983-84 had on average been allocated almost half as much again as those that had operated in the fishery in 1983-84. The New South Wales and Western Australia boats exiting the fishery, which had not been used to fish for SBT in 1983-84, had received only half as much quota as those that had fished.

Although boats taken out of the fishery constituted two-thirds of the boats to which quota had been allocated, these boats were only allocated 4500t, or one-third of the total available quota. On average, boats for which quota were sold and taken out of the fishery, were allocated 55t of quota compared with the average of 176t allocated to boats which remained in the fishery. On average, the former Western Australia boats received 36t, which was substantially less quota than the quota allocated for the New South Wales and South Australia boats leaving the fishery, which were respectively 71t and 116t.

3.2.3 Impact on adjoining fisheries

The effect of restructuring on other fisheries in the area differed between states. According to the estimates based on survey results (Campbell and Wilkes 1988), by July 1986 almost half the departing boats operating in the SBT fishery in 1983-84 had been sold (Table 4). Other than two boats from South Australia, all of the boats that had been sold were from Western Australia. The latter were sold for use in other local fisheries such as wet fish, shark, for use outside of fishing, or into fisheries outside of southern Western Australia. Owners of some of the Western Australian boats which had been sold replaced them with other boats for use in another local fishery aside from SBT. That is, of the 41 boats in Western Australia which had operated in the SBT fishery in 1983-84, 12 boats, or nearly 1/3, were removed from local fisheries.

All of the boats sold into another fishery either moved into one in which constraints on entry existed (therefore to replace existing boats) or into a fishery that was considered to be under-developed. This, in addition to the decrease in the number of fishing boats, implied that there is likely to have been an overall decrease in fishing effort rather than an increase of fishing effort.

In New South Wales and South Australia most departing boats that had operated in the SBT fishery in 1983-84 moved to a local fishery (Table 4). In many cases these boats had already operated in these same local fisheries on a part-time basis prior to leaving the SBT fishery. This most likely resulted in increased fishing effort in New South Wales and South Australia, although the lack of availability of SBT off New South Wales would have resulted in the movement of New South Wales vessels to full-time operations in the south-east trawl fishery, irrespective of the change in management.

Table 4
Use of 'leaver' boats which had fished in 1983-84 [a]

State	Sold to other local fishery	Sold to non local fishery or non-fishing use	Total sold	Replaced and used in local fishery	'Leaver' boat or replacement in local fishery	Left SBT fishery
New South Wales	0	0	0	0	7	7
South Australia	1 (81)	1 (81)	2 (57)	0	4 (23)	7
Western Australia	9 (16)	13 (12)	22 (10)	10 (8)	29 (6)	41
Total	10 (16)	14 (13)	24 (10)	10 (8)	40 (5)	55

[a] As at the end of June 1986.
Numbers in parenthesis are relative standard error. *Source*: Campbell and Wilkes (1986)

3.2.4 Economic impact
3.2.4.1 Change in fleet capitalisation

Based on the 1984 assessed market value, the value of the boats exiting from the SBT fishery was approximately $A17.4 million. However not all of this amount is attributable to the 1984 management changes as one-third of the boats for which quota was sold did not operate in the fishery in 1983-84. Most of these boats, especially the 60% of the boats from New South Wales, would probably not have returned to the fishery, as a result of the decline in SBT availability on the east coast. Excluding these 27 boats, the value of the 55 boats exiting the fishery that had operated in the fishery in 1983-84 was $A9.15 million.

There were two other difficulties in estimating how much of the capital-value of the exiting boat's value should be apportioned to the SBT fishery. Other than in South Australia, the boats used to catch tuna were regularly used to catch other fish species. Therefore, the costs of the vessels were jointly shared between the different fishing operations. To measure the change in capitalisation following the exit of vessels it would be necessary to apportion the total boat-costs between the SBT and the other fisheries in which the boats were operated. However, while the quota for ten of the South Australia 'leaver' boats was sold, the capital-value of an additional six boats taken out of the South Australian fleet as a result of owners amalgamating their quota, was not included in the survey.

Of the 55 exiting boats used to fish for SBT in 1983-84, 48 did so on a part-time basis in conjunction with other fishing activities. The remaining seven boats exiting the fishery that had operated full-time in the fishery were from South Australia.

3.2.4.2 Economic impact on leavers

Only two boats were sold in South Australia and none in New South Wales, and for reasons of confidentiality, it is not possible to properly assess any change in market value of exiting boats in South Australia. In Western Australia, the average value of exiting boats apparently dropped by $A32 000 or 40% (Table 5). While a number of elements may have been important in explaining this fall, the sudden increase in the number of boats on the market, and the decline in demand for vessels arising from the cutback in the Western Australian SBT fishery, would have been important.

In Western Australia most exiting vessels had fished for SBT in the year prior to the introduction of the quota-based management-plan. By comparison, most of the New South Wales boats had stopped fishing prior to the 1983-84 season, as a result of the reduced availability of SBT off the east coast.

Table 5
Change in average asset-value of Western Australia exiting boats - 1984-85 to 1985-86

Assessed value of boat $A	Sale value of boat $A	Sale value of quota SA	Sale value of boat plus quota $A	Increase in asset value $A
82 223	50 049 (10.25)	65 022 (7.65)	115 071 (10.02)	32 848 (35.55)

Source: Campbell, Brown and Battaglene (2000). Note: values in brackets are standard errors.

There appears to have been some variation in the price received for quota between the various state fisheries. In Western Australia, the estimated average price received was approximately $A1770/t, while in New South Wales and South Australia the respective average prices were around $A1600 and $A1300/t. For the fleet as a whole, the estimated average payment received per departing boat was approximately $A87 600. Anecdotal evidence indicates that the price received for quota early in the first weeks of the new management plan was much lower than the average price.[8] The fall in boat-value was more than offset by the sale of the associated SBT quota. Thus, the estimated average change in asset-value per boat was an increase of $A33 000, or 40%. Because most of the boats were sold within twelve months of their assessed valuation, no allowance is made for depreciation; otherwise the estimated improvement in asset-value would have been greater.

The impact on the income of those owner-operators, who had left the SBT fishery after the introduction of individual transferable quota, but had not retired, was estimated by comparing the 1983-84 combined taxable income of the leavers and their spouses with their 1985-86 income (Table 6).

The average taxable family-income for the New South Wales and Western Australia owner-operators was estimated to have increased in real terms, on a per boat basis, by 13%. The average family-income per boat in New South Wales was estimated to have increased by 23%, and that in Western Australia by 8%. In this latter state, not only is the average increase in taxable-income less, but the standard error is also greater. These two statistics indicate that, when compared with New South Wales, a higher proportion of Western Australian owner-operators were likely to have been worse off following the introduction of ITCQ.

3.3 Further reductions in global and national quota

Research by Hampton and Majkowski (1986) indicated that in spite of the restrictions on allowable catch in 1984-85 and 1985-86 the parent-stocks continued to fall. Further national quota-reductions were introduced in the 1986-87 season for both Australia and Japan. The Japanese quota was reduced from 23 150t to 19 500t (Table 7). While the nominal Australian quota was retained at 14 500t, the Japanese industry provided funds to enable the Tuna Boat Owners Association of Australia to lease and withhold from capture 3000t of Australian quota. This placed an effective catch limit of 11 500t on Australian operators for the three-year period to 30 September 1989.

---

[8] Some banks, uncertain of the affects of the new management plan, withdrew loan facilities to operators in the SBT fishery in Western Australia. This apparently led to forced sales by some quota holders. The Western Australian Government, by acting as guarantor to these loans, provided short-term stability to the Western Australian quota market.

Table 6

Change in family income of owner-operators who left the fishery [a]

State	Income per boat		
	1983-84 A$	1985-86 A$	Difference [b] A$
New South Wales	51 318(48)	62 897 (41)	11 579 (34)
Western Australia	13 870 (20)	14 937 (14)	1 067 (171)
Total	19 331 (22)	21 931 (19)	2 600 (64)

[a] Survey estimates for exiting boats that operated in the SBT fishery in 1983-84; 1983-84 values inflated, using the consumer price index, to 1986 equivalent values. [b] For New South Wales, there is a less than 5% probability that the change in income is less than zero. For Western Australia there is a less than 30% probability that the change in income is less than zero. For both states together there is a less than 10% probability that the change in income is less than zero.

Note: Numbers in parentheses are relative standard errors, expressed as a percentage of the estimates.

Source: Campbell, Brown and Battaglene (2000). Note: values in brackets are standard errors.

Despite the further reduction of 6650t in the global quota, concern that the parent stocks would continue to fall below an acceptable level remained (Caton and Majkowski 1987). This led to a further reduction of 17 500t reduction to 15 500t in October 1988. The Australian quota was reduced to 6250t and the termination of the leasing agreement with Japan (although funds were retained) the Japanese quota was reduced to 8800t and the New Zealand quota to 450t. In the following year the global quota was reduced by another one-third to 11 750t with national quota of: Australia 5265t, Japan 6065t and New Zealand 420t. Australia, Japan and New Zealand, through the Commission for the Conservation of SBT (CCSBT), retained these quota-levels up until 1996, when the members of the Commission were unable to reach agreement on an acceptable global quota.[9]

Table 7

Catch limits and landings

Season	Australia		Australia		New Zealand	
	Catch limit t	Catch [a] t	Catch limit t	Catch [a] t	Catch limit t	Catch [a] t
1983-84	21 000	15 483	no limit	23 328	10 000 fish [b]	132
1984-85	14 500	13 486	no limit	20 396	10 000 fish [b]	93
1985-86	14 500	13 237	23 150	15 182	1 000	94
1986-87	11 500 [d]	11 308	19 500	13 964	1 000	60
1987-88	11 500 [d]	10 976	19 500	11 422	1 000	94
1988-89	6 250	5 984	8 800	9 222	450	437
1989-90	5 265	4 849	6 065	7 056	420	529
1990-91	5 265	4 316	6 065	6 474	420	165
1991-92	5.265	4 894	6.065	6 137	420	60
1992-93	5 265	5 212	6 065	6 320	420	216
1993-94	5 265	4 937	6 065	6 064	420	277
1994-95	5 265	5.080	6 065	5 866	420	435
1995-96	5 265	5 188	6 065		420	140
1996-97	5 265	4 978	6 065	5 588	420	333
1997-98	5 265	5 087	6 065	6 038	420	331
1998-99	5 265	5 232	6 065	na	420	457

[a] Australian catches include Australia-Japan joint venture and real-time monitoring programme longline catches.

[b] Number of fish.

[c] The Japanese and New Zealand seasons do not correspond with the Australian season. For purpose of this table calendar year data for the first year of the calendar year is given; *e.g.* Japanese catch for 1993-94 of 6 320 is the calendar year catch for 1993.

[d] The nominal national quota remained at 14 500t, however, the leasing of 3000t meant the amount available was 11 500t.   na: Not available.

Sources: Bureau of Resource Sciences; Australian Fisheries Management Authority.

---

[9] In an interim ruling, the International Tribunal for the Law of the Sea (1999) required that, without the presence of a new agreement, the members of the Commission should maintain the previously agreed quota-levels. Since then, the five-member international arbitral tribunal of the International Centre for Settlement of Investment Disputes (2000) has ruled that it lacks jurisdiction to decide the merits of the dispute and has revoked the provisional measures.

3.4 Longer-term adjustments

Geen and Nayar (1989) examined the likely economic performance of the fishery in order to estimate the effect that the management changes may have had on it. They assumed that the fishery management-system in force in 1983-84 was maintained, *i.e.* with the total Australian quota of 14 500t managed under a competitive total allowable catch (TAC) management-system, rather than under an individual transferable quota (ITQ) management-system. They concluded that by 1986-87, fishing in New South Wales would have ceased, operations off South Australia would have fallen substantially, and that SBT operations would have been concentrated on the smaller age-classes off Western Australia. This would have resulted in an annual economic profit to the Western Australia fleet of $A1.6 million, which Geen and Nayar believed would be unsustainable because of decreasing stock numbers.

By contrast, it was estimated that by operating under ITCQ, the SBT operators would have earned an annual sustainable economic profit of $A6.5 million. In addition to the increased profitability, between 1983 and 1986 the average size of fish taken had increased by 11%, thus promoting conservation objectives through a reduction in the number of fish per tonne of catch (Geen and Nayar 1989).

This increase in the size of landed SBT is likely the result of the incentive provided by the higher price per kilogram received for larger fish in the canning and Japanese sashimi markets. This incentive to take larger fish, the shortage of SBT off New South Wales, and the limited alternatives for operators in South Australia appear to have been important factors in the concentration of quota in South Australia (Table 8).

Further apparent increases in resource-rent returns have occurred in the Australian SBT fishery through increasing long-line catches, careful handling of the pole-and-line and purse-seine catches, returns from an Australian/Japanese industry joint-venture and an increasing proportion of Australian catch being directed to the ranching operation off Port Lincoln. These operational changes have been accompanied by marketing changes with nearly all of the Australian production being directed to the Japanese sashimi market.

In 1982-83, the landings of 21 300t of SBT were estimated to be worth $A14.3 million (Table 10). In 1983-84 the catch fell to 15 800t, worth $A12.6 million, yet, by 1994-95, the 5200t of landed quota was valued at $A86.3 million. Much of this increase in the value of landings was the result of Australia-Japan industry-to-industry cooperation, including the joint venture operation in which a proportion of the Australian quota is leased to the Japanese joint-venture partners to take tuna within the Australian fishing zone. The importance of the joint-venture operations to provide returns to Australian SBT quota-holders was highlighted when joint-venture catch fell from 1684t in 1994-95 to 431t in 1995-96. As a result, from 1994-95 to 1995-96, the value of the SBT fishery fell by nearly 50% (Table 9). This was caused by the failure of Australia, Japan and New Zealand, as members of the CCSBT, to arrive at an agreement on a global quota. As a result, since 1996, Japanese fishers have been excluded from fishing in Australian waters.

Table 8
Changes in the distribution of individual transferable quota

Jurisdiction	Season							
	1984		1986		1994		1998	
	t	%	t	%	t	%	t	%
New South Wales	2 022	14	520	3	185	4	123	2
South Australia	9 271	64	12 563	88	4 596	88	4 762	92
Western Australia	2 752	19	1 249	9	424	8	267	5
Queensland	-	-	-	-			48	1
Commwealth Govt.	454	3	-	-	-	-	-	-

Note: this is for boats having quota in excess of 5t.
Source: AFS and AFMA.

The longer-term effect appears to be the increasing proportion of the Australian quota which is being used to capture fish for the production of farmed-tuna. While such operations are feasible without ITCQ, the existence of ITCQ appears to have facilitated this operation. Boats that would otherwise have been engaged in the 'race-for-catch' are now directed to catching 4- to 5-year-old SBT by purse-seine in the Great Australian Bight as well as the time-consuming operation of towing them to grow-out cages in the waters off Port Lincoln. This operation normally occurs in the last quarter of the calendar year, with farmed-fish being sold into the Japanese sashimi market six months or more after capture. Most is sold by October of the following year. During this time individual fish can increase in body weight by one-third, and there is improved flesh-quality.

The total gross value of production from the farm-sector was estimated at $A38 million in 1994-95 (Table 10) with a gross value of SBT for the year of $A124 million. In April 1996 the production of farmed tuna suffered as a result of a storm that killed up to 75% of the caged fish. Most of this kill was replaced with additional quota caught in the 1995-96 financial year and sold in the 1996-97 financial year. The value of farmed tuna was maintained through 1995-96 to 1996-97, but it more than doubled in 1997-98, and doubled again in 1998-99, to $A167 million.

Table 9
Changes in gross value of the Australian southern bluefin tuna fishery[a]

Year	Wild caught		Farmed		Gross total[a]	
	Value $A m	Quantity t	Value $A m	Quantity t	Value $A m	Quantity t
1982-83	14.3	21 300			14.3	21 300
1983-84	12.6	15 800			12.6	15 800
1984-85	16.2	13 500			16.2	13 500
1986-87	23.0[b]	11 800			23.0[b]	11 800
1987-88	18.1[b]	10 100			18.1[b]	10 100
1988-89	20.3	6 000			20.3	6 000
1989-90	60.4[c]	5 000			60.4[c]	5 000
1990-91	55.5	4 300			55.5	4 300
1991-92	63.4	5 100	1.8	100	65.2	5 200
1992-93	98.0	4 900	10.2	500	108.2	15 100
1993-94	80.3	4 700	24.2	1 300	104.5	6 000
1994-95	86.3	5 200	38.0	1 900	124.0	7 100
1995-96	45.9	5 100	39.9	2 000	85.8	7 100
1996-97	40.9	5 900[d]	40.2	2 100	81.1	8 000
1997-98	40.9	4 800	87.2	5 100	128.1	9 900
1998-99	56.8	5 600[d]	166.7	6 400	223.5	12 000

a No allowance is made in these figures for wild caught catch that goes into the cage-raising operation at Port   Lincoln.
b Includes $A7.57 million paid by Japanese industry to the Australian industry to reduce the annual catch of SBT by 3000t for three years. As the Australian quota, along with the global quota was substantially reduced in 1988-89, and there was no additional annual 3000t decrease of Australian catch in that year, the payment is distributed over two rather than three years.
c The Australia-Japan joint venture commenced this year. The value of Australian quota taken by the joint venture is based on the price received in Japan less transport and marketing cost.
d The apparent discrepancy between catch and quota limits is because the fishing season is not synchronous with the financial year.
Source: ABARE (2000) and earlier editions.

3.5   International co-operative fishing arrangements under ITCQ
In 1986 an agreement was made between Japanese industry and South Australia quota-holders for the South Australian industry to forgo catching 3000t of its quota in the 1986-87, 1987-88 and 1988-89 seasons. In return, those forgoing catching quota received a total compensation of $A7.57 million from the Japanese industry. The pay-off to the Japanese tuna long-liners was: that tuna not taken by Australian operators would later become accessible to Japanese long-line operators.

In 1988-89, continuing concern over parent stock levels led to a reduction in global and in national quota levels. As a result the Australian quota was reduced to 6250t and the industry-to-industry agreement was suspended. Australian operators did not forgo catching any of their quota in 1988-89, the third year of the agreement, although the initial payment was retained.

The 1988-89 season saw the introduction of a new element into the fishery, namely the use of Australian quota by Japanese owned and crewed long-line boats to take SBT in the Australian fishing zone in a joint-venture with Australian industry. This operation peaked in 1992-93, when more than 50% of quota was taken by the joint-venture operators. A further change was the use of Australian quota by the Japanese fleet operating on the high-seas beyond the Australian fishing zone. This was part of a programme using Japanese as well as Australian quota to take catch as part of a real-time monitoring-programme.[10] As a result of these two programmes, Japanese vessels took 65% of the Australian quota in 1992-93.

---

[10] The Real Time Monitoring Programme (RTMP) was used to provide data on catch, effort and size composition for a small component of the Japanese long-line fleet operating on the high-seas. The availability of this programme did improve the data-access and timeliness in provision of the data.

These changes resulted in a substantial increase in the net-returns enjoyed by the Australian quota-holders involved in these operations.[11] The conditions of this agreement also included $A500 000 over three years for SBT research, an annual levy of $A1.4 million for monitoring and enforcement, training of Australian crew in long-lining, and funds for research on SBT farming. In addition to the Australian quota used in the joint-venture, some additional Australian quota was used by Japanese vessels to catch SBT outside the Australian fishing zone for purposes of real time monitoring of fish stocks. The value of the catch taken under the Joint Agreement and the Real-Time Monitoring-Programme is presented in Table 10.

The last year of real-time monitoring and joint-venture operations was 1994-95, when Japanese boats took 37% of the Australian quota. The reason for the cessation of these co-operative operations was the failure of Japan and Australia to come to an agreement on the global quota. In the meantime, Australia has continued to operate according to the Australian national quota arrived at within the CCSBT.

Table 10
Catch under Joint-venture and the Real-Time Monitoring programmes

Operations		1988-89	1989-90	1990-91	1991-92	1992-93	1993-94	1994-95
Joint	t [a]	684	400	881	2057	2735	2299	1295
venture	$A [b]	37 620	22 000	28 190	37 643	74 726	17 128	35 379
RTMP [c]	t [a]			300	800	650	270	650
	$A [b]			9893	14 640	17 759	7377	17 759
Total	t [a]	684	400	1181	2857	3385	2569	1945
	$A [b]							

Source: [a] Landings are provided by Agriculture, Forests and Fisheries Australia.
[b] Unit values used to estimate total values are from ABARE (1996) and previous years. Note that as ABARE volume values are for financial year not fishing season, they have not been used in the table.
[c] Real Time Monitoring programme.

While not directly related to the SBT fishery ITCQ management programme, Japanese fishers have been able to take SBT within the Australian fishing zone (Table 11). This access has been particularly important as it has enabled Japanese operators to target large fish off south-east Tasmania.

Table 11
Catch and value of Japanese quota taken by Japanese operators in the Australian fishing zone

	1986-87	1987-88	1988-89	1989-90	1990-91	1991-92	1992-93	1993-94	1994-95	1995-96
Catch (t)	1134.4	1026.6	2308.2	1693.1	660.6	386.8	804.4	436.3	410.7	358.8
Value ($m)	62.5	61.7	126.3	72.9	25.3	12.2	31.6	17.9	15.7	13.1

Source: Hogan, Van Landegham and Topp (1997).

Japanese fishers were excluded from fishing within the Australian fishing zone in 1984-85 as well as since 1995-96 as a result of the failure to agree on the amount of global quota. Acting outside of the Commission for the Conservation of Southern Bluefin Tuna, the Japanese fishery took an additional 2000t of SBT in 1999 and as a result, Australia and New Zealand took Japan to the International Tribunal for the Law of the Sea. In 1999, the tribunal placed an interim injunction on Japan taking additional SBT, even if the catch were to be part of an experimental fishing programme. The Tribunal ruled that the members of the CCSBT ensure they maintain their annual catch at the quantities last agreed to within the Commission. The Tribunal ruling requires Japan to treat the 'experimental' catch as part of its national quota of 6065 t. The 714t from the 'experimental fishing programme' was counted against the 1999 allocation of 6065t, with the remaining 1484t counted against the 2000 allocation of 6065t.

On 7 August 2000, a five-member international arbitral tribunal invoked the interim provisional measures imposed on Japan by the International Tribunal for the Law of the Sea. The Tribunal noted that the successful settlement of the dispute would be promoted by the parties abstaining from any unilateral act that may aggravate it (International Centre for Settlement of Investment Disputers 2000). Since the interim decision, the members of the CCSBT have moved closer to resolving the issues on which the original dispute was based.

---

[11] It is also likely that these transfers in access to quota resulted in an improvement in the economics of the over-capitalised Japanese fleet.

4   CONCENTRATION OF OWNERSHIP
4.1   Status prior to the programme

In 1981-82, the Australian SBT fishery consisted of 117 operators, although, on the 1 October 1984, 136 boats were allocated SBT quota allocation in excess of 5t. Of these 136 quota-holding boats, 26 (19%) were in New South Wales, 40 (29%) in South Australia, and 70 (52%) in Western Australia. In the three years prior to the programme the average seasonal catch by the Australian fleet was 18 700t, with an average global catch of 43 600t.

Without action being taken in Australia, the focus of operation for the Australian SBT fleet is likely to have moved to targeting of the 2- to 3-year-old (juvenile) fish off Western Australia This would have likely lead to declining catch, reduced gross returns and the eventual collapse of the fishery. At the global level, it was recognised that the global catch would have to be reduced, and researchers proposed that the parent stocks be allowed to recover to their 1980 levels by the year 2020.

4.2   Restrictions on transfer of ownership

While there are no restrictions on the lease or sale of Australian SBT quota within Australia, quota can only be used by a boat operator who either holds, or is operating on behalf of the holder of, a Commonwealth fishing permit. However, the Australian Fisheries Management Authority (AFMA) has placed a requirement on the long-liners (operating in waters off southeastern Australia during those periods of the year when southern bluefin tuna are likely to be taken) to hold a minimum SBT-quota of 500kg (AFMA 2000).

4.3   Effects of the programme

A number of changes occurred following the introduction of ITCQ into the SBT fishery. These included structural changes to the national fishery, to fishing technology, and changes in market provision and product. While many of these changes may not have been the direct result of the introduction of ITCQ, it is likely that the management-changes facilitated these outcomes. In addition, without changes in the management of the fishery, the fishery would have collapsed. Without an ITCQ programme, the benefits received by leavers on exiting the fishery would not have been possible.

The major structural changes observed in the fishery were the immediate concentration of quota-ownership and fishing operations to South Australia, the cooperation between Australian and Japanese fishing operations, and the later return of SBT to waters off the coast of New South Wales. The cooperation between Australia and Japan included the leasing and withdrawal from the global fishery of Australian quota by the Japanese fishing industry in the 1986-87 and 1987-88 Australian fishing seasons, the use of Australian quota by the Japanese high-seas fleet for the purpose of real-time monitoring of catch and the Australia-Japan joint-venture long-line fishery. The closure of the real-time monitoring and the Australia-Japan joint-venture programmes has seen most of Australia's quota shift to the SBT grow-out ranching operations off Port Lincoln in South Australia. These farm operations were assisted by Japanese finance and technology.

Changes in the use of fishing technology included the use of long-line fishing methods off the New South Wales coast, with the quantity of landings taken by long-line increasing from 15t in 1990-91, to 648t in 1997-98 (Table 12), or 12% of SBT quota. The use of purse-seine to capture and transport live fish to the fish-farms off Port Lincoln, and the use of sea cages to grow out and improve the quality (intra-muscle marbling) of juvenile fish. These changes in fishing technology included changes in product and in marketing. Wild-caught landings are no longer being directed to canning in Australian, or shipment to Italy, rather now to the Japanese sashimi market either directly, or following grow-out activities.

In addition to a gross increase in the value of SBT product (Table 9), there has been an increase in realised rent-returns to the fishery to the benefit of fishers, tuna farmers and society as a whole. The expected rent-returns at the introduction of the management plan were important in affecting the price paid to those owner-operators leaving the fishery, and assisted the structural adjustments that occurred in the fishery. While there has been a recovery of SBT stocks off eastern Australia, the rate of recovery in fishing operations is not as great as was expected. The degree to which landings increase off eastern Australia under the current management plan will depend on the relative profitability of east coast operations compared to elsewhere. To-date, SBT quota-holdings and operations continue to be concentrated off South Australia. A possible factor in that this is the high growth-rates obtained from 4- to 5-year-old SBT in the grow-out cages more than compensates for any improvement in the return on quota-holdings from taking larger fish.

5.   DISCUSSION
5.1   Reduction in fleet-capacity
5.1.1   Objectives

The purpose of the 1984 ITCQ-based management plan was to (a) conserve fish stocks while (b) meeting an economic efficiency objective and to move closer to optimal resource. The introduction of the 1984 management plan appears to have moved the Australian SBT fishery towards meeting both objectives. There has been a reduction in the fleet-capacity and the amount of catch, while the overall value of Australian SBT production has increased several fold, particularly with the inclusion of farmed production.

Table 12

Distribution of landings and catch methodology used to take Australian quota 1988-89 to 1997-98

Quota year	Domestic operators [a]							Joint-venture long-line	Real-Time Monit. Prog [b]	Total landings with Austr. quota	National quota
	Jurisdiction				Method		Total				
	West-ern Austr.	South Austr.	New South Wales	Tas-mania	Long-line	Surf-ace					
1988-89	425	4 872	1	2	1	5 299	5 300	684	0	5 984	6 250
1989-90	230	4 199	6	14	6	4 443	4 449	400	0	4 849	5 265
1990-91	220	2 588	15	57	15	2 865	2 880	881	300	4 061[c]	5 265
1991-92	17	1 781 (138)[d]	124	56	124	1 854	1 978	2 057	800	4 835	5 265
1992-93	0	1 506 (722)	254	67	350	1 477	1 827	2 735	650	5 212	5 265
1993-94	0	1 970 (1294)	286	112	446	1 922	2 368	2 299	270	4 937	5 265
1994-95	0	2 864 (1954)	157	113	268	2 866	3 134	1 295	650	5 079	5 265
1995-96	0	4 809 (3362)	117	262	351	4 837	5 188	0	0	5 188	5 265
1996-97	0	4 498 (2498)	236	244	471	4 507	4 978	0	0	4 978	5 265
1997-98	0	4 403 (3487)	388	694	648	4 403	5 087	0	0	5 097	5 265
1998-99		(4991)			216	5 016	5 207	216	0	5 232	5 265

[a] Involving domestic operators.
[b] Real time monitoring programme
[c] 700t of quota was not allocated in 1990-91.
[d] The values in brackets show the amount of catch taken to be grown out in farm cages.
Source: AFMA.

While the reductions in Australian, Japanese and New Zealand catch has resulted in conservation benefits, the long-term conservation benefits are less certain. However, over this time, the amount of catch by others, outside of the CCSBT, has increased. This has impacted on the conservation objective of the CCSBT and on the incentive for Commission members to cooperate.

5.1.2 Initial impact of programme

In the first two years following the introduction of the new management plan, fleet numbers as well as the number and capital value of boats active in the fishery were reduced, as too were landings. At the same time fleet profitability increased (Table 13). The expected increase in returns under ITCQ helped to finance the restructuring of the fishery, while the realised economic returns have benefited operators in the industry and the Australian public as a whole.

Table 13

Summary of immediate impact following the introduction of ITCQ

	Landings 1000t		Gross value $A m		No. of Boats	Quota alloca-tion tonnes	Gross reduction in capital-isation $A m	Changes from 1983-84 through 1985-86 financial years ($A)		
	83-84	86-87	83-84	86-87				Rent	Family Income	WA capital value
Fleet	21.0	-	12.6	-	136	14 046	-	-	-	-
Stayers	-	11.5	-	23.0	54	9 520	-	+6.5 m	-	-
Leavers	-		-		82	4 526	-17.4	-	+2 600	+32 848

Source: from previous tables.

Another element in the use of individual, transferable catch-quota in restructuring the fleet was the average annual earnings of those who left the fishery increased, and (for Western Australia at least) the capital value of their fishing assets also increased (Table 13). The results indicate that on a state fishery basis, the state-by-state differences in the rates at which quota-holders left the fishery, is consistent to the relative differences in the opportunity-cost of remaining in the fishery. It could be expected that if the same economic incentive would have existed for operators within each state fishery and, as a result, the same relative behavioural responses would have been observable within each state. The increase in the average family-income observed for Western Australian 'leavers' is consistent with this.

In spite of these benefits, the management changes introduced in 1984 did cause stress and disruption to individuals and their families involved with the SBT fishery. However, compared to the possible impact of other alternatives, such as input-controls or use of a total-quota (see Section 3.4.1), the disruption and economic loss to operators and their families is likely to have been less. In Western Australia, the Commonwealth government also provided financial assistance for restructuring.

5.1.3 Ongoing effects

There have been a number of changes in the Australian SBT fishery following the introduction of ITCQ on 1 October 1984, in particular the restructuring and the reduction of annual landings to around 5000t and the substantial increase in returns to individual operators and to the fishery as a whole. This has come about due to: a bilateral agreement between Australian and Japanese operators in the use of Australian quota both within the Australian fishing zone and on the high seas; the development and growth of tuna-farming; and changes in marketing. The institutional structure provided by ITCQ is likely to have been important in facilitating these outcomes. The fall of the Australian dollar over this time against the Japanese yen has also been important in the improvement of the value of SBT product.

In addition, under their bilateral agreement, Japan provided expert input to the Australian long-lining operations and to the growing-out, or farming, of juvenile SBT. The withdrawal of the Japanese SBT fishing industry from the joint Australia-Japan SBT operations in 1996, saw the value of the SBT production fall from $A128 million in 1995-95, to $A85.8 and $A81.1 million in 1995-96 and 1996-97. However, by 1997-98 and 1998-99, the value of Australian SBT production had increased to $A128.1 and $A223.5 million respectively. The reason for this recovery has been the transfer of over 90% of the quota-catch to the grow-out ponds in South Australia. As a result, the assumed value of SBT landings were, respectively, $A40.9 million and $A56.8 million.[12]

At the global level, there were further reductions in the global quota in 1986-87 from 38 650t to 19 500t, and finally to 11 750t (the current level) in 1989-90. Questions remain in regard to whether the CCSBT's conservation objective, that parent stocks be allowed to recover to the 1980 level by 2020, will be achieved. While the substantial catch reductions by CCSBT members would have had a positive long-term effect on SBT stocks, it is questionable whether the stock will achieve the agreed performance objectives. While there is a general agreement among the members of the CCSBT on the current status of stocks, the members disagree on future stock-projections, with projections ranging from significant stock-rebuilding to further declines in stocks under current harvest-levels. Australian estimates indicate that a 15% reduction in catch would provide a 50% probability of recovery, and a 35% reduction would give a 75% probability of recovery (Hayes 1997, pp. 16-17).

Globally, there was a two-thirds reduction in reported southern bluefin tuna landings to 17 700t between 1983 and 1987. However, the build-up in stock numbers and the limitation of the make-up of the membership of the CCSBT has provided an incentive for others to enter and to increase their effort in the SBT fishery. As a result, the estimated global catch of SBT in 1998 was 19 240t, with an additional 1460t catch by the Japanese fleet beyond their national quota. This is a one-third increase in the global quota set by the CCSBT.

Kennedy (1999) found there is little incentive for Australian, Japan and New Zealand to co-operate as members of the CCSBT in setting global TAC-levels as long as SBT fishers outside of the CCSBT are taking catch in addition to the CCSBT global quota. Bioeconomic modelling by Kennedy, Davies and Cox (1999) indicates that reductions in present catch can lead to an increase in future catch and an increase in the future value of the fishery.

5.2 Concentration of ownership

The introduction of ITCQ into the Australian SBT fishery in October 1984 has been followed by a number of structural and operational changes. In October 1984, 64% of the quota allocation went to operators in South Australia, with 19% to those in Western Australia, and 14% to those in New South Wales. In 1998, South Australia operators had increased their proportion of quota by almost 50% to 92% of the national quota. In spite of the increasing availability of catch off the New South Wales coast and the expected shift in quota to the Australian east coast, only 2% of the quota was held in that State. The retention of quota appears to be the result of the quantum jump in the value of sales of farmed-tuna over what would have been received for quota-catch.

---

[12] This raises an interesting question in regard to the distribution of resource-rent, whether it is attributable to the scarcity of SBT or the scarcity of suitable farming areas. It is probable that ABARE's assumed value is an under estimate.

In operational terms, there has been a reduction in landings to one-third of what they were in the season prior to the introduction of quota, there has been a reduction in the capitalisation of the fleet and there have been substantive change in the technology used to catch and produce sashimi-quality SBT products (Table 14). These changes on the production-side were also accompanied by a shift on the demand-side with the diversion of product from canning to the high value Japanese sashimi market.

Table 14
The distribution of Australian SBT quota at October 1988, November 1999

	New South Wales	South Australia	Western Australia	Queensland
Quota distribution October '84	2 022t (14%)	9 271t (64%)	2 752t (19%)	-
Quota distribution November '99	123t (2.4%) number of owners 4	4762 (91.6%) number of owners 33	267 (5.1%) number of owners 13	48 (0.9%) number of owners 3

Note, this is for quota-holders holding in excess of 5t, seven boats having received 5t or less. While the statistics presented in this paper have been on the basis of those holding in excess of 5 tonne of quota, at November 1999, 44 owners held 5t or less of quota, which accounted for 65t.
Source: AFMA

6. LITERATURE CITED

ABARE - Australian Bureau of Agricultural and Resource Economics 1994. Commodity Statistical Bulletin, ABARE, AGPS, Canberra.

ABARE - Australian Bureau of Agricultural and Resource Economics 1996. Fishery Statistics 1996, ABARE, AGPS, Canberra.

ABARE - Australian Bureau of Agricultural and Resource Economics 1999. Commodity Statistical Bulletin, ABARE, AGPS, Canberra.

AFMA - Australian Fisheries Management Authority 2000. New management arrangements for east coast tuna to minimise the incidental catch of SBT, AFMA News, 4, issue 3, pp 1 and 8.

BAE - Bureau of Agricultural Economics 1983. BAE warning on 'commercial extinction' of tuna, Australian Fisheries, 8, pp. 12-13, AGPS, Canberra.

BAE - Bureau of Agricultural Economics 1986. Southern Bluefin Tuna Survey 1980-81 and 1981-82, Project no. 62300, Bureau of Agricultural Economics, Canberra.

Campbell, D. 1984. Individual Transferable Catch Quotas, their Role use and Application, Fishery Report no. 11, Northern Territory Department of Primary Production, Darwin.

Campbell, D. and L. Wilkes 1988. Report to the Australian fisheries Service on those leaving the southern bluefin tuna fishery. ABARE, Canberra.

Campbell, D., T. Battaglene and D. Brown 1996. Use of individual transferable quotas in Australian fisheries. Paper presented at the International Institute of Fisheries Economics and Trade International Workshop on Assessment and distribution of Harvest Quotas in Fisheries, Aalesund, Norway, 8-11 July, 1966.

Campbell, D., D. Brown and T. Battaglene 2000. Individual transferable catch quotas: Australian experience in the southern bluefin tuna fishery, Marine Policy 24(2):109-117.

Caton, A. 1987. Australia fills SBT quota in 1986-s87 season, Australian Fisheries 46(12):12-14.

Caton, A. and J. .Majkowski 1987. Warning issued on global catch limits, Australian Fisheries 42(3):278-91.

Geen, G. and M. Nayar 1989. Individual Transferable Quotas and the Southern Bluefin Tuna Fishery, ABARE Occasional Paper no. 105, AGPS, Canberra.

Grainger, G. 1988. Australian Fisheries and Administrative Appeals Tribunal, Department of Primary Industries and Energy, AGPS, Canberra.

Hayes, E.A. 1997. A Review of the Southern Bluefin Tuna Fishery: Implications for Ecologically Sustainable Management, Traffic Oceania, Sydney.

Hampton, J. and J. Majkowski 1986. Scientists fear SBT problems worsening, Australian Fisheries, 45(12):6-9.

Hogan, L., K. Van Landegham and V. Topp 1997. Access Fee Arrangements for Japanese Fishing Vessels in the Australian Fishing Zone, ABARE report to the Fisheries Resources Research Fund, Canberra.

Industries Assistance Commission 1984. Southern Bluefin Tuna, Industries Assistance Commission Report, AGPS, Canberra.

International Tribunal for the Law of the Sea 1999. Southern Bluefin Case (New Zealand v. Japan; Australia v. Japan), Requests for provisional measures (nos. 3 and 4). Rendered under Annex VII of the United Nations Convention on the Law of the Sea.

International Centre for the Settlement of Investment Disputes 2000. Southern Bluefin Tuna Case (Australia and New Zealand v. Japan), Awarded on jurisdiction and admissibility. Rendered under Annex VII of the United Nations Convention on the Law of the Sea; www.worldbank.org/icsid.

Kennedy, J. 1999. A dynamic model of co-operative and non-co-operative harvesting of southern bluefin tuna with an open access fringe, 1999 World Conference on Natural Resource Modelling, Saint Mary's University, Halifax, Canada, 23-25 June.

Kennedy, J.O.S. and J.W. Watkins 1984. Optimal Quotas for the Southern Bluefin Tuna Fishery, Economic Discussion Papers no. 4/84, Latrobe University.

Kennedy, J.O.S. and J.W. Watkins 1985. Modelling the economic returns of SBT, Australian Fisheries, 44. no. 7, pp. 37-40.

Kennedy, J.O.S., L. Davies, and A. Cox 1999. Joint rent maximisation and open access competition in the southern bluefin tuna fishery. ABARE paper presented at the Combined 43rd Annual Australian and 6th Annual New Zealand Agricultural and Resource Economics Society Conference Christchurch, New Zealand, 20-22 January.

Majkowski, J. and A. Caton 1984. Scientist's concerns behind tuna management, Australian Fisheries, 43(3) 371-82.

Smith, P. 1986. Tuna. *In*: Haynes, J. and P. Smith. Market Trends for Australian Fisheries Products, BAE (now ABARE) report no.86.3, AGPS Canberra, pp. 54-64.

Wesney, D., B. Scott and P. Franklin 1985. Recent developments in the management of major Australian fisheries: Theory and practice. In Proceedings of the Second Conference of the Institute of Fisheries Economics and Trade. Vol.1: Christchurch, New Zealand, August 20-23, 1984. pp. 289-303.

CHANGES IN FLEET CAPACITY AND OWNERSHIP OF HARVESTING RIGHTS IN NEW ZEALAND FISHERIES

R. Connor
Centre for Resource and Environmental Studies
Australian National University, Canberra, ACT 0200 Australia
<rconnor@cres.anu.edu.au>

1. INTRODUCTION

In 1986, New Zealand introduced management by Individual Transferable Quota (ITQ) for 26 of the country's most economically significant fishery species. The system has since been expanded and extensively modified, and remains the most comprehensive and perhaps most robust example of ITQ-management in the world today.

The system was born of dual motivations: concern for the stress on stocks being imposed by rapidly expanding effort in inshore fin-fisheries; and the desire for a mechanism to allow the domestic industry to capture rents and build capacity in the offshore sector, dominated in the past by foreign fleets. These objectives implied restructuring of the fleet to better cater to resource-limitations on the one hand and opportunities for a great expansion of domestic effort on the other. ITQs do not dictate fleet structure nor distribution of catch entitlements, but provide a mechanism for such decisions to be taken at the level of the individual firm. It is important that the progress of adjustment of fleet-capacity and ownership of harvesting-rights is assessed so that other jurisdictions may learn from the early experience with ITQ-management.

The objectives of this paper are to analyse, to the extent possible with readily available data, the impacts of the introduction of ITQ-management on the structure of the New Zealand commercial fishing fleet, and on the distribution of "ownership" of the catch. Such an appraisal has not previously been undertaken for New Zealand, and only limited assessments of other quota-managed fisheries have been published. Together with the scale of the New Zealand Quota Management System (QMS) and data availability, this has limited the analysis to a broad assessment that is useful in itself, but which also raises issues for further investigation. These include methodological issues as well as questions requiring more detailed information on the structure and dynamics of the New Zealand fishing industry.

An analysis of fleet structure is presented following a brief description of the historical context and the New Zealand ITQ-management reforms. This stratifies the fleet by length-class and uses gross registered tonnage (GRT) as a proxy for vessel fishing-capacity. Trends are identified and discussed. Catch figures are reviewed to provide a basis for assessing capacity trends. An analysis is made of the age-structure of the fleet to aid understanding of fleet structural dynamics and the impacts of quota-management, which are then discussed. The next section undertakes analysis of the trends in distribution of quota and catch among quota-holders. A range of indices are generated from quota registry and catch data. These are used to assess the degree of, and trends in, concentration of ownership of quota and how this differs from concentration of end-of-year holdings and actual catch. The pre-ITQ baseline distribution used for comparison in this analysis is that of the provisional maximum individual transferable quota (PMITQ), issued by the Ministry as part of the ITQ-allocation process, based on assessment of individual fisher catch histories. The final section of the paper provides some discussion of the overall results of the analyses and potential for further research.

2. BACKGROUND
2.1 The historical context[1]

In 1978 New Zealand declared a 200-mile Exclusive Economic Zone around its coasts encompassing some 4.1 million square kilometres. About one-third of this area is fishable by modern demersal methods, and the zone has an estimated sustainable yield in excess of half a million tonnes. As in other coastal states, the declaration of the EEZ was in part a response to fishing of the zone by distant-water fleets of other nations, in particular by Japan, the Republic of Korea and the former USSR. This foreign exploitation of New Zealand fish-stocks had begun in the 1950s when the domestic industry was highly regulated. The government response was to completely deregulate fishing in 1963 and to provide subsidies and other encouragement for the domestic industry to compete for a larger share of the catch. The industry responded with a vessel-building boom and a rapid increase in catches from the inshore fisheries. However, the foreign fleets also increased their efforts, and by 1977 were taking nearly 90% of the known 476 000t catch of fin-fish from the area (Sharp 1997).

---

[1] A more complete description of the historical context of New Zealand commercial fisheries is given in Connor (2001).

Responsibility for the management of New Zealand's fisheries lay with the Ministry of Agriculture and Fisheries (MAF).[2] Initially, following the declaration of the EEZ, the fisheries outside the twelve-mile Territorial Sea were managed separately. Total allowable catches (TACs) were struck for the offshore species, and these were first allocated preferentially to the domestic industry, and second to the foreign fleets under licence and government bilateral agreements. The policies offered the foreign fleets less of the prime species and areas than they had previously been fishing, changing the economic balance, and resulting in a much reduced total catch for the next few years (OECD 1997).

Government policies also proved incentives for domestic companies to invest in onshore processing-plants and vessels for offshore fishing, but the main initial domestic involvement was developed through joint-ventures with foreign companies and foreign vessel charters. Joint ventures brought local crew onto the big vessels, and direct involvement of domestic companies in the management of fishing operations and marketing, paving the way for further domestic expansion. Foreign vessels began delivering large catches to onshore processing. By about 1982 the local companies had learnt what they needed to know from joint ventures, and arrangements with foreign vessels moved to simpler contracts to charter fishing-capacity to catch against domestic company-quotas. Foreign vessel charter has remained an important part of offshore fishing in New Zealand since that time, gradually diminishing as domestic companies have invested in large freezer-trawlers. Both arrangements brought greatly increased cash-flow to the domestic industry, foreign exchange from exports and employment in processing.

It was at this time that the inshore fisheries began showing signs of stress and management gradually moved into crisis mode. New powers to declare controlled fisheries were introduced in 1977 and a moratorium on scallop and rock lobster permits followed in 1978. Alarming fluctuations in catches of the most economically important inshore species, snapper (*Pagrus auratus*), and rapidly increasing catches of vulnerable species of sharks and gropers, brought a total moratorium on new fishing permits in 1982. Both management and industry had recognised that there were economic as well as stock problems in the inshore fisheries (Riley 1982). Five per cent of the fleet was taking two-thirds of the catch, and there were large numbers of part-time operators.

During 1983 a consultative policy review process was initiated by the Ministry for the inshore fisheries, and a trial "enterprise allocation" (EA) quota-scheme was introduced in the offshore fisheries. After several rounds of consultation and a change of government, a decision was made in 1985 to adopt a near-comprehensive ITQ-based management system for both inshore and offshore sectors. For the offshore, existing EA quotas were converted to ITQ directly. For the inshore, a complex process of assessment and allocation was undertaken. Initial allocations of entitlement were based on catch-histories from the best two of three qualifying years and a tendering process was undertaken for reduction of total allocations through a government-funded quota buy-back. Where reduction targets were not met for critical species, administrative reductions were made to establish the required total allowable catches[3,4].

2.2   The nature of the harvesting-right

Prior to the introduction of ITQs in New Zealand, rights to engage in commercial fishing were focused primarily on the fishing permit. Permits were issued by the Ministry and were subject to conditions elaborated on the permit, as well to the provisions of fisheries legislation and regulations. From 1963 to 1982 (1978 for rock lobster and scallops) there was no restriction on the number of permits available. The system was that of regulated open-access. From March 1982 there was a moratorium on the issuance of new permits for the inshore fin-fish fisheries. Part of the system of permit conditions was the nomination of the species being fished and methods to be used. The moratorium was intended to prevent further such endorsements to existing permits, and this was at least partially successful. In effect, the permit-moratorium limited the participants in each fishery and therefore the eventual recipients of ITQ allocations.

The passing of the *Fisheries Act 1983* introduced a new definition of "commercial fisherman" that permanently excluded an estimated 1500-1800 part-time fishers from renewing their permits. This significantly reduced the population of fishing entitlements (by around 40%) and the potential for expanded effort, but did not significantly reduce the actual fishing-effort applied. Also in 1983, the introduction of administrative "enterprise allocations" (EAs), modelled on the system used in the Canadian Atlantic offshore trawl-fishery, established individual catch-entitlements for the first time. These were issued to seven large companies and two consortia active in the offshore sector, and additional amounts of the TACs were fished competitively by other smaller

---

[2] The Ministry of Agriculture and Fisheries was reformed in 1994 as the Ministry of Fisheries (MFish). In this document this agency will be referred to as MAF or the Ministry.

[3] Under the New Zealand quota management system (QMS) TACs are set for the overall take of a fish stock, including recreational and indigenous customary fishing. The commercial catch limit is a subset of the TAC and is termed the total allowable commercial catch or TACC.

[4] The basis of the QMS is described in Clark and Duncan 1986, and Clark et al. 1988.

participants. These quotas were a trial scheme, were valid initially for ten years, and covered the nine key mid-depth and deep-water species. They were applicable to the company, not to individual vessels or skippers, so that the most efficient use could be made of catching-capacity. Transfer was possible subject to approval of the Ministry. These off-shore quotas became the trial for the later wider adoption of ITQs.

New Zealand ITQs came into effect in October 1986 applying to 153 management stocks of 26 species - the nine off-shore species under EAs, plus 17 inshore species. Catches of these species at the time comprised about 83% by weight of the total commercial fin-fish catch. Allocations were subject to appeal to a quasi-judicial Quota Appeal Authority, but this did not affect the full operation of the management-system or quota-trading. ITQs were created as a perpetual right to a part of the fish harvest, designated in absolute weights of whole fish (in tonnes) for a particular species or species group to be taken annually from a specified quota-management area. These rights were allocated free of charge to existing participants in the fisheries, and were to be fully compensated in the event of TAC reductions. Free transferability and lease was subject to reporting of all transactions with prices to the Ministry, and to aggregation limits of 20% for inshore and 35% for deep-water stocks. The ITQ allocated rights to utilise the resources, but the fishing permit remained as the right of access. Under the QMS legislation, a fishing permit was to be granted to anyone who fulfilled the minimum quota-holdings requirement of five tonnes for finfish.

Responsibilities attached to quota-ownership included legal obligations to land all catch of quota species, unless under minimum legal size; to submit monthly quota-monitoring reports in addition to completing catch and landing returns and catch-effort logs; and to pay resource-rentals on all quota held, whether caught or not. Some flexibility was built into the system by allowing the carry-over of up to ten percent of uncaught quota to the following year, or for up to 10% over-catch of holdings to be counted against the following year's entitlements.

These characteristics established the character of the ITQ as private property in the right to harvest fish from a given stock - not in the fish stocks themselves - and a clear understanding of this character has become generalised in New Zealand since 1986. There was no legal impediment to the use of ITQ as security for bank loans, but the Ministry did not make provision for the registration of liens or caveats against the title to ownership and this in many cases prevented such use.

The nature of the ITQ-right underwent a major change in 1990. The original specification of ITQ in tonnes of fish required the government to enter the quota-market to buy or sell quota when it wished to alter the total allowable catch. When faced with potential for stock collapse in orange roughy[5] and the need to reduce this valuable quota by large percentages, the system was changed so that ITQ were denominated as a percentage of the TAC, rather than as a specific tonnage. Adjustment then implied merely the automatic *pro rata* adjustment of all ITQ-holdings at the beginning of each season to match the TAC.

A typical inshore trawler moored in the Iron Pot, Napier. These vessels all became part of the QMS on the introduction of ITQs in New Zealand

3. ASSESSMENT OF FLEET-CAPACITY
3.1 Characterising fleet-capacity
3.1.1 Introduction

The New Zealand fisheries to which quota management has been applied cover a full range of species, habitats and methods. The fleet subject to quota-restrictions thus covers every possible size and configuration from 3m dinghies to freezer trawlers above-100m. Methods include single and pair trawling, seining, drop line, pelagic and bottom long-line, pole-and-line, trolling, set-net, potting, and trap. More than 100 finfish species caught ranging from estuary flounders to deep-sea oreos, and shellfish include scallops, clams, dredge oysters, abalone (paua) and rock lobster. Few vessels within the fleet pursue a single species. In such high-value specialist fisheries as rock lobster and abalone, fishers may be content with one target, but most exploit other species in other seasons, and/or take some bycatch. Some 32 species were managed under ITQ in 1996 when

---
[5] For scientific species names see Appendix I.

new fisheries legislation was passed mandating the gradual inclusion of all commercial species in the system. By 1998 there were 42 species under ITQ and another 40 species considered commercially significant, with more being moved into the system each year.

Detailed study of capacity-utilisation under these conditions would require specific data collection and would need to be confined to a small sample of the fleet. This study takes a broad descriptive approach: data available from the agency vessel-registry is used to examine trends in a vessel-capacity proxy (gross registered tonnage) by vessel size-class, and to assess the age structure of the fleet. Separation by method or target species was not possible with the data used, but would be possible using vessel logbook data. The fleet has been stratified by length over-all (LOA) into size-classes of 3m, so as to match data summaries of earlier years (*e.g.* King 1985).

3.1.2 Methods and data

In data supplied from the New Zealand Ministry of Fisheries' Quota Management System (QMS), vessels are categorised as domestic, chartered or foreign. Vessel dimensions include gross registered tonnage (GRT), which is used here as a proxy for fishing-capacity. GRT by length-class is summarised for domestic vessels, and GRT by flag-state for charters from 1987 to 1998. Data on the dimensions of the foreign-licensed vessels were not available.

The available data contained many errors and duplications, with some significant gaps, although the error rate decreased in later years. For vessels with no recorded GRT, the averages of recorded GRT for the relevant length-class from the same year were used as estimates. After elimination of gross entry-errors these averages are quite consistent, with any changes being smooth trends. Pre-QMS data on vessels was only available for a few years (1984-88), was in a different format and was extremely patchy. For 1974 to 1982, GRT was estimated as the product of vessel numbers in each length-class published by King (1985) and mean GRT for each class calculated from QMS data for the years 1989-1991. Identified gaps have been filled by interpolation - particularly over 1983-4 - when the basis for data collection changed with the change in official fishing-year to begin in October. The vessel data for 1984-88 were the worst for data-entry errors, missing data, duplicates and so on, and the results for these years should be treated as indicative only.

The analysis of capacity is focused firstly on the inshore fleet, which was viewed as over-capitalised at the time that the QMS was introduced. An assumption is made in the analysis of GRT by length-class that vessels over 33m are not primarily part of the inshore fleet. This is somewhat arbitrary and, in part, is an artefact of the pre-QMS data that grouped together all vessels over 33m. However, it is considered reasonable that most vessels greater than 33m in length would be primarily deployed fishing in deeper waters. In fact many vessels smaller than 33m are likely to be deployed in fishing for offshore species.

The New Zealand Fishing Industry Board reported details of fleet composition and catch by method for 1984, 1986 and 1987 (FIB 1987; FIB 1989). These data are used to increase understanding of structural change in the fleet. In addition, QMS data recording the year of construction of vessels are used to examine the age structure of the fleet by size-class.

3.2 Results of capacity-assessment
3.2.1 Inshore fleet

Changes were expected in the fleet across the boundary where the QMS was introduced in 1986, particularly for the inshore where total allowable catches were reduced considerably through a quota buy-back scheme. In addition, the *Fisheries Act 1983* introduced a new definition of commercial fisherman that effectively excluded many smaller vessels prior to the implementation of quota, on the grounds that they were only fishing on a part-time basis.

The results of the analysis of trends in GRT for the inshore domestic fleet (less than 33m LOA) produce a natural stratification into three groups: under 12m; 12-24m; 24-33m. The general pattern has been for the small-boat sector to decline in

The fate of small multipurpose vessels such as these (shown in Westport), and their crews, under quota systems, is one of the socio-economic areas requiring further research

both numbers and capacity; the mid-size fleet has remained fairly constant; and the larger vessels have increased in number. The result has been no net change in capacity from the pre-QMS peak in 1982 to 1998, although a higher peak was reached in 1994. Figure 1 illustrates the trends. The mean total GRT for vessels under-33m for the years 1976-85 was 31 301, and for 1987-1998 it was 34 219: an increase of 9%. The 1998 figure was 33 352 GRT.

A number of other features are evident in this data. Some of the variation in the period 1982 to 1989 is likely due to data problems, but the impacts of policy-change can be clearly seen. The enactment of the *Exclusive Economic Zone Act 1977* stimulated expansion in the number of both small (<12m) and large (24-33m) vessels. From 1976 to 1982 small-vessel capacity increased by some 27% while that for large vessels increased by 400%, each of these groups contributing about the same amount of additional tonnage over the period. The 1983 exclusion of part-timer fishermen hit the small-vessels harder than the other classes, as expected. Numbers of vessels less-than-12m LOA have dropped substantially from 4800 at their peak in 1978, to just over 2000 following the implementation of the exclusion policy and ITQs. This represents a 54% drop in numbers and a 34% reduction in capacity, indicating that a greater number of smaller-boats from the group exited (see Figure 2). A brief resurgence in numbers of small-boats in the late 1980s was followed by further declines, to around 1300 in 1998 - representing 28% of the peak number and 54% of peak capacity for this sector. This reduction represents 14% of the 1978 total capacity of the inshore fleet less-than-33m.

The data trend settled down following the transition to ITQs in 1986, and the whole fleet experienced steady growth until 1994. From this date (the year that full cost-recovery was implemented in New Zealand) over the next four years the capacity in the small-boat sector declined by one-third, and by about seven percent in the other two classes. Further detail is provided in Figures 3 and 4. Notable in the data are the large numbers involved and the volatility in the smallest size-classes, the static nature of capacity and numbers in the 12-24m classes (this might be regarded as the core inshore fleet), and the high growth rate in the large-vessel classes. The 24-33m class increased in number and capacity by a full order of magnitude over two decades before growth was halted in the mid 1990s. As will be seen below, this indicates for these vessels 24-33m in length greater affiliation with the offshore-sector than the inshore.

Figure 1
New Zealand domestic fleet (under 33m LOA) total GRT for three vessel length classes - 1974 to 1998

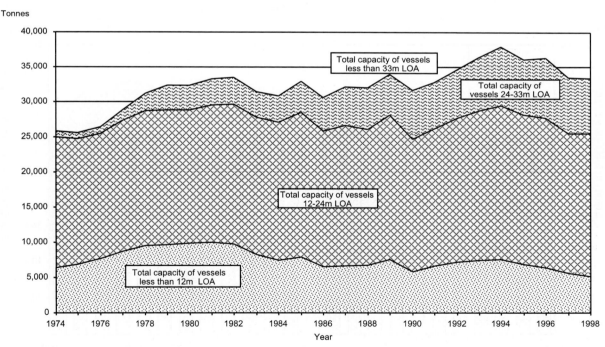

Data sources: King 1985; FSU data; QMS data.

Figure 2

New Zealand small boat sector - Total capacity (GRT) and count of registered vessels less than 12m length

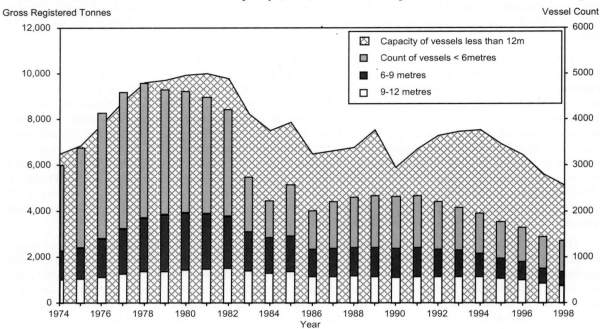

Data sources: King 1985; FSU data; QMS data.

Figure 3

Total capacity (GRT) and count by length class of registered fishing vessels of 12 to 24ms length - 1974 to 1998

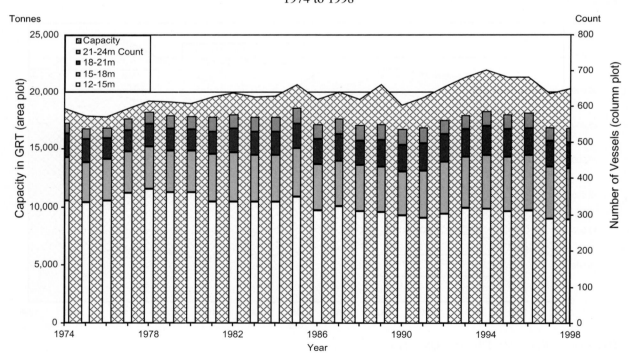

Figure 4

Total capacity (GRT) and count by length-class of registered fishing vessels of 24 to 33m length - 1974 to 1998

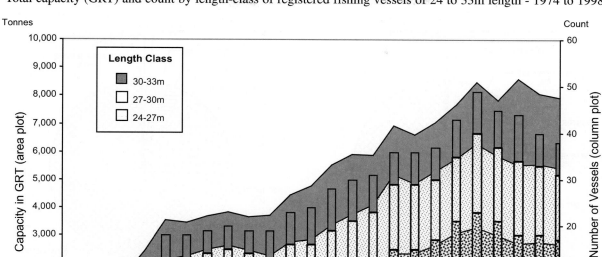

Data sources: King 1985; FSU data; QMS data.

Overall, although the total capacity of this under-33m fleet has not been reduced over time, policy changes have certainly had some impact in halting expansionary phases. Further, close examination of the activities of the larger vessels (24-33m) included in the analysis may significantly change the picture, as a significant proportion of the capacity in these classes is undoubtedly applied to offshore fisheries (such as hoki and the pelagic species). This would imply that the capacity being applied in the inshore fisheries has effectively declined from its pre-QMS peak in 1982. If the vessels greater-than-24m are excluded as being part of the inshore fleet, then capacity in 1998 is back down to pre-EEZ (1974) levels. The aggregate catch for the main inshore species has not changed between peak-years before and after the QMS, although the catch composition has changed through quota, effectively constraining the catch of particular species, while others have increased. Catch trends are considered in more detail below.

Because of the data-sources used here detailed analysis of the distribution of capacity by species and method has not been possible. However, Figures 5 and 6 show data published by the Fishing Industry Board (FIB 1987, 1989, 1990, 1994, 1996) covering the whole domestic fleet. Figure 5 indicates the proportion of catch taken by method, for the 1984-5 and 1987-8 fishing years. The total catch taken by the domestic fleet declined from 150 000t to 130 000t between these years with the implementation of ITQs. Catch from single-trawls is constant for the two years and so increases as a proportion of the total. The other mobile gear, purse seining, expanded its catch of jack mackerels and kahawai rapidly in this period. The catches from all other methods declined. Set-netting was particularly hard hit due to a targeted reduction in catches under quota, of shark species, snapper and groper, and most of the vessels exiting from this sector were from the under-12m classes. Figure 6 shows the change in numbers of vessels by method, and indicates that by the mid-1990s set-netting was making something of a comeback. Of note also is the similar recovery of hook-and-line methods, and a gradual but steady decline in numbers of rock lobster boats.

Figure 5
Proportions of total New Zealand Domestic catch by method - 1984/5 and 1987/8

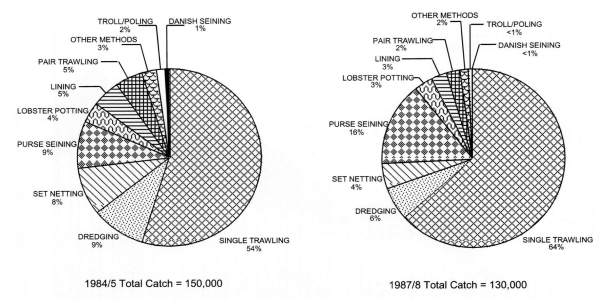

1984/5 Total Catch = 150,000   1987/8 Total Catch = 130,000

Data source: FIB 1987, 1989.

3.2.2 Offshore fleet

In the length-classes of vessels suited to fishing the offshore species, more dramatic trends are apparent. Figure 7 shows the growth in capacity and numbers for the over-33m fleet compared with that of the large inshore classes already discussed. While the 24-33m class has expanded to some 8000GRT of the 33 000GRT inshore fleet, numbers of over-33m domestic vessels have increased from two in the late 1970s to 49 vessels in 1998, totalling some 45 000 gross registered tonnes. This large fleet of offshore vessels has been gradually replacing the capacity provided in the past by foreign vessels chartered by New Zealand fishing companies. After some earlier use of very large vessels, the dominant vessel length-classes are now 60-70m and 40-45m (see Figure 8).

This bimodal configuration is due in no small measure to a regulation (the 43m rule) that limits the use of vessels over 43m long in the inshore and in designated areas important for the huge hoki fishery. Vessels have been custom built for this fishery to conform to the regulation, and these "fat boats" have increased the average tonnage for this length-class dramatically. In the 40-45m range, there were 4 vessels in 1987 and 18 in 1998. Average tonnage of these vessels has increased by 50%, so total tonnage has climbed from 1600 to over 11 000GRT. The 60-70m class are now the largest vessels in the domestic fleet: there was 1 vessel in 1987, and 12 in 1998. Again, average tonnage has increased by 50% and total GRT has increased from about 1400 to 23 000. The data shows that up to ten vessels in the length-classes greater-than-70m working in the years since the implementation of the QMS, but all have now gone. This capacity has been more than accounted for by the expansion in the two classes described (40-45m and 60-70m).

Figure 6

Numbers of full-time* domestic vessels by method for selected years

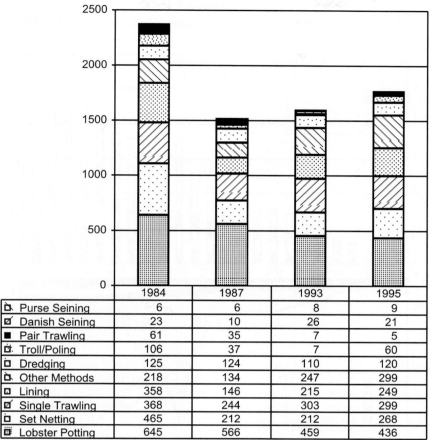

	1984	1987	1993	1995
Purse Seining	6	6	8	9
Danish Seining	23	10	26	21
Pair Trawling	61	35	7	5
Troll/Poling	106	37	7	60
Dredging	125	124	110	120
Other Methods	218	134	247	299
Lining	358	146	215	249
Single Trawling	368	244	303	299
Set Netting	465	212	212	268
Lobster Potting	645	566	459	436

* Full-time vessels are defined as those landing more than 7t of fin-fish or molluscs, or 2t of rock lobster. Data sources: FIB 1987, 1989, 1994, and 1996.

Trends for registered charter-capacity are shown in Figure 9. The charter fleet is still important to the New Zealand fishing industry, with vessels with a combined 125 000 GRT active during 1998. This compares with less than 80 000 GRT for the total domestic fleet, but the charter vessels do not generally spend all-year fishing in New Zealand waters. The 1998 charter tonnage was within 4% of the total in 1987, with 1997 being the lowest total since the start of the QMS. In the interim, a huge peak of 288 000 GRT was registered in 1990. The majority of these vessels (176 000 GRT) was Russian, possibly reflecting difficulties in the administration of the fleet following the collapse of the Soviet Union. The Japanese charter fleet was already declining from its peak the previous year. From a traditional base in the Russian, Korean, and Japanese distant-water fleets that have fished New Zealand waters since the 1950s and 60s, the flag status of the charter-capacity has diversified substantially since 1992, with some 20 nations now represented. Russian- and Ukrainian-flagged vessels still provide about 45% of chartered tonnage.

3.3 Catch

Catch figures over the period have been reviewed in order to put fleet changes in context. Figure 10 shows the catches of the main inshore species from 1974 to 1998, and Figure 11 compares catches immediately before ITQs with the 1998 result. As mentioned above, the aggregate catch for these species has not changed dramatically, although the catch mix has changed. Red cod catches are volatile due to variable recruitment, but among the other 17 species examined, catches of four species were sharply reduced on the introduction of the quota system and have been kept down since. These were the specific targets of the quota buy-back programme implemented in 1986. Effort seems to have shifted to other species whose total allowable commercial catches (TACCs) were under-caught, but as mentioned above, there are interactions between vessel-size, fishing-method and species. Those fishing the four key buy-back species included small-boats using set-nets and long-lines, and production by these methods was reduced sharply by the reforms. Groper is the only one of the four species not currently constrained by aggregate quota, but this is an artefact of the distribution of quota across areas. The areas of highest historical catch are fully fished against quota, while more remote areas are not. Another feature is the dip in total catches in the first year of the quota system. The reported catch for many species was well under the TACC-levels in 1987, but rose again the following season as the fishers adjusted to the new system.

Figure 7

Domestic fleet - numbers and capacity of large length-classes 1974 to 1998

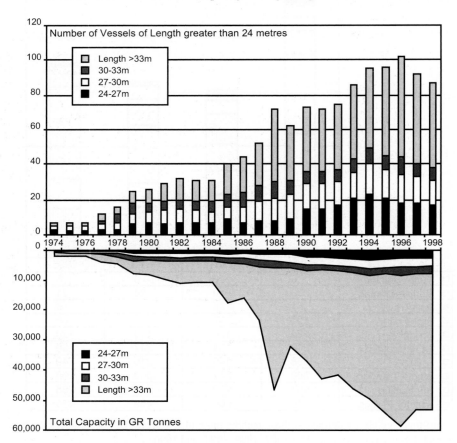

Data sources: King 1985; FSU data; QMS data.

The relevance of catch data from offshore species to the issues of domestic-capacity is perhaps to point out the 'blue-sky' opportunity for capacity-development in this sector. From the declaration of the EEZ in 1978, the New Zealand fishing industry worked to develop and domesticate the offshore fisheries. In 1977, foreign fleets took almost 90% of the 500 000t catch from the zone. This foreign catch was cut right back in 1978 and the fisheries were gradually redeveloped under joint venture and charter arrangements with foreign vessels. Some foreign-licensed fishing also continued. Figure 12 shows the split of total production from the zone between foreign licence, charter/joint-ventures, and domestic vessels for the period 1985-95. This indicates the domestic fleet's increasing share of an expanding total catch. The gradual switch from foreign- to domestic-fishing accelerated in the 1990s as a consequence of the development of the large-vessel capacity described above.

Total catches of offshore quota species are shown in Figure 13.[6] The plot shows the cuts in catch in 1978 with the implementation of the EEZ, particularly for the rapidly expanding foreign catches of hoki, hake and ling. Orange roughy then came to dominate this sector in value terms until the major expansion of hoki catches from 1986.[7] Figure 12 shows the use of charter vessels that took this catch and the response of the domestic fleet, and Figures 8 and 9 indicate the massive investment in domestic capacity taking place in the offshore sector from 1987.

---

[6] These figures include foreign-licensed and charter catches. Squid has been omitted, as data prior to the QMS were not available.

[7] Orange roughy is a high value species for a trawl fishery, with an indicative port price of $NZ2/t in 1988 compared with $NZ350/t for hoki. By 1995 these port prices were $NZ3500 and $NZ500/t respectively.

Figure 8
New Zealand domestic fleet capacity trends (GRT) - Large vessels by length-class 1987 to 1998

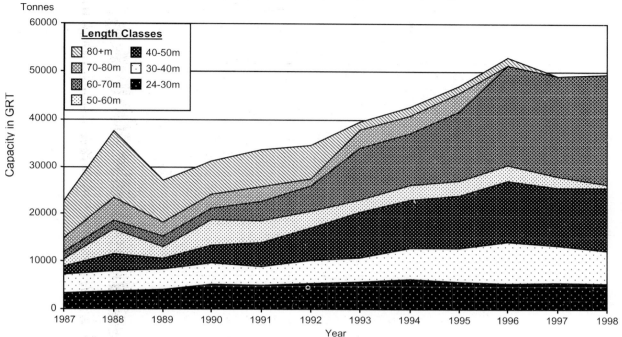

Data source: QMS data.

Figure 9
New Zealand foreign charter fleet capacity (GRT) by major flag-State
Capacity in GRT

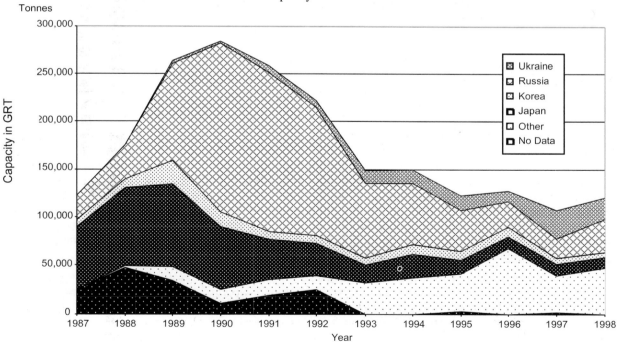

Data source: QMS data.

Catches of orange roughy catches have declined markedly from the mid-1980s, with the 1998 landings being only 39% of 1987 totals, at just over 20 000t. The several roughy fisheries had had TACCs initially set high on biological assumptions that proved overly optimistic. Since 1990 the TACCs have been progressively reduced to what are thought to be more sustainable levels, and these changes have brought catches down. This has been a highly-charged process and there have been overt political and economic trade-offs made against scientific recommendations for large cuts in the TACC. To some extent reductions in the TACC may have been tracking falling catches, and explicit deals with the industry have delayed overall reductions by creating new divisions of quota-management areas, with industry investing in exploration of new areas to maintain their catch. This 'serial discovery' of new sea-mounts and other aggregation sites has protected the industry from what otherwise would have been major economic impacts, but knowledge of the impacts on stocks and their ecology is uncertain.

Figure 10
Total catch for NZ inshore finfish species: 1974-1998

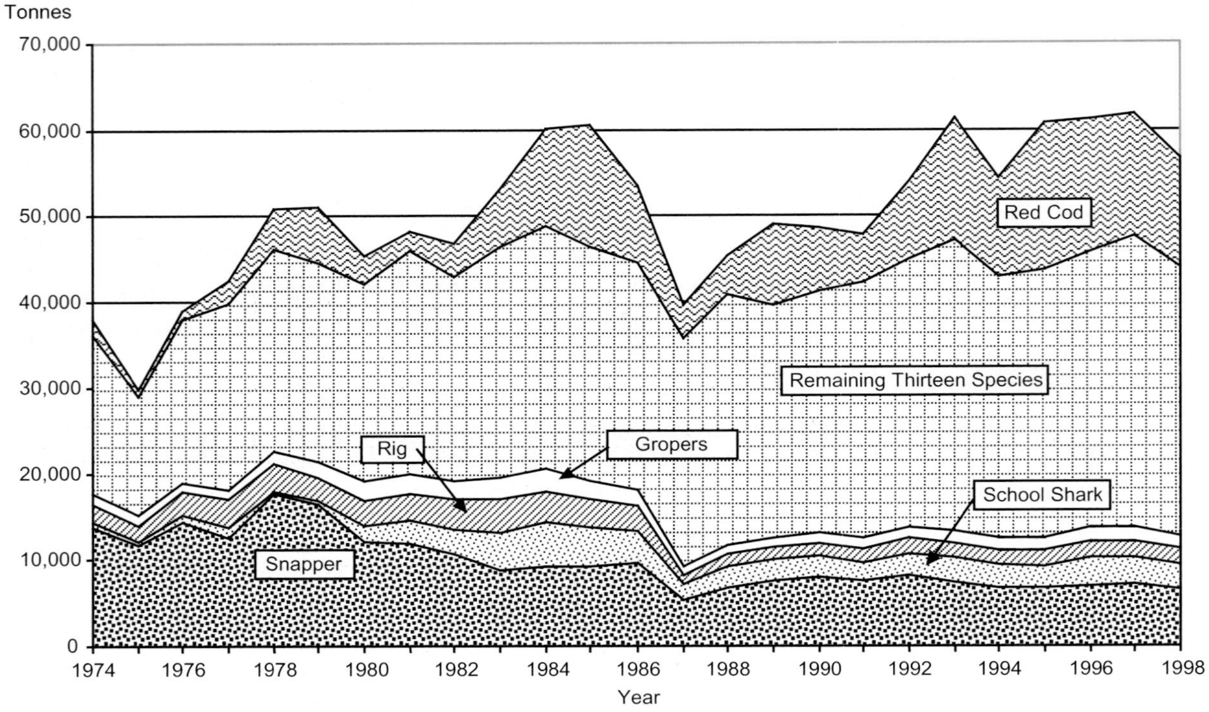

Data sources: King 1985; FSU data; QMS data.

3.4 Age profile of the domestic fleet

An analysis was made of the age structure of the domestic fleet from the vessel-registry records. The most complete data set was for vessels registered in 1996 and this is used for the main assessment. A further limited analysis was conducted of 1987 and 1984 data, which were less complete but allow some comparisons to be made. Histograms were plotted by length-class showing vessel numbers and total GRT by year built (see Appendix II).

Regular cycles of vessel-building activity are clear from this analysis. In general, peaks in numbers of vessels built occur at approximately decade frequency, but these cycles are also strongly correlated with the major regulatory changes. A decade-long boom in vessel-building occurred in New Zealand following the deregulation of fisheries in 1963, and another occurred with the declaration of the EEZ. The bulk of the capacity in the offshore fleet was built between 1986 and 1992, following the implementation of the QMS, but the established fleet did not experience another boom at this time.

These vessel cohorts are evident in the 1996 registry data for the 12-24m vessels, where vessels built in the 1960s remain the dominant cohort in numbers and capacity (Table 1). Earlier cohorts are also still evident with peaks immediately post-WWII and again in 1956. In this range about 25% of the vessel numbers and the total capacity fishing in 1996 was built before 1960, with the oldest vessel built in 1906.

The general picture for this core inshore fleet is that it is aging, with 80% of capacity more than 15-years old. Replacement since the implementation of the QMS has been at a low rate, with 80 vessels totalling 2400GRT built between 1987 and 1996. This has maintained the aggregate capacity of the fleet, but this pattern does not reflect any great incentive provided by ITQs to increase productive efficiency by investing in new vessels. This sector of the fleet warrants closer study to determine the impacts of the quota system on exit and entry incentives, and could provide a good opportunity to examine strategic decision making of vessel owners under uncertainty.

Table 1
Cohorts of 12-24m vessels

Year vessel built	Number still registered in 1996	Total GRT 1996
Pre-1963	152	4 970
1963-1974	201	7 780
1975-1985	134	5 050
1986-1996	89	2 700
All vessels	576	20 500

Figure 11
Long-term impacts of institutional change on catch of inshore species

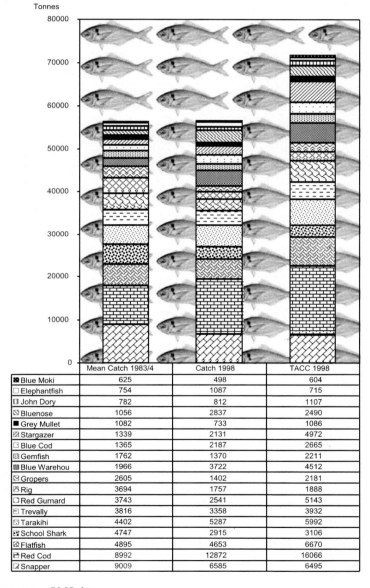

	Mean Catch 1983/4	Catch 1998	TACC 1998
Blue Moki	625	498	604
Elephantfish	754	1087	715
John Dory	782	812	1107
Bluenose	1056	2837	2490
Grey Mullet	1082	733	1086
Stargazer	1339	2131	4972
Blue Cod	1365	2187	2665
Gemfish	1762	1370	2211
Blue Warehou	1966	3722	4512
Gropers	2605	1402	2181
Rig	3694	1757	1888
Red Gurnard	3743	2541	5143
Trevally	3816	3358	3932
Tarakihi	4402	5287	5992
School Shark	4747	2915	3106
Flatfish	4895	4653	6670
Red Cod	8992	12872	16066
Snapper	9009	6585	6495

Data source: QMS data.

There were few vessels larger-than-24m in the New Zealand fleet before the 1970s. The 1996 records show 7 vessels in the 24-33m range built between 1967 and 1975, but the bulk of the capacity in this class was built between 1977 and 1981 (the EEZ cohort). Twenty-seven vessels totalling 4350 GRT from these 5 years were still fishing in 1996 - around 60% of the total tonnage in this range. The origin of these vessels is unknown, but it is certain that a good number of them were imported second-hand as 1984 records show fewer than 20 registrations for the class at that time (refer Figure 4).

Figure 12
Proportion of total catch from New Zealand EEZ caught by foreign-licensed, charter and domestic vessels - 1985 to 1995

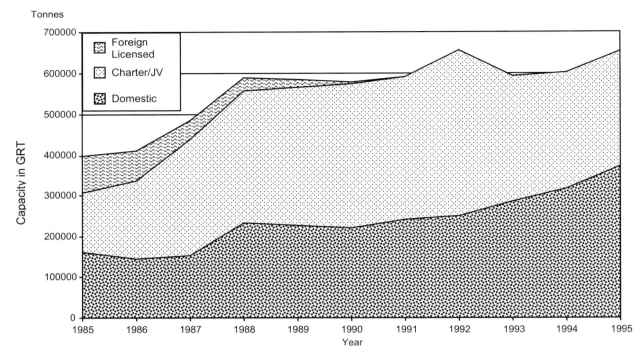

Data sources: FIB 1987, 1989, 1990, 1994 and 1996.

Figure 13
Total New Zealand catch for selected deep-water species - 1974 to 1998

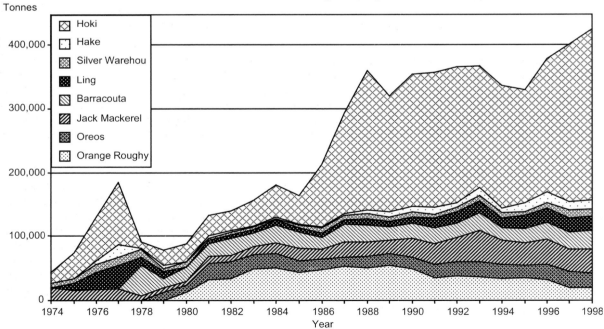

Data sources: Annala and Sullivan 1997; QMS data.

For very large vessels (greater than 33m) a similar strong cohort is present from the 1970s with 27 of 58 vessels registered in 1996 built between 1971 and 1981. However, another 25 vessels were built between 1986 and 1992, and these have an average tonnage of more than twice that of the earlier group. Again many of these vessels were imported into New Zealand from service in other countries' fisheries. In 1984 there were only 12 vessels of over-33m in length registered, but records show, that of vessels in this range registered in 1996, 30 were built before 1984. Similarly for the second cohort, there were 20 additional registrations in this class between 1992 and 1996, but only two of these vessels were built in that period. Hence, the majority of the capacity added to the New Zealand fleet as a consequence of the domestication of the offshore fisheries seems to have been sourced from existing foreign fleets, although some vessels have been purpose-built.

In the smaller vessel classes (less-than-12m LOA), the 9-12m vessels show the two earlier cohorts as dominant - the first peaking in 1969 and the second in 1978, with no sign of a resurgence of building after the QMS implementation. The replacement rate has been low - an average of 8.6 vessels (60t) per year since 1987 for a fleet of 500 vessels, and the average tonnage of these replacements is significantly down (28%). Total capacity was maintained from 1987 to 1996, but has fallen nearly 20% through to 1998.

In the 6-9m vessel-length range there are, as expected, fewer older vessels with only a handful built before 1976 still operating. The EEZ fleet build-up is the dominant feature here with 35% of 1996 capacity built between 1976 and 1980. Since 1980 the replacement-rate has been fairly constant at about 30 GRT/year, or 2.7% of the 1996 fleet size. This low rate has no doubt contributed to the decline of the class in recent years as boats built in the late 1970s are retired. One possible explanation for the decline in this class is the replacement of the older, permanently-moored vessels with smaller (5-6m) aluminium vessels, moved on trailers, so as to improve productive efficiency and reduce maintenance costs.

Most boats under-6m LOA in the fleet in 1996 had been built since 1984 (70%) at an average of 37.5 boats (5% of the 1996 total) per year. This seems to be a rate high enough to maintain the fleet but numbers have still fallen about 25% since 1987. Within this period the building-rate and total number was higher from 1989 to 1994, but has since tapered off. This sector is flexible, due to the low cost and relatively short life of boats. There will always be a place for these small boats, but it is unlikely under quota-management that they will ever again reach the numbers of the late 1970s, when almost 3000 boats under-6m LOA were registered for commercial fishing.

3.5   Summary of fleet-capacity trends

In summary, the total capacity of the New Zealand domestic fishing fleet grew by a net 43% from 1987 to 1998. This is accounted for by growth in the offshore (>33m) fleet to replace charter vessels and increase specialisation. The notable expanding classes include the 43m vessels and the 60-70m freezer trawlers. The inshore fleet has changed little in aggregate capacity although it has undergone significant restructuring. This restructuring has important dimensions not covered in the current analysis, such as vessel replacement: changed ownership patterns; gear configurations; and changed targeting. A core inshore fleet of 20 000 GRT in the range 12-24m LOA has been maintained as a constant capacity component since the mid-1970s. This section of the fleet is aging and there is a suggestion in the data that the replacement rate may be increasing through the 1990s.

The reduction in fleet-capacity by a drop in numbers of more than 70% in small-boats (<12m) has brought the capacity for this sector down below that of the early 1970s, and seen a shift to larger average vessel-size. The 24-33m class has developed rapidly from a few boats in the mid-1970s to become a significant sector of the domestic fleet. Many of these vessels will be deployed in fishing for species other than the inshore ones, indicating an overall decline in the capacity dedicated to the inshore waters since the introduction of quota. At the same time the overall catch of inshore species has been maintained, although the proportions of the species-mix have changed significantly, as catch levels for some over-fished stocks have been brought down while others have increased.

Insufficient data were available to describe in detail the impact of charter-capacity. Charter vessels do not operate in New Zealand all-year-round, and a fuller assessment of the relationships and trends in capacity and catch will require data on how long vessels are in New Zealand waters. However it is clear that the domestic fleet is taking an increasing proportion of the catch, and the early 1990s were a time of significant expansion in domestic capacity. Some of the large vessels added to the fleet had previously operated as chartered vessels and have been purchased by New Zealand companies. Others have been purpose-built or imported from overseas. The numbers of foreign licensed vessels fishing New Zealand waters, as well as their catches, declined steeply as New Zealand companies increased charter-operations following the QMS implementation. In 1984, for example, the foreign-licensed vessels took a total of 120 000t of a range of species from New Zealand waters. By 1994 this had been reduced to about 30t of bluefin tuna.

3.6   Effects of the introduction of transferable property-rights on fleet-capacity

Without more detailed study, including interviews with vessel owners, it is not possible to fully separate the effects of the introduction of ITQs from those of other regulatory measures, changes in export markets, and other factors. However, some observations are possible.

First, the introduction of quota did not provoke a new vessel-building boom for the inshore fleet as other major regulatory changes had done in the past. The historical context is important, in that the bulk of the core inshore fleet constructed in the late 1960s was still serviceable at this time. Estimates made by the Fishing Industry Board in 1983 suggested the inshore fleet was over-capitalised by about 20% (NAFMAC 1983). This was highlighted at the time as a signal that the regulatory framework required reform, but the key issues for the inshore were the over-exploitation of a few valuable and vulnerable species. Through the setting of TACCs under the QMS, these catches were brought down effectively, while the permitted-catch levels for other species were increased over historical levels. This allowed the fleet to adjust without an urgent need to shed capacity. However, because of the implications of the targeted cuts for particular gears used by the small boats, particularly lining and set netting, a substantial number of smaller vessels exited at the implementation of the QMS. Pressure has continued on this sector and a significant decline in numbers of small-boats has occurred through the 1990s. The 6-9m class in particular seems to be disappearing, and this may be a product of both efficiency considerations for the size-class, and the age of many of these vessels.

Some vessels built elsewhere in the universal boom of the late seventies were imported into New Zealand after 1986 both for the core inshore fleet and the offshore sectors, but some new vessels have also been built. Whether the new constructions constitute any more than an efficient replacement-rate for the fleet or not, given that the offshore catches were being domesticated from foreign charters, is an unanswered question at this stage. In general, it seems that replacement of vessels in most classes has become more regularised than it has been in the past, with a similar number of new vessels brought-in each year, rather than all at once, in response to a policy change.

The increased security offered to businesses by quota ownership, as part of a credible commitment from government regarding access to resources, has undoubtedly promoted the massive investments required in vessels and shore processing-operations required to patriate the catching and processing of fish taken from the NZ EEZ. ITQ provided the means for repatriation of the flow of resource-rents, previously captured by the distant-water fishing nations, through charter and joint venture operations. Local companies used these cash flows to back investment to expand their own capacity, turning the fishing industry into a major contributor to GDP and export earnings for New Zealand.

The significant negative impact on capacity-growth trends for the inshore fleet from 1994 suggests that the new cost-recovery regime may be driving out marginal operators. Although "cost recovery" may be easier to justify in political terms than taxes or "resource rentals", any increases in charges under transferable quota will affect less well-capitalised businesses harder. Higher charges will flow through into lower quota prices eventually, but for those who have paid top prices to get into the fishery, or have borrowed against high quota values, new charges could cause trouble. If operators have high debt-equity ratios and are not high-liners, they may be driven out by increases, where they could have survived had the charges been in place before they bought their quota, and thus been factored into the price.

In conclusion, although ITQ systems are often advocated where over-capacity is a problem, they do not act directly to regulate fleet-capacity. By limiting catches of individual species independently, quota systems can establish conditions for structural change within the fleet, and they provide the mechanism of transferability for catching-rights to allow autonomous adjustment. Whether a net decrease in capacity occurs in a particular fishery or sector of a fleet depends on a range of factors, including the degree of over-capacity initially present, opportunity-costs of holding vessels, labour and quota, and perceptions about the future of fishing and quota prices. In the case of New Zealand, the almost complete coverage of the quota system precludes movement of small vessels to other non-quota fisheries, and some may have been locked in while their vessels still had a useful life. As the fleet ages some small-boat capacity seems to be dropping-out, without being replaced, and this trend is likely to continue. At the other end of the scale, the QMS has provided the conditions for large-scale development of the offshore industry and the use of charter vessels has allowed the domestication of capacity to occur in ways most advantageous to New Zealand companies. It is unclear whether the future fleet will ever be totally domestically owned.

4.   CONCENTRATION OF QUOTA-OWNERSHIP
4.1   Status prior to programme

The decade prior to the introduction of the general ITQ-regime in 1986 was one of rapid change for the fishing sector in New Zealand, as it was for the industry all over the world. The introduction of EEZs precipitated major changes in patterns of global fishing and catch distribution and shifts in fleet structure. The

UNCLOS[8] negotiations and the EEZs were themselves a response to expanding international fishing activity and capacity that was causing conflict between distant-water fishing nations and the coastal states. It was this historical process of change, promoting rapid expansion of domestic fishing activity in New Zealand during the 1970s, that brought the need for new methods of regulation.

New Zealand fisheries management was liberalised in 1963 after four decades of conservative management that severely constrained growth of the industry and the activity of fishers. For the following period up to 1978 management was under a regime of regulated open-access. Fishing operators were required to be licensed and vessels registered, but most fisheries were able to be exploited by anyone who applied for access, and the industry expanded rapidly. Permits carried endorsements nominating methods, areas and species to which they applied. In 1978 entry to fisheries began to be restricted, and a complete moratorium on new permits was implemented in 1982. The moratorium was applied to existing permits by not allowing further endorsements to be added.

It is at this point of the virtual closure of fishing to new entrants, that it could be said that exclusive rights to access were first established. When allocation of ITQ was eventually undertaken in 1985-86, it was to be based on the three fishing-years immediately following this closure. Hence the set of fishers who might be eligible for ITQ-rights was fixed in 1982. This group was subsequently greatly reduced in number with the removal of part-time fishers by the *Fisheries Act 1983*, but this did not affect the distribution of the vast majority of the catch. The part-timer fishermen removed before quota was introduced were estimated to have been responsible for between 2 and 5% of the reported domestic catch at the time[9].

The outcome of a process of policy review and consultation with the industry between 1983 and 1985, was a government decision to implement the quota-management system. Allocation of quota was undertaken in a series of steps. The first was for the Ministry to assess the recorded catch-histories of all commercial fishers over the qualifying three fishing-years and notify these to the stakeholders. Following an objections-process the fishers were then issued provisional maximum ITQ (PMITQ) - reflecting the assessed catch history - and a guaranteed minimum ITQ (GMITQ) - reflecting the TAC levels desired by the regulator. The sum of the PMITQs for a species or fish-quota stock were in many cases higher than the desired TAC level for two reasons. First, each fisher nominated the best two of the three catch-history years which were then averaged to obtain their PMITQ allocations. Hence the sum of these averages was often greater than the actual total catch in any one year. Second, some stocks were considered over-fished and in need of catch reductions from pre-QMS levels.

The desired TAC reductions were made through a government-funded buy-back of PMITQ and lastly involved some *pro rata* administrative adjustments to reach the final ITQ allocations. Some fishers sold all of their quota back to the government, while others sold some so as to adjust the balance of their holdings and to avoid administrative reductions. For species where large reductions in allocations were made from PMITQ to final ITQ during the implementation (such as snapper, rig, school shark and gropers) the buy-back may have significantly altered the distribution and therefore concentration of catching-rights.

The relevance of PMITQ to this analysis is that it provides a convenient baseline for judging the impact of the QMS on the distribution of catching rights. The purpose of the extended process leading to the allocation of PMITQ was to produce a fair and agreed assessment of the distribution of catch in the qualifying years. It therefore provides a useful, and the most accessible, baseline data series for the analysis.

4.2   Restrictions on transfer of ownership

The implementation of "management stocks" under the QMS - where species are divided into between two and ten stocks by area - to some extent restricts the geographic concentration of fishing. Each quota can only be fished in its specified area, preventing companies buying-up quota being fished in one area of the New Zealand's marine area in order to fish against it in another. There are exceptions to this, depending on management requirements: for example, hoki has only one management stock and one ITQ for the whole fishery. The initial 26 species brought under ITQ management were divided into a total of 153 stocks.

---

8 UNCLOS – the United Nations Convention on the Law of the Sea – was negotiated between 1958 and 1982, and introduced internationally agreed standards for the jurisdiction of coastal states over adjacent ocean resources. Within this institutional framework the 12 mile Territorial Sea and the 200 mile Exclusive Economic Zone (EEZ) have been adopted for the management of fishing.

9 However, these excluded fishers may have been disproportionately responsible for unreported catch. From a total of 5184 fishing vessels licensed in 1978, 2942 are reported to have earned less than $NZ500 from fishing (National Research Advisory Council 1980, cited in Wallace 1997). This may indicate widespread under-reporting of catch, possibly to evade income tax, and hence these exclusions may have been more significant economically as well as for fish stocks than indicated by the records. In this case the changes in distribution of actual catches may also have been more significant than believed.

Several provisions were made in the QMS restricting the ownership of ITQ and thereby its transfer. First, quota could only be held by permanent residents of New Zealand, or by companies with less than 25% foreign ownership. Second, maximum holdings were restricted as a percentage of the total quota (or TACC) for a stock or species. For a range of mainly deep-water and mid-depth stocks, listed in Schedule 1 of the *Fisheries Amendment Act 1986*, a maximum holding of 35% of the species over all areas was permitted. For other stocks, a maximum of 20% of the total by QMA was allowed, or 10% for rock lobster. Hence, for inshore stocks local concentration was restricted to 20%, whereas for off-shore species the restriction applied to the proportion of the aggregate for the whole country held by one interest, and could thus allow individual stocks of these species to be dominated by one or two interests.

Both the maximum-holdings and the foreign-ownership provisions are difficult to enforce strictly due to complex cross ownership of companies and other relationships that can be established to avoid these restrictions. The Minister of Fisheries retained the power to exempt companies, or individuals, from the aggregation limits and to allow them to accumulate further quota up to a stipulated level. During the 1990s, the government modified the foreign-ownership provisions to allow exemptions on the judgement of the Overseas Investment Commission with concurrence of the Ministry of Fisheries. This remains a poorly-defined and discretionary area of the fisheries law[10].

Minimum-holdings and transfer-restrictions also applied, being generally a total from all areas of at least one tonne for shellfish and 5 tonnes for fin-fish that must be held before fishing can be undertaken.

The *Fisheries Act 1996* lifted the aggregation-limits for a schedule of 14 species to 45% of the New Zealand-wide total for each (most of these species had been under the 35% provision). It also changed minimum-holdings to 3 tonnes of rock lobster from the same area, 5 tonnes of southern scallops or a 3 tonne total from all areas for other shellfish species.

4.3 Concentration of catching rights

From an economic point of view there are two reasons for monitoring the concentration of ownership in fisheries. The first is that where production has been inefficient due to over-capitalisation and likely accompanied by stock-depletion and higher-than-necessary variable catching-costs, the introduction of quota-management often has as the explicit objective of reducing the number of vessels in the fishery. Assuming that 'grand-fathering' is used in allocation, if rationalisation of the fleet is to occur, then some degree of quota-concentration is to be expected as a result. Hence increasing quota-concentration can indicate that the system is achieving one of its objectives. In addition, economies of scale in catching, processing and marketing, and the commercial advantages of secure supply, may mean that higher net gains are possible from more concentrated quota-ownership. In this way efficiency of resource utilisation may be improved even without a reduction in fleet size.

On the other hand, there are limits beyond which concentration can be negative. Should a small number of owners control the large majority of quota, monopoly-type market-power effects may be manifest. These can include the manipulation of prices for both fish and quota so as to capture rents and to facilitate further accumulation of quota, and undue dominance in the labour market affecting wages and conditions of fishers (National Research Council 1999). The development of such monopoly-power may be guarded against by the specification of aggregation limits for quota. In practice such limits have been set at levels ranging from 0.5% to 45% of a fish stock. Alternatively, anti-trust legislation or quasi-governmental regulatory commissions may be relied upon to guard against monopoly power without having to specify an arbitrary limit on individual holdings. However, some means to estimate market-concentration is required. To date there have been few published studies estimating quota-concentration in fisheries (for examples see Gauvin, Ward and Burgess 1994; Hogan, Thorpe and Timcke 1999).

4.4 Methods and data
4.4.1 Concentration indices

Standard measurements of concentration of such factors as market share, assets, physical output, or employment among competing industrial companies are considered by Scherer (1973). Should data on marginal costs of production be available (which is usually not the case) a direct comparison with price can estimate the degree of monopoly-power present in a market. The most straightforward index for commonly available data is the Concentration Ratio (CR), which is the proportion of the factor chosen represented by a selected number of the largest firms. The top four firms are commonly used (CR4), and often a table is presented with a range of values (*e.g.* CR4, 8, 20 and 50).

---

[10] In May 2000, the newly elected Labour led coalition Government intervened to prevent the sale of a 50% share in New Zealand's largest fishing company and quota owner to foreign interests.

An extension of this comparison is to use percentiles, or to construct a Lorenz curve, which is a plot of the cumulative total proportion of the factor represented, against the proportion of the firms represented, sorted in rank-order for the factor. For example, if quota-owning firms are ranked by the amount of quota held, then the cummulative values for the proportion of the total held by each firm are plotted. In the case that all participants hold the same amount (for instance 100 firms with 1% of the quota each) a 45-degree straight line is the result. Where there are differences in holdings, the curve will sag below the 45-degree line but start and finish at the same points (0 and 100% of quota owned by 0 and 100% of owners). Figure 14 plots two Lorenz curves, showing the distribution of allocations of Provisional Maximum Individual Transferable Quota (PMITQ) for inshore species (reflecting catch-history assessment), and ownership of ITQ for the same stocks after one year of quota-trading. This illustrates the initial impact of ITQs in the sector through significant consolidation of ownership.

Figure 14
Example of Lorenz curves using inshore PMITQ and 1987 ITQ ownership data

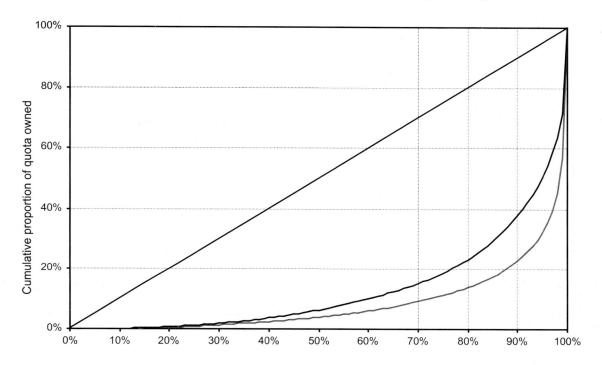

Data source: QMS data.

The Lorenz curve can be converted to a single number, the Gini Index (GI), by comparing the area between the 45-degree line and the curve with the total area under the 45-degree line. For any number of firms all with identical shares, the curve will be the line, and the GI is zero. As the distribution of shares becomes less equal, the index approaches unity. The GI is a measure of inequality in shares, that does not take into account the number of participants. It is therefore not ideal for measuring competitiveness where firms are likely to have similar sized shares, as it will give the same answer for 2 firms with 50% of quota each and for 100 firms with 1% each. However, it may be useful in indicating inequality when used with other indices.

The Gini Index (GI) has a range from zero to one and is calculated as follows:

$$GI = 1 - \frac{\sum\limits_{i=1}^{N}\left(\sum\limits_{j=1}^{i} q_j \Big/ Q\right)}{(N+1)/2}$$

where:

q = quota or catch amount
Q = total of all quotas or catches
N = number of quota holdings or catches and
all q are ordered, 1 to N from smallest to largest.

The GIs for the distributions illustrated in Figure 14 are: 0.74 for PMITQ, and 0.84 for 1987 ITQ-ownership.

An index more useful for estimating potential market-power, the Herfindahl-Hirschman Index (HHI), which sums the squared proportionate shares of all firms. This takes account of both the number of firms and inequality in quota shares, weighting the larger firms quadratically. The HHI is calculated as follows:

$$HHI = \sum_{i=1}^{N}(q_i/Q)^2$$

where:

q = *quota or catch amount*
Q = *total of all quotas or catches and*
N = *number of quota holdings or catches*

The index as calculated has a range of zero to one. In this study, both the GI and the HHI are expressed as percentages. Hence a calculated index of 0.75 is written as 75%.

4.4.2 Data and indices used

Data supplied by the New Zealand Ministry of Fisheries included PMITQ/GMITQ allocations (1986), and balances of quota-ownership, leased-holdings and catches by quota-holder on the final day of the fishing-year for each year from 1987 to 1998. These data were processed to generate a range of indices:

i.   Total quota/catch
ii. Number of owners/holders/catchers
iii. HHI
iv. GI
v.   CR1; CR3; CR4; CR10
vi. Number and percentage of owners with 95% share
vii. Percentage share controlled by top 5% of owners and
viii. Number with less than minimum holdings for the class.

The fisheries were split into three general classes of fin-fish species: inshore, mid-depth, and deep-water. The species in each group are listed in Appendix I. Jack mackerel, squid, and paua have been omitted from the analysis as they were added to the quota-system in the second year and some data problems remain unresolved. Rock lobster, which was added to the QMS in 1990 is included as an example of a more sedentary species suited to small operators and is treated separately.

4.5   Results of concentration analysis
4.5.1 Presentation of results

The analysis is presented in several parts. First, a short section discusses some issues with respect to interpretation of the indicators. Then, each sector is examined more closely in two steps:

i.   immediate restructuring on the introduction of the system through the quota buy-back and quota market-trading during the first twelve months and
ii. trends over the period 1987 to 1998.

The detailed results data are presented in Tables 6 to 10 and in Figures 15 to 18.

Figure 15
Percentage of total inshore fin-fish quota owned by top 1, 3, 4 and 10 owners and by top 5% of owners

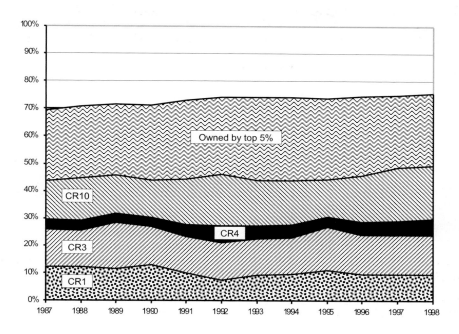

Figure 16
Percentage of total mid-depth fin-fish quota owned by top 1, 3, 4 and 10 owners and by top 5% of owners

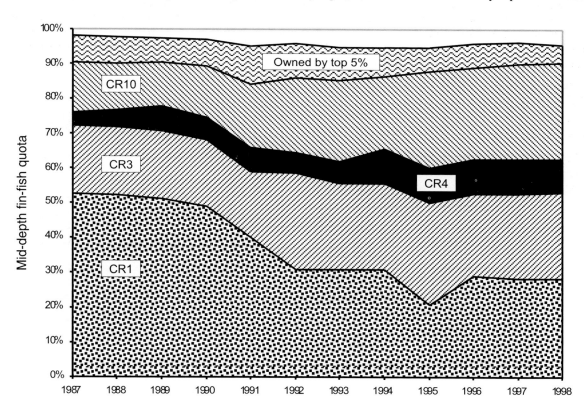

Figure 17

Percentage of total deep-water fin-fish quota owned by top 1, 3, 4 and 10 owners and by top 5% of owners

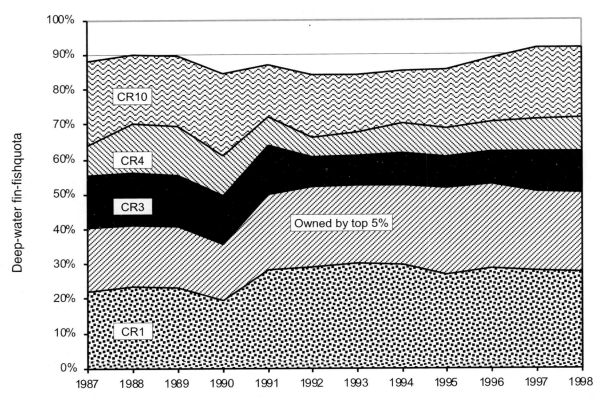

Figure 18

Percentage of total rock lobster quota owned by top 1, 3, 4 and 10 owners and by top 5% of owners

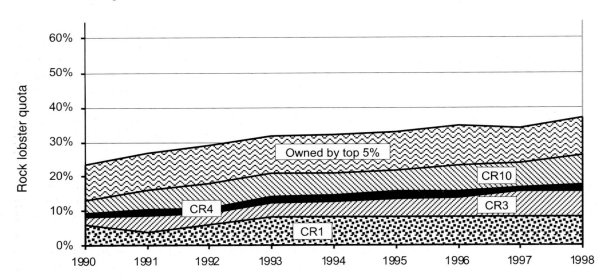

4.5.2 Factors in interpretation of results

As will be seen in the results, no single index gives a clear indication of the nature of the concentration of rights. However, taken together and compared across sectors, the set of indices used here can provide a richer insight. The GI tends to be quite high for all fin-fish sectors, in the 75% to 97% range, reflecting the diversity of operator-size to be found within these groupings. In Rock Lobster, the GI is much lower reflecting the nature of the fishery, which dictates relatively small tonnages per operator. However, for the same GI, the HHI can range widely. For example, the holdings-data in Table 2 show the same degree of inequality in distribution (GI) with very different indications of market concentration (HHI).

Table 2

Disparate concentration with similar inequality

Sector:year	GI	HHI
IS:1988	83.4%	1.8%
DW:1994	83.3%	15.3%

The similar GI reflects a similar distribution of quota by proportion of holders - something also indicated by the proportion of quota held by the top 5%. However, the HHI, in taking into account the number of holders (for the inshore being some 20 times the number for the deep-water), and thereby the absolute proportion held by the largest players, indicates the inshore structure is considerably less subject to potential problems from concentration.

Many year-classes of quota have a long tail of small holdings that are less than the minimum for a fishing permit to be issued. If these are eliminated from the index calculations, the GI decreases slightly as the degree of inequality has decreased, but the HHI increases as the number of firms has decreased and thereby the concentration of ownership has increased. This illustrates the utility of the two indexes. In fact, neither is very sensitive to large numbers of small holdings being eliminated.

On the other hand, all the indices are sensitive to the presence of single large owners at the top of the scale. Relevant here are the large government quota-holdings in the first years of the system. On initial allocation of the fixed-tonnage ITQs, any part of the TACCs that was in excess of private entitlements was allocated to the government. At the end of the first year of the QMS, the New Zealand Government retained ownership of around 196 000t of quota, of which just over 150 000t was hoki, out of a total TACC for all quota species of 467 000t. Some of the offshore quota was sold by auction in the first years of the system, and the remainder was leased out to operators. With the transition to proportional quota in 1990 the government eventually got out of quota-ownership with some of the assets being used for settlement of Maori claims. But at the start of the QMS the government was the largest owner of quota, and this distorts the ownership-concentration indices to some extent. This is most marked in the mid-depth ownership figures for 1987 to 1991 - HHI and CRs. In the deep-water species the government held some 22% of quota in 1987, but here this has the opposite affect on the indices after 1990. Concentration of ownership in this sector increases when the government sold its holdings as the top few private owners increased their holdings.

After 1989 the government began transferring some of its quota-holdings to a Commission established to receive assets to be held in trust as part of the settlement of Maori claims to fisheries resources. Further quota was purchased by the government on the quota-market to fulfil these obligations until approximately 10% of all quota was held by the Treaty of Waitangi Fisheries Commission (TOKM)[11]. This has created a new large owner in all sectors, and the Commission is the largest owner in fisheries such as rock lobster. Again this distorts the picture in the short-term as this quota is being held in trust and will (eventually) be distributed in small holdings to the 78 tribal groups (Maori *iwi*) recognised by TOKM. The Commission, like the government as owner before it, differs from other large quota-owners in not being a commercial fish buyer, and is therefore not in a position to exert monopolistic influence on fish prices itself. However, TOKM does own, or have controlling, shares in several large fishing companies. It also leases large amounts of quota with discounted rates for companies controlled by Maori interests and this may affect demand and prices in the wider quota-market.

Another data issue that should be borne in mind with respect to interpreting indices, is that some interests in the industry effectively control more than one fishing company. For the major players these connections are fairly well known. In most sectors the same three principal interests effectively control at least two companies each of the top ten. Cross ownership by larger interests in the industry also effectively concentrates control over quota, but this is even more difficult to take into account.

The number of quota-holders with less than 5t of fin-fish must be interpreted with caution. Some of these quota-owners may be targeting non-quota species primarily and be catching quota-species in small amounts as bycatch. Some are rock lobster fishers who, after 1990, would qualify for a quota permit if they held more than one tonne of rock lobster quota. In 1990, for example, 180 amongst the 455 quota-holders with less than 5t of fin-fish quota, also owned rock lobster quota.

---

[11] In 1989, following several years of litigation, the New Zealand Government agreed to provide Maori with 10% of all quota over four years as an interim position while negotiations for a final settlement of fisheries claims was in process. The final settlement was reached during 1992. The body established in 1989 to receive what has become known as the pre-settlement assets was the Maori Fisheries Commission. This was reconstituted in 1992 following the final settlement as the Treaty of Waitangi Fisheries Commission.

4.5.3 Inshore species
4.5.3.1 Effect in the first year of the QMS

The inshore is the most complex of the three fin-fish sectors considered. The quota buy-back applied to 16 of these 17 species (see Appendix I for species lists with scientific names), but was targeted to substantially reduce the fishing mortality of six of them. The others were bought largely to allow fishers to offer up all of their allocated PMITQ - that is, to quit the fishery entirely. Table 3 shows the combined impact of the buy-back and the first year of quota-trading. A 25% reduction in the overall number of quota-owners understates the proportions exiting from the species under pressure, which approached 40%. However, the largest numbers of fishers remain in these high-value fisheries, particularly snapper, rig and school shark.

The top 25 companies in the inshore fisheries are more-or-less the same as those in the other sectors and are heavily dominant in ownership of quota, with some 53% of privately-owned inshore-quota at the end of the first year, up from 33% of PMITQ. While the total privately-owned tonnage in 1987 was 30% less than PMITQ levels, the top 25 owners actually increased their aggregate tonnage by 10%. This represents a substantial concentration of ownership and financial restructuring of the sector in a single year. Five of the top eight quota-owners were either new companies or had substantially increased their holdings by acquisitions in the first year, with another five similar companies further down the order in the top twenty-five. The 156 largest owners had 60% of the PMITQ, but at the end of the first year this proportion of quota was owned by just 48 companies.

Table 4 tests statistics by quintile for the inshore sector. The 80:20 rule applies here, with the percentage of quota owned by the top 20% of owners moving from 76% to 85% over the first year, while the last 20% of owners have declined from having 0.7% of quota to just 0.4%.

Table 6 shows the jump in the indices between PMITQ allocations and the balances at end of 1987 for owned-quota. If the government holdings are taken out, the steps for the CR indices are significantly reduced, but the difference fades fairly quickly up the order. Table 5 shows the CR indices for 1987 with government-holdings excluded.

Large tonnages of quota have been bought up during the first year of trading by highly-capitalised companies. It seems that much of this quota has come from the medium-sized operations (in the 100 to 200t range) selling out or amalgamating. Many small holders have also sold up with a total of 440 quota-owners dropping out of the inshore in the first year.

Table 3
First year impacts of ITQ on number of owners and TACs - Inshore species

Species Code[1]	No. of PMITQ owners	1987 ITQ owners	Change %	Change %	PMITQ tonnes	1987 ITQ tonnes	Change %	Change %	Gvt holdings %
BCO	649	500	-149	-23	1 968	2 328	360	18	6
BNS	237	159	-78	-33	1 825	1 351	-474	-26	8
BYX	52	41	-11	-21	1 984	1 780	-204	-10	10
ELE	299	183	-116	-39	1 037	471	-566	-55	17
FLA	796	546	-250	-31	6 170	6 055	-115	-2	5
GMU	197	170	-27	-14	1 125	972	-152	-14	5
GUR	938	593	-345	-37	4 828	4 302	-526	-11	9
HPB	831	538	-293	-35	3 353	1 845	-1 508	-45	18
JDO	406	261	-145	-36	1 072	863	-210	-20	10
MOK	518	330	-188	-36	862	268	-594	-69	24
RCO	528	344	-184	-35	13 862	15 311	1 448	10	15
SCH	1 122	688	-434	-39	6 156	2 601	-3 555	-58	10
SNA	1 007	643	-364	-36	12 468	6 609	-5 860	-47	1
SPO	988	601	-387	-39	4 508	1 448	-3 060	-68	5
STA	305	194	-111	-36	2 123	4 162	2 039	96	54
TAR	713	454	-259	-36	5 985	5 390	-596	-10	9
TRE	629	403	-226	-36	4 692	3 261	-1431	-30	2
Totals	1 749	1 309	-440	-25	74 017	59 015	-15 002	-20	12

---

[12] For species names see Appendix I.

Table 4

Inshore quota owned by quintile - PMITQ and 1987 ITQ

Owner percentile	PMITQ (t)	PMITQ %	1987 ITQ (t)	1987 ITQ %
20	503	0.68	196	0.38
40	2 231	3.01	1 047	2.02
60	4 680	6.32	2 082	4.02
80	9876	13.34	4 399	8.49
100	56 728	76.64	44 079	85.09
Totals	74 018		51 803	

Table 5

Concentration ratio indices for inshore species - PMITQ and 1987 ITQ

Quota	CR1	CR3	CR4	CR10
PMITQ	8%	13%	15%	23%
Ex-Govt ITQ	9%	20%	23%	38%
All ITQ	12%	26%	30%	44%

4.5.3.2 Trends under the QMS: 1987 to 1998

While total allowable commercial catches (TACCs) and therefore quota owned for these species increased by 15% between 1987 and 1998, the number of quota-owners decreased by a further 26%, from 1309 to 963 (see Table 6). The HHI has increased marginally, which would be predicted by falling numbers of owners, but there is more happening than just this. While the top-ten owners have steadily increased their share of the quota from 43% to just under 50%, some jockeying has been going on among the top-three or top-four owners. In 1987 the Government was the largest quota-owner in this as in the other sectors. This changed with the move to proportional quota in 1990 and the agreement in 1989 to provide 10% of all quota to Maori in interim recognition of their claim to the resource. The Maori Fisheries Commission subsequently acquired from the government 2.5% of all quota per year for four years, to take its place as the largest inshore quota-owner. Figure 15 shows the trends in concentration for the largest owners.

Table 6
Quota owned by sector

Fishery	Year	Total quota owned (t)	No. owning quota	HHI %	CR1 %	CR3 %	CR4 %	CR10 %	No. owning 95%	% owning 95%	% owned by top 5%	No. owning < 5t	Gini index %
Inshore	PMITQ	74 017	1749	1.1	8	13	15	23	957	55	49	428	73.5
	1987	59 015	1309	3.1	12	26	30	44	563	43	69	454	84.1
	1988	61 198	1289	3.2	12	25	29	45	533	41	71	445	84.8
	1989	64 715	1312	3.5	12	28	32	46	525	40	72	452	85.3
	1990	67 026	1320	3.4	13	27	30	44	512	39	71	459	85.6
	1991	66 666	1291	2.9	10	23	28	44	491	38	73	439	86.1
	1992	68 391	1244	2.8	7	21	27	46	457	37	74	411	86.7
	1993	67 694	1187	2.8	9	23	28	44	426	36	74	398	86.9
	1994	67 858	1161	2.8	10	23	28	44	411	35	74	386	86.9
	1995	68 057	1110	3.3	11	27	31	45	391	35	74	364	86.9
	1996	68 444	1077	3.0	10	24	29	46	366	34	75	350	87.3
	1997	69 071	1023	3.2	10	24	29	49	343	34	75	323	87.5
	1998	67 958	963	3.3	10	24	30	49	322	33	76	292	87.6
Mid-depth	PMITQ	160 178	708	10.6	21	48	55	81	32	5	95	451	97.1
	1987	325 175	493	30.2	53	72	76	91	15	3	98	327	98.2
	1988	326 081	478	29.9	52	72	77	90	17	4	98	317	98.1
	1989	329 119	475	28.9	51	71	78	91	16	3	97	312	98.1
	1990	333 569	466	26.6	49	68	74	89	18	4	97	291	97.9
	1991	308 161	457	19.0	40	59	66	84	23	5	95	290	97.3
	1992	290 266	449	14.8	31	58	65	86	20	4	96	278	97.3
	1993	290 694	424	14.1	31	55	62	85	22	5	95	259	97.0
	1994	290 770	417	14.6	31	55	65	87	22	5	95	247	97.0
	1995	311 864	409	11.4	21	50	60	88	21	5	95	236	96.9
	1996	331 514	396	13.6	29	53	63	89	18	5	96	224	97.0
	1997	340 668	378	13.5	28	53	63	90	17	4	96	217	97.0
	1998	338 242	360	13.6	28	53	63	90	17	5	96	210	96.9
Deep-water	PMITQ	64 559	54	11.9	22	52	60	86	24	44	52	1	77.4
	1987	83 010	44	12.4	22	55	64	88	19	43	40	1	76.3
	1988	85 216	38	13.8	23	56	70	90	16	42	41	1	76.0
	1989	86 623	43	13.6	23	56	70	90	16	37	41	3	78.2
	1990	70 353	42	10.9	20	50	61	85	20	48	36	2	72.7
	1991	64 110	41	15.8	28	64	72	87	18	44	50	4	76.9
	1992	63 076	49	15.5	29	60	66	84	19	39	52	2	78.3
	1993	61 700	47	15.9	30	61	68	84	18	38	52	4	78.1
	1994	61 534	49	16.1	30	62	70	85	18	37	52	6	79.8
	1995	56 233	47	15.6	27	61	69	86	18	38	52	5	79.1
	1996	50 050	41	16.5	29	62	71	89	14	34	53	6	79.3
	1997	50 474	39	16.2	28	62	71	92	13	33	50	7	79.9
	1998	50 474	40	16.1	28	62	72	92	13	33	50	8	80.2
Rock lobster	1990	3 726	686	0.6	6	8	9	13	503	73	23	411	48.7
	1991	3 597	656	0.6	4	9	10	16	479	73	27	418	51.2
	1992	3 286	598	0.7	6	9	11	18	438	73	29	382	52.6
	1993	2 936	554	1.1	8	12	14	21	404	73	32	379	54.5
	1994	2 932	525	1.1	8	13	15	21	383	73	32	350	55.0
	1995	2 915	517	1.2	8	14	15	22	372	72	33	342	55.5
	1996	2 968	512	1.2	8	14	16	23	358	70	35	336	57.5
	1997	2 894	490	1.3	8	15	17	24	344	70	34	315	57.1
	1998	2 954	470	1.4	8	15	17	26	329	70	37	293	58.1

Data source: QMS data.

Given that there are almost one thousand owners (1998), having 50% of quota owned by 10 interests (1% of owners) might seem concentrated, but these same companies own greater proportions of quota in the other fin-fish sectors. The top 5% of owners have 75% of the inshore quota, up a little from 69%, and the proportion of all owners holding 95% of all quota has dropped from 43 to 33%. The HHI is low at around 3.3%, the GI has moved from 84% to 88%, indicating a moderately high degree of inequality in holding sizes, and the number with less than minimum holdings has dropped by one-third.

The figures for end-of-year holdings (Table 7), which take into account effective redistribution of access through leasing, show a decrease in concentration with respect to the ownership figures, systematically expressed in all indicators. Numbers of holders are about 10% up on numbers of owners, HHI is down slightly, all concentration ratios are down and so on. The numbers with less than the minimum holdings are up slightly, presumably because some have leased almost all their quota.

For catch, Table 8 shows the HHI and GI indicators are lower again than for holding and ownership. The CR1 and CR3 are the same as holdings indicating the top owners are getting all their fish, but CR4 and CR10 drop off slightly. The proportion caught by the top 5% in 1998 was two-thirds the total catch, whereas the top 5% of owners owned three-quarters of all quota. This indicates a small shift in the effective share of owners to smaller operators.

4.5.4 Mid-depth species
4.5.4.1 Effect in the first year of the QMS

The mid-depth species-group comprises seven of the nine species (or species groups) included in the enterprise allocation quota scheme from 1983 (refer Appendix I). PMITQ allocations reflected either existing enterprise allocation quotas held. For the competitive TAC fishers in this sector, were based on assessed catch-histories. All but one of these species had their TACCs set higher than the sum of PMITQ, with the government taking ownership of the increases as well as some ling- and gemfish-quota surrendered as part of the inshore buy-back. With a total holding of 53% of this sector the government was by far the biggest quota-owner.

The adjustment in the first year seen in this sector is similar in character, although not as extensive, as in the inshore - catch in the sector was already highly concentrated. The same four new large business entities who were thus present in the top-25 owners, with the basic order in this group unchanged from PMITQ allocations at the end of 1987. However, below this about a dozen of the next 25-biggest entities have disappeared, their quota having been taken up through mergers and acquisitions by the new larger players. There were 215 departures during the year, reducing the total number of quota-holders by 30% from 708 to 493, with the species caught by smaller vessels showing the greatest reduction in numbers.

The distribution of rights in this sector is the most highly skewed of all. The top quintile of owners had over 99% of both PMITQ and subsequent privately owned ITQ. The majority of fish in these quota are caught by large factory freezer-trawlers, but almost all of the species are also available to relatively small boats as targeted species or bycatch. Hence the distribution-plot has a long tail. There were only 86 of the 493 owners with greater than 20t of mid-depth quota in the first year of the QMS. Tables 6 and 7 tell the story. Note the numbers owning or holding 95% of the quota. Ownership is significantly more concentrated than holdings indicating large amounts of leasing, some of which will doubtless involve conditions as to where fish is to be landed and processed. The number of firms holding (rather than owning) 95% of quota in 1987 is in fact not too different from the PMITQ figure.

4.5.4.2 Trends under the QMS: 1987 to 1998

The figures in Table 6 for the first four years of the quota system are confounded by the fact that the New Zealand Government owned large amounts of quota that was leased out. Hence in this early period the concentration ratio is 53% which would have been illegal for any owner other than the government (under the prevailing aggregation limits of 20% for inshore fin-fish and 35% for deepwater and mid-depth species). In 1992, the concentration ratio reached 31%, as the government sold quota with the move to proportional quota, and has stayed about there since. The number of quota-owners has dropped by another 25% since 1987 and those with less than minimum holdings (5t) by a third, accounting for most of the overall reduction. The HHI is about 14% and the GI is 97%; both indices have remained stable since the exit of the government from ownership. The concentration ratios are all high with 17 of 360 quota-owners having 95% of the quota in 1998 (see Figure 16).

The relation of 'end-of-year holdings' to 'owned-quota' is similar to that of the inshore sector. The number of holders are about 10% up on the number of owners, and all the indicators show the quota is spread around a little more, with CR10 dropping from 90 to 82% indicating about 30 000t of quota is been leased out from this group of owners. The HHI drops from 14% for owned to 10% for held. The effect of government ownership on the indices can be seen in the contrast between owned and held figures for the early years. The HHI for held quota in 1987 was only 11%, almost the same as in 1998, as are all the concentration ratios. The dominance of the top few owners in the hoki fishery can be seen in the change in numbers-owning-95% column of Table 6 for the years 1991 to 1995, reflecting the 20% quota reduction for hoki in place during that time.

Table 7
End of year quota holdings by sector

Fishery	Year	Total quota held (t)	No. holding quota	HHI %	CR1 %	CR3 %	CR4 %	CR10 %	No. holding 95%	% holding 95%	% held by top 5%	No. holding < 5t	Gini index %
Inshore	1987	60 272	1408	1.9	7	17	21	37	625	44	68	444	83.1
	1988	61 871	1432	1.8	7	17	20	35	618	43	68	459	83.4
	1989	65 566	1473	2.2	9	20	23	37	616	42	70	448	84.2
	1990	67 575	1502	1.9	8	18	21	36	597	40	69	485	84.9
	1991	66 750	1567	1.6	6	15	20	32	658	42	68	449	83.7
	1992	68 395	1608	1.5	6	16	19	31	666	41	70	458	84.4
	1993	67 815	1503	1.6	6	16	20	31	604	40	70	439	84.8
	1994	67 867	1436	1.9	7	19	22	35	557	39	70	431	85.2
	1995	68 065	1276	2.4	11	21	24	36	471	37	70	413	85.6
	1996	68 444	1155	2.9	10	24	29	44	381	33	75	390	87.6
	1997	69 071	1168	2.5	8	21	26	42	426	36	72	345	86.2
	1998	67 958	1091	2.8	8	22	28	44	395	36	73	322	86.3
Mid-depth	1987	330 583	565	11.4	25	51	58	81	27	5	96	336	97.1
	1988	329 063	567	10.5	24	49	56	78	27	5	95	322	97.1
	1989	330 331	542	11.0	25	51	58	77	27	5	95	296	96.9
	1990	341 261	543	9.2	20	46	56	74	31	6	94	304	96.6
	1991	308 166	554	7.0	16	36	45	71	36	6	92	323	96.1
	1992	290 266	583	10.9	25	49	55	79	34	6	93	334	96.8
	1993	290 751	547	8.7	20	45	51	75	41	7	91	298	96.1
	1994	290 770	544	9.7	24	45	52	75	45	8	90	257	95.9
	1995	311 864	589	13.0	27	57	64	80	40	7	93	260	96.8
	1996	331 514	464	10.7	22	47	57	87	21	5	96	226	97.0
	1997	340 668	455	10.5	23	48	55	81	29	6	93	242	96.4
	1998	338 242	400	10.3	22	48	54	82	27	7	92	219	96.1
Deep-water	1987	83 469	61	9.7	18	45	54	83	27	44	45	2	77.6
	1988	85 416	43	11.6	22	53	61	86	19	44	38	0	74.4
	1989	87 519	48	10.5	20	48	58	84	19	40	35	3	76.1
	1990	70 353	53	8.9	19	41	50	80	22	42	41	2	74.4
	1991	64 160	66	12.9	24	58	62	78	28	42	58	4	78.9
	1992	63 076	55	14.6	25	59	65	84	21	38	59	11	80.0
	1993	61 850	66	14.9	30	61	68	86	20	30	61	12	83.8
	1994	61 534	74	15.3	26	65	68	83	27	36	68	12	83.3
	1995	56 233	76	15.8	26	64	69	84	22	29	69	14	85.1
	1996	50 050	45	15.6	27	60	69	87	16	36	51	5	79.8
	1997	50 474	55	16.1	27	65	74	89	15	27	65	11	84.0
	1998	50 474	54	15.1	25	63	72	89	16	30	63	9	82.9
Rock lobster	1990	3 757	679	0.6	6	8	9	13	498	73	23	398	48.0
	1991	3 603	686	0.3	2	4	5	8	514	75	20	417	45.2
	1992	3 288	658	0.3	1	3	4	8	510	78	19	416	42.5
	1993	2 936	599	0.3	2	4	5	10	466	78	21	395	43.5
	1994	2 932	573	0.3	1	3	4	8	451	79	17	338	39.6
	1995	2 915	548	0.4	4	5	6	11	438	80	20	328	40.5
	1996	2 968	556	0.4	3	6	7	12	423	76	21	326	44.1
	1997	2 894	507	0.5	4	7	8	13	397	78	21	288	43.0
	1998	2 954	482	0.5	5	8	8	13	376	78	21	250	43.7

Data source: QMS data.

Table 8
Total catch by sector

Fishery	Year	Total Catch (t)	No. reporting catch	HHI %	CR1 %	CR3 %	CR4 %	CR10 %	No. catch-ing 95%	% catch-ing 95%	% caught by top 5%	No. catching < 5t	Gini index %
Inshore	1987	33 972	1489	1.8	10	17	19	30	659	44	61	746	81.1
	1988	40 291	1390	1.6	8	15	18	32	571	41	65	649	82.9
	1989	44 927	1446	1.8	7	16	19	36	520	36	70	731	85.9
	1990	44 360	1350	1.6	7	16	19	33	502	37	69	684	85.0
	1991	43 414	1317	1.7	7	17	21	33	523	40	66	623	83.6
	1992	44 214	1254	2.1	8	21	25	36	478	38	69	602	85.1
	1993	50 529	1238	2.1	8	20	25	37	464	37	69	559	85.3
	1994	49 820	1260	2.1	8	20	25	36	450	36	70	598	86.0
	1995	58 352	1196	2.1	8	20	23	37	404	34	70	535	86.4
	1996	59 295	1132	2.0	7	19	23	36	390	34	70	486	86.2
	1997	58 407	1035	2.1	7	20	23	37	390	38	69	383	85.0
	1998	54 272	925	2.5	9	23	26	41	348	38	67	349	84.6
Mid-depth	1987	210 283	569	11.7	25	53	59	83	22	4	97	402	97.5
	1988	275 358	541	12.7	26	56	64	83	24	4	96	379	97.5
	1989	231 637	525	11.6	24	53	60	81	26	5	95	377	97.3
	1990	263 867	488	9.2	18	46	56	76	28	6	94	339	96.6
	1991	275 908	476	9.0	19	44	53	76	31	7	92	308	96.3
	1992	243 232	482	12.6	28	51	59	85	24	5	95	318	97.1
	1993	231 871	485	11.5	24	51	60	84	29	6	94	320	96.9
	1994	241 837	481	9.9	25	44	51	75	30	6	93	306	96.3
	1995	245 265	480	9.3	21	43	53	79	37	8	91	286	96.1
	1996	287 706	497	8.4	17	42	51	76	39	8	90	310	96.0
	1997	327 712	470	7.5	13	37	47	75	35	7	90	275	95.8
	1998	351 521	450	7.0	14	36	45	72	30	7	91	278	95.8
Deep-water	1987	65 725	76	9.1	19	42	51	80	29	38	51	14	80.2
	1988	65 565	51	12.2	21	54	64	86	20	39	54	8	78.7
	1989	69 946	52	11.4	22	50	58	83	20	38	50	5	77.4
	1990	66 496	49	10.7	20	48	57	83	21	43	38	7	75.5
	1991	56 389	60	13.9	25	61	65	80	26	43	61	8	78.8
	1992	56 799	48	17.0	28	65	72	90	15	31	54	8	81.9
	1993	54 919	42	19.7	32	73	77	90	14	33	57	5	81.6
	1994	52 109	61	16.2	24	68	72	88	17	28	68	9	84.3
	1995	40 752	57	14.8	24	64	69	88	18	32	64	10	82.7
	1996	43 718	62	15.0	26	62	72	90	14	23	62	23	85.9
	1997	40 369	50	17.3	27	69	74	92	13	26	52	14	84.4
	1998	37 425	50	15.1	26	63	69	90	13	26	47	16	83.0

Data source: QMS data.

The catch indices show a further dilution of concentration with respect to holdings and ownership, with the biggest holders taking the losses - that is, the largest proportion of uncaught quota was held by the big companies. However, the top 5% of those companies reporting catch still controlled 90% of the catch.

4.5.5 Deepwater species
4.5.5.1 Effect in the first year of the QMS

This group comprises orange roughy and the oreo species. These were part of the EA quota scheme and EAs were converted directly to PMITQ in 1986. TACCs were increased by some 29% overall in moving to ITQ, and again the government retained ownership of this share for leasing and sale. This share was similar in size to that of the next biggest owner, so the concentration ratio does not show any movement from PMITQ to 1987 (refer Table 6).

Table 9
Quota owned: all fin-fish sectors combined

Fishery	Year	Total quota owned (t)	No. holding quota	HHI %	CR1 %	CR3 %	CR4 %	CR10 %	No. owning 95%	% owning 95%	% owned by top 5%	No. owning < 5t	Gini index %
IS+MD	PMITQ	298 759	1756	7.0	17	40	46	66	434	25	85	412	92.1
+DW	1987	467 201	1357	20.7	42	63	68	83	62	5	95	457	97.2
	1988	472 495	1333	20.6	42	63	70	83	57	4	95	448	97.3
	1989	480 458	1353	19.9	41	61	70	83	65	5	95	452	97.2
	1990	469 696	1362	18.0	39	59	63	81	71	5	95	455	97.1
	1991	412 272	1332	14.4	34	53	59	76	82	6	94	430	96.7
	1992	421 726	1292	10.6	26	47	53	77	76	6	95	392	96.7
	1993	420 079	1230	10.4	26	46	51	75	69	6	95	380	96.7
	1994	420 023	1203	10.8	26	46	55	77	67	6	95	370	96.7
	1995	433 347	1151	9.5	18	46	56	79	67	6	94	349	96.7
	1996	446 246	1113	10.8	25	46	55	80	57	5	95	334	96.9
	1997	454 783	1054	10.9	25	46	55	82	50	5	95	301	97.0
	1998	450 631	992	11.0	25	46	56	83	48	5	95	276	97.0

Data source: QMS data.

Table 10
Quota held: all fin-fish sectors combined

Fishery	Year	Total quota held (t)	No. holding quota	HHI %	CR1 %	CR3 %	CR4 %	CR10 %	No. holding 95	% holding 95	% held by top 5%	No. holding < 5t	% Gini index
IS+MD	1987	473 657	1461	8.5	21	43	49	74	97	7	94	450	96.5
+DW	1988	476 350	1524	7.8	20	40	46	71	88	6	95	442	96.6
	1989	483 416	1552	8.3	21	42	49	69	92	6	94	437	96.7
	1990	477 937	1584	6.7	18	37	45	65	96	6	94	458	96.5
	1991	412 412	1648	4.7	12	28	35	60	137	8	92	407	95.7
	1992	421 731	1687	6.6	18	36	43	67	126	7	93	409	96.0
	1993	420 407	1576	5.7	14	34	41	65	120	8	93	385	95.9
	1994	420 032	1518	6.4	17	35	44	66	122	8	92	375	95.9
	1995	433 355	1425	10.3	23	51	57	71	108	8	93	375	96.2
	1996	446 247	1255	8.7	20	40	50	79	65	5	95	368	96.9
	1997	454 783	1216	8.6	21	42	51	74	78	6	94	333	96.5
	1998	450 631	1128	8.4	20	42	50	75	72	6	94	308	96.5

Data source: QMS data.

The total number of owners in the group was reduced by 19% from 54 to 44 by the end of the first year, with the quota being acquired by a few new large business entities. Most of this quota came from those fishers operating under the competitive TACC section of the fishery, who received 100 and 200t ITQ allocations and who sold out.

4.5.5.2 Trends under the QMS: 1987 to 1998

The same set of large companies controls the bulk of this sector as is the case in the mid-depth stocks. The nature of the fishing limits participation to large vessels and the total of around 40 owners has remained relatively stable over the period. However, ownership has become more concentrated among the top four as a result of the government relinquishing quota-ownership, with CR4 moving from 64-72% while CR10 has increased less than four percentage points. The HHI reflects the movement of the government share to the big three private owners in shifting from 12.5 to 16% over the period. Ninety-five percent of deepwater-quota is held in 13 accounts. Despite this high concentration, the deepwater-quota is more evenly spread among the owners than the other categories according to the GI, due to the limited scope for smaller players. The top 5% of owners (2) had 50% of the quota in 1998 with a GI ranging from 72-80% over the period (see also Figure 17).

Holdings show somewhat different patterns to the other groups relative to owned-quota. Numbers of participants are 25-30% higher, but the HHI barely moves, and the concentration ratios are all about the same as for ownership. This pattern is likely to be a result of the leasing-out of small amounts of quota to cover incidental catches in mid-depth fisheries, rather than any attempt by non-owners to target these species.

Both catches and TACCs have fallen by about 40% over the period. Thirteen accounts reported 95% of the catch in 1998, and the other indicators are almost identical to ownership.

4.5.6 Rock lobster

This fishery has a relatively large number of participants with small tonnages, as it is a high-value, small-boat, near-shore fishery. Rock lobster was introduced into the QMS in 1989-90, and TACCs had fallen 20% in total by 1998. PMITQ data for rock lobster was not investigated.

Total participants fell by one-third from 686 to 470 over the nine year period. Along with the entry of TOKM as the largest owner, this has contributed to a doubling of the HHI, but it is still comparatively low at 1.4%. The CR10 is only 26% in 1998, but this has also doubled since 1989-90 (see Figure 18). The proportion of all quota owned by the top 5% has increased from 23 to 37%, and the GI has ranged from 49% in 1990 to 58% in 1998, indicating a relatively low, but increasing, inequality in parcel sizes among quota-owners. The average holding for rock lobster is around six tonnes.

The number holding quota at the end of year is almost identical to ownership, but indices of holdings show large amounts of leasing by the big owners. This reflects both the fact that TOKM is the largest owner, and the nature of the fishery, as it would be a busy lobsterman that brought in 250t in a season. The holding concentration ratios are all about half the ownership values and have remained stable over the period. A full 78% of holders are included in the group with 95% of the quota at year's end.

4.6   Summary and discussion of quota concentration

This assessment has shown significant restructuring in the New Zealand fishing industry as a result of the implementation of the QMS. Dramatic changes were seen in the first year of the system, particularly in the inshore fisheries.   Some adjustment was due directly to the government-sponsored quota buy-back, to administrative cuts to quotas as part of the allocation process, and part of the remainder can be assumed to have occurred as a knock-on effect from that process. However, a large amount of quota-trading in the first year seems to have occurred through aggressive acquisition by larger trawling companies bolstering their portfolios in the face of the TACC cuts and the formation of new highly-capitalised commercial entities.

A total of 400 (23%) of 1750 quota-owners sold out in the first year of the system. All of these held PMITQ for inshore species and about half of them held mid-depth species. It is not possible to tell from the available data how many left completely through the buy-back, and how many through trading-down during the first year of the system.

The retention and subsequent disposal of large amounts of ITQ by the government, the conversion to a proportional quota system and the settlement of Maori claims complicate the interpretation of the analysis, but all seem to have been handled without great disruption to the system. Changes in indices through the transitions show consolidation of the dominance of the top 10 to 15 companies.

Since the initial adjustment over 1986-87, all the fin-fish sectors have gradually seen an increase in the degree of concentration of ownership. The total number of quota-owners in 1998 was 57% of the number who were issued PMITQ in 1986. The top-10 owners in the inshore have moved from owning 23% of PMITQ in 1986 to 38% of quota a year later, to 49% in 1998. In the other two sectors they own in excess of 90% of the quota. Taking into account the ownership relationships between companies, the dominant three interests in the New Zealand industry have controlled between 65 and 80% of all quota over the life of the system.

Consistently, end-of-year holdings and catch have been less concentrated than ownership, with the exception of deep-water species. The major holders in the inshore and mid-depth sectors are leasing out about 10% of their quota to others. Where catches fall short of TACCs in the inshore, the largest owners have generally leased out surplus quota, whereas in the mid-depth fisheries the major players bear the losses associated with catch shortfalls.

The lowest concentration of ownership found was in the rock lobster fishery where small parcels of quota are comparatively evenly distributed reflecting the practicalities of the fishery. A couple of large holders (including TOKM which holds quota in trust for Maori) distort the indices somewhat, with half the quota of the top-10 owners leased out of the group. When ownership of TOKM-held quota is distributed to tribal groups, indices in all of the sectors will fall.

The ownership patterns observed over the four sectors examined conform to the characteristics of each fishery. Many owners of mainly small parcels of rock lobster quota makes sense in a small-boat high value per unit catch fishery. Specialised central processing is not required and large rents are available for capture by the individual fisher. Although processors and export companies may wish to purchase rock lobster quota, fishers have good reason to hold on to this profitable asset. The diverse inshore fin-fishery exhibits moderately high, and gradually increasing, concentration following a burst of adjustment after several years of anticipation. The stock-specific restrictions on quota-concentration for these species provides some protection against severe market-power developing, but it is reasonable to postulate that a degree of monopoly-power may be exercised in some fisheries to control port-prices for fish and consequently the price of quota.

The high concentration of ownership in the mid-depth and deep-water sectors also reflects the nature of these predominantly trawl fisheries which require large capital investments in vessels and processing plant. Some cause for concern may arise with the mid-depth species, in that many small operators still hold quota in these fisheries but the vast majority of the quota is controlled by a few entities. Identification of the specific problems associated with market-power would require detailed research at a local level.

What is clear is that the predominant patterns of catch-ownership, particularly in the offshore sectors, were in place before the QMS came along. The biogeographical and technological issues in fishing these species, the degree of capitalisation required to harvest and process, and the nature of international fish-marketing are the key factors in the economic equation that dictates highly concentrated ownership of catching-rights. ITQs have provided increased security of fish supply that improves competitiveness in marketing. They also provide a means to optimise operations and maximise returns from New Zealand's fisheries resources, but they do not change the fundamental economics of the fishery.

For the inshore, the major adjustment in distribution of catch-rights in the first year of the QMS tends to support the decision to introduce a means such as ITQ through which rationalisation could take place. Prior management regimes had allowed the development of an inefficient configuration in the industry and of over-exploitation of vulnerable high-value species. The permit moratorium and several years of reform policy consultation contributed to a build-up of tension and expectation in the industry, that was released when quotas were implemented. The large cash-flows and expectation of further large-scale development in the offshore fisheries no doubt contributed to a spree of deal-making and quota buy-outs of inshore quota by the corporate sector. There may be ongoing cause for concern in the relations between large quota-owners and contracted catching-capacity as well as the exercise of undue influence over prices for fish and quota in the inshore fisheries, but its elucidation would require detailed study of local conditions.

Unfortunately New Zealand lacks the well-developed checks and balances in the area of anti-competitive commercial behaviour, and the vigilance, compared to, for example, the Australian Competition and Consumer Commission, or the anti-trust watchdog groups and concerned authorities in the United States. Further independent research in this area may be necessary to pinpoint any current abuses of market-power and potential hazards for the future due to high concentration in ownership of catching-rights.

5. CONCLUSION

The general discourse concerning individual transferable quota often includes assumptions about what might, or does, happen under quota-management in regard to ownership-concentration and fleet-change. A common understanding is that the key economic argument made in support of ITQ-type management is to reduce over-capitalisation in the harvesting sector through rationalisation of fleet-capacity. This in turn implies some concentration of quota-ownership.

Although the analysis in this paper has been developed at a broad level, and only from available agency data, it gives some indications as to the veracity of the above understanding in practice under New Zealand conditions. At best the argument is incomplete. The data indicate that the restructuring of ownership of catching-rights under quota is the predominant change, rather than being a consequence of fleet-rationalisation. This supports a view that significant gains in efficiency were available to the industry elsewhere than in the catching sector itself. Returns to scale in processing and export marketing, security of supply, and synergies between inshore and expanding offshore operations saw the emergence of new large companies and significant rationalisation among existing large and medium enterprises, but without significant impact on fleet capacity.

The patterns of vessel registrations support the view that boats tend to remain in use as long as possible. With no alternative application, vessels may need to decline in value to that equivalent to their salvage value before being withdrawn from fishing. The fleet appears to be adjusting and reducing through the attrition of old vessels rather than exit of working boats. This is seen in the reduced numbers of the smallest vessels in the fleet during the 1990s, but significant reduction in total capacity of the inshore fleet may not come until the largest cohort of 12 to 24m vessels reach the end of their economic life. In addition, the over-capitalisation argument may have been somewhat oversold in relation to the pre-QMS fleet. Modelling at the time suggested the fleet was about 20% larger than it needed to be (NAFMAC 1983; Sissenwine and Mace 1992). This is not a great deal of surplus capacity, and given the shifts in target species and increases in TACCs and catches for many stocks under the QMS, the inshore fleet may not require significant reduction. On the other hand, technological change has undoubtedly increased fishing-power per registered tonne (GRT) over the past two decades.

Data examined here suggest that the QMS successfully checked and contained expansion of the inshore fleet in areas that were over-capitalised, and provided the means to redirect effort and existing capacity away from over-fished and vulnerable stocks, toward those capable of higher production levels. Given the levels of capacity existing in the mid-1980s, the use of ITQs undoubtedly produced a more efficient economic outcome than could the alternatives of input regulation or competitive TAC-management, and took pressure off increases in the fishing of vulnerable stocks.

At the same time the QMS provided a significant boost to the rapidly-expanding export-oriented industry led by the major companies. Ownership of exclusive rights to harvest provides greatly enhanced security of supply - an essential component for success in international marketing of fish products. New Zealand is a relatively small player in the world market, but it has been remarkably successful in selling into distant and competitive markets such as Japan, the Republic of Korea, the United States, and Europe, prospering as a result. Being able to confidently predict forward-supply is an important component of this success. This factor more than fleet-rationalisation may help explain the patterns in ownership of fishing-rights and changes seen on transition to quota-management.

The factors discussed here make New Zealand something of a special case in fisheries management. The country is small enough to make a unified management system and administration of fisheries possible, so that all economically significant commercial fisheries are under quota, leaving nowhere for vessels to go to exit the system. A small domestic market (relative to production levels) has led to export-orientation driving both vertical integration and concentration resulting in firms large enough to succeed in world markets.

However, for other management jurisdictions implementing or considering quota-management, these types of factors are becoming increasingly significant. Many fisheries are managed under some form of limited-entry, reducing the potential for vessel displacement. The concentration of ownership in domestic marketing channels in supermarket chains makes security of supply issues more important and will have implications for transferable catching rights. Vessel lock-in, in that no other options exist for their use, may have implications for levels of discarding and high-grading in some fisheries, and local concentration could develop to the point that independent catchers are exploited by large quota-owners. It is all too easy to believe that under ITQs a generally prosperous industry indicates efficiency and management success. However, an argument can be made that part of the price of such prosperity is an obligation of vigilance to look more closely at the areas of potential market-failure such as the commercial relationships between large and small players and possible undue influence on prices.

Certainly, further study of New Zealand's implementation of ITQs is warranted, given the rather halting progress towards success in fisheries management globally. Closer examination of the issues raised in this paper on a fishery-by-fishery and method-by-method basis with input from the industry could yield further insights into both the detailed outcomes of the New Zealand system and the specific conditions for positive results with ITQs. Research would also prove helpful in the related areas of the fate of exiting boats and fishers, the interaction of property-rights and processes of stakeholder involvement in management, and the changing management cultures in regulation and within the industry under quota-management.

6.   ACKNOWLEDGEMENTS

The data upon which this paper is primarily based were supplied by the New Zealand Ministry of Fisheries from their Quota Management System (QMS) databases. The author gratefully acknowledges the assistance of Ministry staff in the extraction and provision of this data. Special thanks are owed to John Annala, Kim Duckworth and William Emerson.

7.   LITERATURE CITED

Annala, J.H. and K.J. Sullivan 1997. Report from the Fishery Assessment Plenary, May 1997: stock assessments and yield estimates. Ministry of Fisheries, Wellington.

Clark, I.N. and A.J. Duncan 1986. New Zealand's Fisheries Management Policies - Past, Present and Future: The Implementation of an ITQ -Based Management System. Fisheries Access Control Programs Worldwide: Proceedings of the Workshop on Management Options for the North Pacific Longline Fisheries, Orcas Island, Washington, April 21-25, 1986. Alaska Sea Grant College Program, University of Alaska.

Clark, I.N., P.J. Major and N. Mollett 1988. Development and Implementation of New Zealand's ITQ Management System. Marine Resource Economics **5**: 325-349.

Connor, R. 2001 Initial allocation of Individual Transferable Quota in New Zealand fisheries. 220-250. *In:* Shotton, R. (Ed.) Case studies on the allocation of transferable quota rights in fisheries. FAO Fish. Tech. Pap. No. 411. FAO, Rome. 373 pp.

FIB - Fishing Industry Board 1987. Economic Review of the New Zealand Fishing Industry 1986-87. New Zealand Fishing Industry Board, Wellington, New Zealand.

FIB - Fishing Industry Board 1989. G. Bevin, P. Maloney and P. Roberts, (Eds.) Economic Review of the New Zealand Fishing Industry 1987-88. New Zealand Fishing Industry Board, Wellington, New Zealand.

FIB - Fishing Industry Board 1990. G. Bevin, P. Maloney, P. Roberts and N. Redzwan, Eds. Economic Review of the New Zealand Fishing Industry 1988-89. New Zealand Fishing Industry Board, Wellington, New Zealand.

FIB - Fishing Industry Board 1994. The New Zealand Seafood Industry Economic Review 1993. New Zealand Fishing Industry Board, Wellington, New Zealand.

FIB 1996. The New Zealand Seafood Industry Economic Review 1994-1996. New Zealand Fishing Industry Board, Wellington, New Zealand.

Gauvin, J.R., J.M. Ward and E.E. Burgess 1994. Description and evaluation of the Wreckfish (*Polyprion americanus*) fishery under Individual Transferable Quotas. Marine Resource Economics **9**: 99-118.

Hogan, L., S. Thorpe and D. Timcke 1999. Tradable quotas in fisheries management: implications for Australia's south east fishery. Canberra, ABARE.

King, M.R. 1985. Fish and shellfish landings by domestic fishermen, 1974-82. Fisheries Research Division, Occasional Publication, Data Series No. 20. Ministry of Agriculture and Fisheries, Wellington.

NAFMAC 1983. Future policy for the inshore fishery - a discussion paper, August 1983. National Fisheries Management Advisory Committee, MAF, Wellington.

National Research Advisory Council 1980. Commercial Marine Fisheries Working Party Report to the Minister of Science and Technology, 31 October 1980, Wellington.

National Research Council 1999. Sharing the fish: toward a national policy on individual fishing quotas. National Academy Press, Washington D.C.

OECD - Organisation for Economic Co-operation and Development. 1997. Towards Sustainable Fisheries. OECD, Paris.

Riley, P. 1982. Economic Aspects of New Zealand's Policies on Limited Entry Fisheries. *In:* Sturgess, N.H. and T.F. Meany (Eds.), Policy and Practice in Fisheries Management, 365-383. Australian Government Printing Service, Canberra.

Scherer, F. M. 1973. Industrial Market Structure and Economic Performance. Chicago, Rand McNally and Co.

Sharp, B.M.H. 1997. From regulated access to transferable harvesting rights: policy insights from New Zealand. Marine Policy 21, 501-517.

Sissenwine, M.P. and P.M. Mace 1992. ITQs in New Zealand: the era of fixed quota in perpetuity. Fishery Bulletin, U.S. b (1): 147-160.

Wallace, C. 1997. New Zealand's Fisheries Quota Management System Assessed. Creating a green future: 1997 Conference of the Australia New Zealand Society for Ecological Economics, Melbourne, ANZSEE.

APPENDIX I
Species codes, names and groupings

QMS species-code	Common name	Scientific name	TACC 1998 (tonnes)
Inshore fin - fish species			
BCO	Blue cod	*Parapercis colias*	2 665
BNS	Blue nose	*Hyperoglyphe antarctica*	2 490
BYX	Alfonsino	*Beryx splendens; B. decadactylus*	2 727
ELE	Elephant fish	*Callorhinchus milii*	715
FLA	Flatfish (group of 8 species)	*Rhombosolea leporina; R.plebeia; R. etiaria; R. tapirina; Pelotretis flavilatus; Peltorhamphus novaezelandiae; Colistium guntheri; C. nudipinnis*	6 670
GMU	Grey mullet	*Mugil cephalus*	1 086
GUR	Red gurnard	*Chelidonichthys kumu*	5 143
HPB	Groper (2 species)	*Polyprion oxygeneios; P. americanus*	2 181
JDO	John dory	*Zeus faber*	1 107
MOK	Blue moki	*Latridopsis ciliaris*	604
RCO	Red cod	*Pseudophycis bachus*	16 066
SCH	School shark	*Galeorhinus galeus*	3 106
SNA	Snapper	*Pagrus auratus*	6 495
SPO	Rig	*Mustelus lenticulatus*	1 888
STA	Stargazer	*Kathetostoma giganteum*	4 972
TAR	Tarakihi	*Nemadactylus macropterus*	5 992
TRE	Trevally	*Pseudocaranx dentex*	3 932
Rock lobster species			
CRA	Rock lobster	*Jasus edwardsii*	2 927
Mid- depth fin - fish species			
BAR	Barracouta	*Thyrsites atun*	34 233
HAK	Hake	*Merluccius australis*	13 997
HOK	Hoki	*Macruronus novaezelandiae*	250 010
LIN	Ling	*Genypterus blacodes*	22 113
SKI	Gemfish	*Rexea solandri*	2 211
SWA	Silver warehou	*Seriolella punctata*	9 512
WAR	Blue warehou	*Seriolella brama*	4 512
Deep - water fin - fish species			
OEO	Oreos (group of 3 species)	*Allocyttus niger; Neocyttus rhomboidalis; Pseudocyttus maculatus*	25 654
ORH	Orange roughy	*Hoplostethus atlanticus*	21 330

Sources: QMS data; Annala & Sullivan 1997.

CHANGES IN FLEET CAPACITY FOLLOWING THE INTRODUCTION OF INDIVIDUAL VESSEL QUOTAS IN THE ALASKAN PACIFIC HALIBUT AND SABLEFISH FISHERY

M. Hartley and M. Fina
Northern Economics
880 'H' Street, Suite 210
Anchorage, AK 99501
<MarcusH@norecon.com>

1.   INTRODUCTION

In 1995, the federally-managed commercial long-line fisheries for Pacific halibut (*Hippoglossus stenolepis*) and sablefish (*Anoplopoma fimbria*) in the U.S. North Pacific moved from open-access management with limits on the Total Allowable Catch (TAC) to management systems with individual fishing quotas (IFQs). At that time, the halibut fisheries had the greatest number of participants of any fishery managed with individual quotas. This paper examines the halibut and sablefish fisheries and the consequences of the change in management. This is the second of two papers that examine the IFQ system in the U.S. North Pacific halibut and sablefish fisheries. The first paper discussed the circumstances leading up to the IFQ programme, the development of the IFQ programme, and the initial allocation of interests in the fisheries under the programme. This second paper provides a more quantitative examination of effects of the IFQ programme, including participation levels, fleet consolidation, and other changes that have resulted from the IFQ programmes.

The transition from open-access to IFQs was a long, arduous process marked by periods of progress, followed by periods of retreat, eventually leading to approval and implementation of an effective programme. To develop an understanding of the subtleties of the IFQ programme, it is necessary follow a path similar to the path of the IFQ policymakers - taking several steps forward, then stepping back for a fresh perspective. The complexities of the open-access and IFQ-management regimes and the often-conflicting goals and objectives of the involved policy-makers complicate the task of describing programme processes and results. To aid readers in keeping important contextual information in mind as new concepts or data are introduced, this document is repetitive at times. It is hoped that readers already familiar with Alaska's sablefish and halibut fisheries, or the IFQ programme and its implementation, will not be disturbed by this necessary repetition.

2.   BACKGROUND – THE PRE-IFQ FISHERY
2.1   Brief history of the halibut fishery

Alaska halibut and sablefish fisheries are regulated by similar IFQ programmes that were developed as by a single process. The fisheries differ both historically and in the manner of prosecution but in general have a high degree of overlap. This section describes the two fisheries before IFQ programme implementation, the development of the programme, and the initial allocation. The section begins with a brief historical description of the two fisheries and then describes management of the fisheries leading up to the IFQ programme, including some of the conditions that led to the IFQ programme.

**Pacific halibut *(Hippoglossus stenolepis)*
can grow 260cm in length and 350kg. Oldest reported age is 42 years.**
Photo: International Pacific Halibut Commision

The IFQ programmes regulate the halibut and sablefish fixed-gear fisheries. Both are primarily long-line fisheries, although some other methods are used[1]. Understanding the differences and similarities between the fisheries are critical to understanding the IFQ programmes.

The halibut fishery developed earlier than the sablefish fishery. Halibut have been harvested commercially since the late 1800s. The fishery occurs relatively close to shore, as halibut can be caught in waters as shallow as 90 feet. Prior to the IFQ programme, the fishery was managed with TAC limits and season limitations. A combination of factors led to the fishery to becoming a part-time, supplemental fishery, drawing fishers from other fisheries during slow seasons. Gear for targeting halibut is relatively inexpensive, allowing fishers to enter the fishery at little cost. Seasons were timed to limit the conflict with other fisheries, so fishers were able to participate without sacrificing time in other fisheries (IPHC 1987).

The increasing number of vessels participating in the halibut fishery resulted in shorter and shorter seasons, even in periods when the TAC of halibut remained steady or increased. Short seasons for halibut limited fishers' ability to target halibut full-time. By the time that the IFQ programme was instituted, the season had been reduced to 24-hour periods in many regulatory areas (Pautzke and Oliver 1997). The inability of halibut fishers to earn their total income from the halibut fishery is reflected in catch statistics from the fishery. Between 1984 and 1990, each year an average of 3275 vessel-owners participated in the halibut fishery. An average of 70% of all vessel-owners who made landings of halibut in these years also made landings of other species. During the same years, less than 25% of the total revenue earned by vessel-owners with halibut landings was attributed to halibut.

Traditional measures of fishing capacity can be inappropriate when applied to a supplemental fishery. Before IFQs, the halibut TAC would have been harvested in a matter of days. This circumstance should not be taken as an indication of over-capacity in the halibut fishery because most of the resources employed were also used in other fisheries when the vessels were not targeting halibut.

2.2 History of the sablefish fishery

The sablefish fishery is farther offshore than the halibut fishery and this species is caught in waters 1200 to 3000 feet deep. Before 1976 (when Exclusive Economic Zone - EEZ - waters were extended to the current 200-mile limit), the Alaska sablefish fishery was dominated by foreign fleets. After extension of EEZ waters, the fishery evolved into a domestic fishery. In 1987, foreign harvests ended by regulation. By that time, domestic harvests had grown to a level equal to historic highs of the foreign fleet. As with halibut, sablefish does not have a history as a full-time fleet supporting a tenured fleet. Unlike halibut fishers, a few fishers have made a living targeting sablefish. Most of these fishers operate vessels 60 feet or more in length, enabling them to fish in less protected areas, particularly the Aleutian Islands and the Bering Sea[2].

The activity of sablefish fishers in other fisheries demonstrates the fishers' reliance on other fisheries. From 1985 to 1990, at least 95% of vessels with commercial sablefish landings had landings of other species and on average, generated 65% of their income from other fisheries. While data indicate that the typical sablefish fisher relied more heavily on sablefish than the typical halibut fisher relied on halibut, sablefish fishers were clearly very active in other fisheries.

The composition of the sablefish fleet was also somewhat different from the halibut fleet, with sablefish vessels tending to be larger on average than halibut vessels. From 1985 to 1990, the number of active sablefish vessels increased from 371 to more than 800. Between 1985 and 1990 an average of more than 20% of the sablefish fleet consisted of catcher-vessels greater than 60 feet in length, compared to an average of 5% in the halibut fleet. In addition, an average of 2.5% of the sablefish fleet consisted of freezer-vessels. During the same years, catcher-vessels greater than 60 feet in length and freezer-vessels harvested a combined average of 53% of the total sablefish harvest, compared to 28.5% of halibut harvests.

2.3 Halibut and sablefish fisheries management

In 1976, the *Fishery Conservation and Management Act* (now known as the *Magnuson-Stevens Fishery Conservation Act*) established the current management-regime for fisheries in U.S. EEZ waters (waters between 3 and 200 miles of the coast). The regime called for creation of the North Pacific Fisheries Management Council

---

[1] In the halibut fishery other hook-and-line methods are used. In the sablefish fishery some pots are used.

[2] Vessels more than 60 feet in length are prohibited from fishing salmon, thus limiting alternative fisheries for large sablefish vessels.

(NPFMC)[3] which is tasked with making management recommendations to the U.S. Secretary of Commerce. Measures approved by the Secretary become binding regulations that are implemented by the National Marine Fisheries Service (NMFS)[4]. Management of sablefish and halibut fixed-gear fisheries in EEZ waters is under this regime[5]. While both fisheries are managed by NMFS and NPFMC, halibut TACs are determined by the International Pacific Halibut Commission (IPHC), a commission created by a treaty between the U.S. and Canada to coordinate regulation of the North American Pacific halibut fishery (Pautzke and Oliver 1997). Further information is given in Hartley and Fina (2001).

3.   THE IFQ PROGRAMME AND INITIAL ALLOCATION
3.1   Introduction

The IFQ system substantially changed rights in the fisheries. At commencement of the programme, NMFS issued quota shares (QS) to fishers for each regulatory area of each fishery. At the beginning of each season, the TAC for each regulatory area of each fishery is determined[6] . The holder of QS is entitled to a portion of annual TAC in the applicable regulatory area. This annual allotment is referred to as the fisher's IFQ and is the weight[7] of the fisher's permitted catch for the year. The amount of IFQ is equal to the area TAC multiplied by the fisher's QS, divided by the total QS pool in the area. IFQs may be fished at any time during the open-season. Under the IFQ programme, an extended season, which begins March 15 and ends November 15, was established. Unused IFQ amounts cannot be retained for use in a future year. On the other hand, recognizing that unintentional overages can occur, over-harvests of up to 10% of a fisher's IFQ are addressed by a reduction in the following year's IFQ without penalty. Fishers with 'overages' in excess of 10% are subject to enforcement sanctions, including confiscation (NMFS 1995).

Holders of QS may sell their rights[8]. Several restrictions on the sale of QS were adopted to avoid excessive consolidation and other changes in character of the fishing fleet. Although the system creates a property-right in the fishery, to avoid costly litigation in the event that the management programme is changed, the programme does not create a permanent interest in the fishery. QS remain valid indefinitely; however, if the programme is discontinued, the QS-holders will not be entitled to compensation (NMFS 1995).

Fundamental to an IFQ programme is the initial issuance of QS in the fishery. Those fishers issued shares receive a right to harvest a predetermined percentage of the TAC. This initial allocation influences both the distribution of wealth among fishers and the character of the fishery. The importance of the initial allocation was increased because of restrictions on transferability of QS. This section examines policy objectives of the initial allocation and the method by which the allocation was determined.

3.2   The initial allocation of quota shares

Several key policy objectives guided initial allocation of QS. Some of the major objectives were to:

i.   Preserve the character of the fleet
ii.   Limit and discourage corporate ownership of the fisheries
iii.   Reward active participants in the fisheries
iv.   Reward long-time participants over relative newcomers to the fisheries
v.   Reward those who invested in the fisheries over those who simply worked in the fisheries
vi.   Limit windfalls, regardless of federal policies precluding any charge for QS distributed in the initial allocation

---

[3] The NPFMC is composed of a panel of 11 voting members (6 from Alaska, 4 from Washington, and 1 from Oregon) and 4 non-voting members. Voting members represent the fishing industry, fish processors, and federal, state, and local agencies (Pautzke and Oliver 1997).

[4] NMFS is an agency of the National Oceanic and Atmospheric Administration (NOAA) under the U.S. Department of Commerce.

[5] Sablefish were under jurisdiction of the NPFMC from its onset. Halibut came under NPFMC jurisdiction only after the resolution of treaty issues in 1982. Halibut is managed jointly by NMFS and IPHC. In general, IPHC is charged with monitoring stocks and setting overall catch limits. Allocation of catch limits within the three separate jurisdictions (Alaska, British Columbia, and the Pacific U.S.) is handled by the federal agencies responsible for management. In Alaska, NMFS and NPFMC determine the allocations and were responsible for developing the IFQ program.

[6] The sablefish TAC is determined by NPFMC, subject to approval by the Secretary of Commerce. The halibut TAC is determined by the IPHC.

[7] Halibut traditionally have been landed in headed-and-gutted form. Halibut TACs and IFQs are set in terms of headed-and-gutted weight. Sablefish TACs and IFQs are set in terms of round weight (NMFS 1995).

[8] The right to sell IFQs (properly considered a lease of QS) was initially limited. Since 1997, sales of IFQs designated for use on catcher-vessels has been prohibited. The sale of IFQs designated for use on freezer vessels is permitted (NMFS 1995).

vii.   Discourage speculative entry into the fishery.

The initial allocation recommended by NPFMC embodied these objectives. The over-riding theme was to preserve the size and character of the fleet (NMFS 1994). By allocating QS to current participants, the initial allocation served this objective and also rewarded fishers who had been active in the fisheries. By 1990, it was apparent that, in the future, entry to the fishery would be limited in some manner. Consequently, NPFMC decided that activity after 1990 would not be used in determining QS, in order to prevent fishers from entering the fishery simply to obtain QS, in essence gaining a property-right in the fishery. Knowing that seasons were short and that different fishers entered and exited the fishery each year, NPFMC chose to allocate shares to all fishers who were active between 1988 and 1990, and to base shares on catches in additional previous years, 7 for halibut and 6 for sablefish (NMFS 1995; Pautzke and Oliver 1997)[9] .

At the time of programme development, the *Magnuson-Stevens Act* prohibited any charge on issuance of QS. The absence of a charge had the added effects of preserving fleet character and deterring opposition to the programme. Requiring payment for the initial allocation might have excluded some traditional participants from the fleet simply because of their inability to afford QS. Sale of the initial allocation might have allowed corporations, banks, or owners of large industrial vessels to purchase a large part of the initial allocation. Not charging for QS, however, created potential windfalls.

To limit the windfalls, NPFMC used a broad, inclusive policy that provided an initial allocation to many more fishers than had participated in any given year. The programme issued QS to all vessel-owners active in the fisheries from 1988 through 1990. Eligible fishers received QS equal to their harvests over an even longer period - for halibut, vessel-owners were allowed to submit the best records for 5 of the past 7 years back to 1984; for sablefish, vessel-owners used the best 5 of 6 years back to 1985.

An ancillary goal of the broad-based initial allocation was to allow fishers to determine the extent of their activity in the fishery. Fishers could choose to exit the fishery by selling QS, increase activity in the fishery by purchasing additional QS or IFQs, or simply fish their initial allocations. With a broad-based allocation of initial shares, market conditions would be more likely to determine activity in the fishery.

One limiting factor in the initial allocation was that only vessel-owners and fishers leasing vessels were allocated QS. Crews were not granted shares (NMFS 1994). The objective of this rule was to reward those parties who had taken the risk to enter the fishery. Boat- and gear-owners were presumed to have invested in the fisheries. Operators who could demonstrate a leasehold interest in vessels were also thought to have invested sufficiently in the fishery to entitle them to QS[10]. Omitting crew members from the initial allocation also made the process workable. Unlike data for vessels and vessel-owners, no official data were available to verify participation of crew members.

A few adjustments were made to QS allocations in determining the final issuance. In regulatory areas in the Aleutian Islands and Bering Sea, the allocations were reduced to develop a Community Development Quota (CDQ) programme. The CDQ programme was designed to assist area communities by allocating them portions of the TAC. Historically, the communities have reaped little reward from the fisheries because of a lack of economic capital. By allocating portions of the TAC to the communities, it was hoped that they would become active participants in the fisheries. Fishers who had been denied QS to accommodate the CDQ programme were granted a proportional amount of QS in regulatory areas that did not have CDQ programmes, and there were proportional decreases in the QS of all other fishers in those areas (NMFS 1994).

4.   RESTRICTIONS ON IFQ OWNERSHIP, USE AND TRANSFER
4.1   Ownership constraints

The development of the IFQ system and the initial allocation of rights in the fishery cannot ensure that programme goals are attained. Trading in the IFQ system allows changes in the interests in the fishery that also can affect the success of the programme. This section describes restrictions on ownership and trading under the programme intended to preclude excessive consolidation and preserve the owner-operator nature of the fishery. In general, these provisions were imposed because of concerns of policymakers that if the transfer and use of

---

[9] Part of the rationale for considering several years was to avoid excluding from the fisheries those fishers who typically participated but could not participate for a year or two due to illness or other uncontrollable circumstances (NMFS 1994; Pautzke and Oliver 1997).

[10] Records such as license purchase, tax records showing deductions of lease and crew payments, and other similar documents could be used to show that a vessel was being operated under a lease.

IFQ were unrestricted, the fisheries would soon be controlled by absentee owners and large corporations. Restrictions were enacted that limit:

i.   Who can own or purchase QS
ii.   The amount of QS persons can own
iii. The number of IFQs that can be used
iv.   The areas in which particular IFQs can be used
v.   The vessel-size (as measured by vessel-length) and vessel-type (catcher-vessels or freezer-vessels) on which particular IFQs can be used.

To fully understand the nuances of many of these restrictions it is important to review the difference between QS and IFQs. QS represent the right to harvest a fixed percentage of the TAC for every year into the foreseeable future, whereas IFQs represent the right to harvest a fixed number of pounds during a specified year.

QS are based on pounds of catch during the qualification period - for example, approximately 13 748 QS would have been issued by NMFS if a participant had landed 13 748lb during the qualification period[11]. The total number of QS issued was approximately equal to the total number of pounds that were landed during each qualifier's best years during the qualifying period (1984 through 1990 for halibut, and 1985 through 1990 for sablefish). Thus, the total number of QS far exceeds the total amount of pounds landed in any given year.

IFQs represent the pounds that can be landed in a specific year, with the total amount of IFQs equal to the TAC for the fishery for that year. The amount of IFQs that each individual receives is calculated based on that individual's holdings of QS as a proportion of the total number of QS outstanding (the QS pool).

Restrictions differed for QS and IFQs. For example, because QS represent a proportion of future TACs, they cannot be physically used or consumed. Therefore, there are no restrictions on the use of QS; restrictions on use are imposed only on IFQs. Similarly, since IFQs create no continuing right or interest but are used or consumed in a single year, the concept of leasing when applied to IFQs would have been nonsensical—once IFQs are used, there is nothing to return to the owner. Thus, leasing restrictions were imposed on QS and not on IFQs. The differences in the rights created by QS and IFQs[12] should be kept in mind when considering the different limitations applicable to each.

4.2   Eligibility to own quota-shares and caps on ownership
4.2.1 Requirements

The IFQ programme uses two basic eligibility requirements for ownership of QS and IFQs to achieve policy objectives. The first restriction is intended to ensure that all interests in the fishery are held by U.S. operators. The restriction limits ownership of QS and IFQ to U.S. citizens or corporations or partnerships formed in the U.S.

The second ownership restriction is a regulatory preference in favor of active fishers. This restriction is intended to serve that preference by defining who, in addition to initial recipients, can become owners of QS through transfers. Eligibility requirements for ownership are designed to assure that most QS are held by people directly active in the fishery. Consequently, corporations or partnerships that did not receive initial allocations can only purchase QS that are designated as freezer-vessel shares[13] many of which are owned and operated by corporations and partnerships. QS designated as catcher-vessel QS can be owned by a corporation or partnership only if the corporation or partnership was the original QS recipient. Eligibility to purchase catcher-vessel QS is also restricted to initial recipients of QS[14] and bona fide crew members. Bona fide crew members are defined by regulations as those individuals who have been directly active in the harvest of fish in a U.S. commercial fishery for at least 150 days (NMFS 1995).

4.2.2 Ownership caps

To prevent fleet consolidation, the system also established caps on the ownership of QS and on the harvest of IFQs. Ownership and harvest restrictions prohibit any single individual or single vessel from owning or harvesting more than a specified percent of the existing QS (the QS pool). In both the halibut and sablefish fisheries, ownership and harvesting caps apply to aggregations of existing regulatory areas.

---

[11] Some quota shares were allocated to CDQ groups without catch-history. These allocations were made by proportionally reducing initial allocations of quota shares based on qualifying catch.

[12] A "lease of QS" is equivalent to the "sale of IFQs"; the two terms are used inter-changeably in the industry.

[13] QS and resulting IFQs are designated for use on either catcher-vessels or freezer-vessels, based on the type of vessel used during the qualifying period.

[14] One exception is a regulatory area in Southeast Alaska, where only individuals (not corporations or partnerships) are permitted to purchase additional halibut QS.

For purposes of ownership and harvesting limits, the halibut fishery is divided into two areas. In the regulatory areas in the Aleutian Islands and Bering Sea, QS ownership and harvesting cannot exceed 1.5% of the total QS of those areas combined. In the Gulf of Alaska and Southeast Alaska regulatory areas, QS ownership cannot exceed 0.5% of outstanding QS. A special restriction caps ownership and harvesting at 1% of the QS in one regulatory area in Southeast Alaska (NMFS 1994). Southeast Alaska generally has more fishing communities and a greater proportion of halibut and sablefish fishers than other areas. This regulation was included at the request of NPFMC representatives from Southeast Alaska who believed that additional protection against consolidation was necessary in that area.

In the sablefish fishery, ownership and harvesting caps are set for two areas, Southeast Alaska and the entire fishery (including Southeast Alaska). The ownership and harvesting cap for these areas is 1% of the outstanding QS pool in the identified area (NMFS 2000b).

4.3   Restrictions imposed on the use of IFQs
4.3.1 Area-use restrictions

Perhaps the most obvious of the restrictions is one that prohibits IFQs resulting from QS issued for one regulatory area from being used in other regulatory areas. In the initial allocation of QS, all QS was allocated to the regulatory area from which the qualifying harvest was taken. Thus, QS generated from catch history in a Southeast Alaska regulatory area results in the issuance of IFQs that are legal for use only in that regulatory area. The same principle holds true for all other regulatory areas. This restriction is intended to maintain the distribution of effort across the range of the halibut and sablefish fisheries.

4.3.2 Vessel-classes

To help preserve the character of the fisheries, the programme created vessel-classes for each fishery. The programme restricts the use of IFQs to the vessel class of the underlying QS. For sablefish, three vessel-classes were created: freezer-vessels, catcher-vessels less than 60 feet in length, and catcher-vessels greater than 60 feet in length. Four vessel-classes were created for halibut: freezer-vessels, catcher-vessels less than 35 feet in length, catcher-vessels between 35 and 60 feet in length, and catcher-vessels greater than 60 feet in length. The additional classification for halibut catcher-vessels was created because of the large number of relatively small boats in that fleet[15]. The initial programme allowed IFQs to be used only on vessels of the same class as the vessel to which the IFQ was initially issued. This restriction is intended to maintain the distribution of the TAC among the vessel-classes (NMFS 1994). An amendment to the programme currently allows most catcher-vessel IFQs to be used by catcher-vessels of a smaller class. The amendment is thought to provide greater flexibility while still limiting any increase in the size of vessels participating in the fisheries (CFEC 1999a; CFEC 1999b).

4.3.3 Restrictions on hired skippers

The regulatory preference for owner-operated catcher-vessels extends to the regulation of who is permitted to fish IFQs. Only catcher-vessel QS-holders who received an initial allocation are permitted to hire skippers to fish their IFQs. All other catcher-vessel QS-holders are required to be onboard the vessel fishing their quota. An amendment to the programme currently requires a corporation or partnership to own at least a 20% interest in any vessel that fishes its QS (NMFS 1995)[16].

The owner-operator requirements do not apply to freezer-vessel QS, which can be fished by hired skippers (NMFS 1994). The exemption was included in the programme because these vessels are generally heavily capitalized and corporate owned, making any owner-operator provisions unreasonably burdensome[17].

---

[15] Since QS were initially allocated by owner, an owner's shares were allocated to the class of the vessel that the owner had used for fishing in the last year used to determine eligibility (1988 1989 1990, or 1991). If multiple vessels had been used in the last year, the catch allocation was apportioned among the vessels used in that year, in proportion to landings.

[16] Corporations and partnerships who used hired skippers before the amendment's adoption may continue to use that hired skipper, the 20% ownership requirement notwithstanding.

[17] Because of the unique character of Southeast Alaska, ownership and transfer restrictions are more restrictive in that area. A special provision prohibits individuals that received initial allocations from hiring captains to fish their QS (only corporations or partnerships are permitted to do so in Southeast Alaska). Individuals receiving an initial allocation (who do not sell their IFQs as otherwise permitted) are always required to be onboard when their IFQs are fished. Although QS-holders may transfer their shares to a solely held corporation, that transfer does not affect the obligation of the QS-holder to be on the vessel fishing the IFQs (NMFS 1995).

4.4   Transfer rights

The ability to transfer QS and the IFQs created by those QS is critical to the rights of QS-holders. Provided that ownership and other restrictions are satisfied, QS may be bought, sold, and transferred. Once QS are sold, the new holder receives the annual IFQs in the same quantity as the original owner.

In general, leasing of QS[18] is prohibited, although there are exceptions: freezer-vessel QS may be leased, subject to ownership cap restrictions. Also during the first 3 years of the programme, owners of catcher-vessel QS were permitted to lease up to 10% of their QS. The decision to permit QS leasing in the first few years of the programme was reached as a compromise to increase flexibility in the fishery (CFEC 1999a, b)[19]. The limitations on QS leasing are designed to reinforce the owner-operator preference in the eligibility requirements.

Restrictions on catcher-vessel QS-leases minimized the number of leases (or equivalently, IFQ sales) during the first three years of the programme. In the halibut fishery, less than 1% of catcher-vessel IFQs have been sold in any year. In the sablefish fishery, less than 2% of catcher-vessel IFQs have been sold in any year. Among freezer-vessels, substantial leasing has occurred in both fisheries. The quantity of freezer-vessel IFQs sold differs from year to year and across the various regulatory areas in both fisheries. In some areas as much as 90% of the freezer-vessel IFQs have been sold in a given year. In both fisheries, the sale of up to 20% of freezer-vessel IFQs, in a few regulatory areas each year has been common (CFEC 1999a; CFEC 1999b).

For any transfers, both the seller and the purchaser must file applications with NMFS. Purchasers of catcher-vessel QS must also be registered as bona fide crew members or be initial recipients. Applications must identify the parties and the type and size of transaction (the amount of QS or IFQs being transferred). Each transaction is recorded by NMFS (NMFS 1995).

4.5   The block programme

An amendment to the original programme (adopted prior to the programme's implementation) placed additional limits on the ability to transfer QS. The amendment, known as the "block programme," restricts the consolidation of small shares by single owners so as to assure that small holdings remain available for part-time fishers (CFEC 1999a; CFEC 1999b). Under the amendment, any initial allocation of QS that would have entitled the owner to less than 20 000lb of harvests under the 1994 TAC is considered a "block." Any block must be transferred as a whole and cannot be consolidated with any other block for purpose of transfers[20]. The amendment prohibits the ownership of more than two blocks, or more than one block and any amount of unblocked QS in any regulatory area (NMFS 1995). The amendment does not directly affect QS that would have entitled the owner to more than 20 000lb of harvest (based on the 1994 TAC). These holdings of shares can be divided and combined, subject only to the other restrictions on ownership and transfer (including the percentage ownership caps) (NMFS 1995).

The prevalence of blocked QS is indicative of the number of vessels that have historically fished in the fisheries. In halibut regulatory areas, between 35 and 70% of all QS are blocked. In the sablefish regulatory areas, between 7 and 20% of all QS are blocked[21]. The amendment is likely to assure that the IFQ programme does not substantially reduce the number of entities with an interest in the halibut and sablefish fisheries (CFEC 1999a; CFEC 1999b).

5.   THE FLEETS - BEFORE AND AFTER PROGRAMME IMPLEMENTATION
5.1   Fleet situation

It is clear that overall capacity in the fleet is still high if judged from the narrow perspective of how much fish the participating vessels could harvest if they were not constrained by their IFQs and by the TAC. There are several reasons why capacity in the sablefish and halibut fisheries has not declined as much as theoretically possible. The primary reason is that the majority of participants has viewed the sablefish and halibut fisheries as

---

[18] Leasing QS, by definition, is equivalent to the sale of unused IFQs since IFQs are good only for the year in which they are issued, and once used, they may not be reused.

[19] "Blocks" of QS must be sold or leased only in their entirety. This restriction, together with the 10% limitation on IFQ sales, precludes almost all leases of block QS.

[20] An exception, known as the "sweep-up" provision, allows consolidation of small blocks into a single block. Sweep-ups (or aggregations of small blocks) are only allowed if the resulting block creates IFQ rights of 3000lb or less of halibut, or 5000lb or less of sablefish.

[21] These figures do not include one halibut regulatory area and one sablefish regulatory area that have a disproportionate number of shares in CDQs. These areas have even higher percentages of blocked QS because the CDQ shares are considered blocked QS.

a means of supplementing income from other major fisheries such as the salmon, crab, and groundfish fisheries, for which many of the vessels were built. These other fisheries are also seasonal, and participants are able to fish sablefish and halibut without disrupting their participation in the other fisheries. For most participants, the sablefish and halibut fisheries are two of several seasonal fisheries in which the vessels participate. A second, very important reason that large-scale fleet reductions have not occurred is that NPFMC, which designed the systems, wished to maintain the existing nature of the fisheries, and to that end, created restrictions that prevent excessive consolidation.

5.2   The pre-IFQ fleets

Before IFQ programme implementation, both the halibut and sablefish fisheries were intensively fished, with large numbers of vessels and very short seasons. The participation levels in the pre-IFQ fisheries were greatly influenced by the ease of entry into the fisheries. Both fisheries are prosecuted with sunken long-line gear that is anchored to the bottom and marked with buoys[22]. Long-line gear is relatively inexpensive and does not require significant investments in specialized equipment or vessels - almost any type of vessel can be outfitted to fish long line gear, including small skiffs and recreational vessels. In addition to the low equipment cost, there are few opportunity costs of participation. Prior to the implementation of the IFQ programme, fishery managers set the opening dates of the seasons to avoid conflict with seasons for other species. Thus, a vessel could participate in the sablefish and halibut fisheries and still participate in the salmon, crab, groundfish, or recreational fisheries for which the vessel likely had been designed. Entry into the fisheries was also significantly influenced by the prospect that fishery managers would soon limit entry with either licences or a quota-based system, and it was believed that demonstrated participation would ensure future participation.

5.3   The halibut fishery prior to the IFQ programme

Participation in the halibut fishery before IFQ implementation is documented in the four tables below. In general, a very large number of relatively small vessels participated in the halibut fishery between 1984 and 1990. Participation was a source of supplemental income for many vessel-owners.

Table 1 shows the number and percentage of vessels that fished halibut participated in other fisheries between 1984 and 1990, the years leading up to the establishment of the IFQ programme[23]. On average, there were more than 3500 vessels participating in the halibut fisheries each year. Of these, an average of more than 2500, or 70%, also participated in other fisheries. Table 3 shows the percentage of total ex-vessel revenue that each participating vessel earned from the halibut fishery and from the other fisheries in which they participated from 1984 through 1990. During these years, the average halibut vessel earned less than 25% from the halibut fishery. Considered together, Tables 1 and 2 demonstrate that the majority of fishers active in halibut fishery

Table 1

Halibut vessels that also had landings in other Alaska fisheries 1984-1990

Indicator	1984	1985	1986	1987	1988	1989	1990
No. of owners active in the halibut fishery	3 472	2 744	3 247	3 777	3 925	3 642	4 206
No. of owners also active in other fisheries	2 036	1 885	2 338	2 912	2 893	2 346	3 178
Percentage active in other fisheries	58.6	68.7	72.0	77.1	73.7	64.4	75.6

Source: NPFMC 1992a.

Table 2

Ex-vessel value of Alaska fisheries for halibut vessels 1984-1990

Fishery	Percent of total ex-vessel value						
	1984	1985	1986	1987	1988	1989	1990
Halibut	18.8	25.6	32.1	23.4	17.7	28.6	23.8
Other fisheries	81.2	74.4	67.9	76.6	82.3	71.4	76.2

Source: NPFMC 1992a.

---

[22] Other hook-and-line gear such as jigs and trolls are legal gears for both species in all areas, but neither are regularly used. Pot gear, which is legal within the IFQ program for sablefish in the Bering Sea and Aleutian Islands, is used only by a few vessels.

[23] These data are for the years considered in determining the initial allocation of QS. The data reflect activity in the fishery prior to the announcement of the change to IFQ management.

were active in the fishery part-time. Thus, while there were clearly more vessels than necessary to catch the halibut available to the fishery, most of the participants were using those vessels in other fisheries in addition to the halibut fishery.

Table 3 shows the number of unique vessel-owners associated with vessel length-classes for the years 1984 to 1990 - also shown are the percentages in each class by year. Table 4 is similar to Table 3, but shows total halibut catch by class rather than the number of owners by class. The tables show that, while more than 50% of participants owned vessels less than 35 feet in length, these vessels collectively harvested less than 15% of the total catch. Less than 6% of the participants used vessels greater than 60 feet in length, but those participants averaged more than 25% of the total halibut harvest.

Table 3
Halibut vessel-owners by vessel length-class 1984 to 1990

Vessel length-class	No. of unique owners						
	1984	1985	1986	1987	1988	1989	1990
≤35 feet	1 951	1 440	1 658	1 924	2 069	1 791	2 018
36-60 feet	988	926	1 199	1 412	1 449	1 429	1 713
> 60 feet	109	112	166	192	178	166	214
Unknown length	64	33	11	11	10	7	4
Freezer-vessels	-	-	-	3	6	5	8
All[a]	**3 077**	**2 479**	**3 001**	**3 489**	**3 649**	**3 346**	**3 883**
	Percent of all owners						
≤35 feet	63.4	58.1	55.2	55.1	56.7	53.5	52.0
36-60 feet	32.1	37.4	40.0	40.5	39.7	42.7	44.1
> 60 feet	3.5	4.5	5.5	5.5	4.9	5.0	5.5
Unknown length	2.1	1.3	0.4	0.3	0.3	0.2	0.1
Freezer-vessels	0.0	0.0	0.0	0.1	0.2	0.1	0.2
All	**100.0**	**100.0**	**100.0**	**100.0**	**100.0**	**100.0**	**100.0**

Source: NPFMC 1992a.
[a] Because some owners have vessels in more than one class, there are actually fewer owners than the sum of owners in each vessel length-class.

Table 4
Halibut catch by vessel length-class 1984 to 1990

Vessel length-class	Volume (000's lb)						
	1984	1985	1986	1987	1988	1989	1990
≤35 feet	4 686	4 998	6 637	8 383	10 066	7 569	6 993
36-60 feet	18 892	25 026	31 884	32 633	34 571	32 923	33 434
> 60 feet	11 319	14 978	18 945	15 038	16 078	14 902	12 007
Unknown length	142	179	318	217	64	70	29
Freezer-vessels	0	0	0	[a]	232	552	483
All	**35 040**	**45 181**	**57 784**	**56 271**	**61 011**	**56 017**	**52 946**
	Percent of total volume						
≤35 feet	13.4	11.1	11.5	14.9	16.5	13.5	13.2
36-60 feet	53.9	55.4	55.2	58.0	56.7	58.8	63.1
> 60 feet	32.3	33.2	32.8	26.7	26.4	26.6	22.7
Unknown length	0.4	0.4	0.6	0.4	0.1	0.1	0.1
Freezer-vessels	0.0	0.0	0.0	[a]	0.4	1.0	0.9
All	**100.0**	**100.0**	**100.0**	**100.0**	**100.0**	**100.0**	**100.0**

Source: NPFMC 1992a.
[a] Due to confidentiality restrictions, this information is included in the >60-foot category.

5.4 The sablefish fishery prior to the IFQ programme

Before the implementation of the IFQ system, the number of participants in the sablefish fishery was only about 20% of the number of participants in the halibut fishery. Table 5 shows the number of vessels participating in the sablefish fishery from 1985 to 1990 - an average of 720 vessels each year. As in the halibut fishery, sablefish participants in general, relied more on income from other fisheries, than on income from sablefish. In the years from 1985 to 1990, more than 95% of the vessels participating in the sablefish fishery had landing of other species. Table 6 shows the relative reliance on income from sablefish compared to income from other species - on average, 65% of the total ex-vessel revenue generated by sablefish fishers came from landings of other species.

Table 5
Sablefish vessels that also had landings in other Alaska fisheries 1985 to1990.

	1985	1986	1987	1988	1989	1990
No. of vessels active in the sablefish fishery	371	606	868	888	768	822
No. of vessels also active in other fisheries	355	576	839	863	746	811
Percentage active in other fisheries	95.7	95.0	96.7	97.2	97.1	98.7

Table 6
Ex-vessel value of Alaska fisheries for sablefish vessels 1985 to 1990

Fishery	Percent of total ex-vessel value					
	1985	1986	1987	1988	1989	1990
Sablefish	37.4	30.0	31.9	41.0	39.0	29.8
Other fisheries	62.6	70.0	68.1	59.0	61.0	70.2

Source: NPFMC 1992a.

The makeup of the sablefish fleet, however, differs from makeup the halibut fleet. Compared to halibut, sablefish are typically found in deeper water farther from shore. Although the two fleets overlap, sablefish fishers overall tend to use larger vessels than halibut fishers do. Table 7 shows the number of vessel-owners who participated in the sablefish fishery in the years 1985 to 1990, classified by the length of their vessels. In these years, between 17 and 30% of the vessels in the fishery were more than 60 feet in length. In addition, freezer-vessels made up between 2 and 4% of the sablefish fleet. The sablefish fleet grew notably in these five years. The number of catcher-vessels less than 60 feet in length tripled from less than 200 to more than 500. During the same period, fewer than 50 catcher-vessels greater than 60 feet entered the fishery, as that portion of the fleet did not even double in size. The number of freezer catchers more than tripled, growing from 6 to 21. Sablefish catch, shown in Table 8, reveals that larger vessels are more active in the sablefish fishery than in the halibut fishery. In only one year between 1985 and 1990 did vessels less than 60 feet in length account for more than 50% of harvests in the fishery, a sharp contrast to the more than 95% of ex-vessel value of the vessels in this class in the halibut fishery.

Table 7
Sablefish vessel-owners by vessel length-class 1985 to 1990

Vessel length-class	No. of unique owners					
	1985	1986	1987	1988	1989	1990
≤ 60 feet	165	339	523	568	502	546
> 60 feet	73	109	136	120	112	116
Unknown length	-	4	10	4	7	9
Freezer-vessels	6	11	14	20	24	21
All vessel-owners [a]	**244**	**460**	**679**	**706**	**642**	**684**
	Percent of all owners					
≤ 60 feet	67.6	73.7	77.0	80.5	78.2	79.8
> 60 feet	29.9	23.7	20.0	17.0	17.4	17.0
Unknown length	0.0	0.9	1.5	0.6	1.1	1.3
Freezer-vessels	2.5	2.4	2.1	2.8	3.7	3.1
All	**100.0**	**100.0**	**100.0**	**100.0**	**100.0**	**100.0**

Source: NPFMC 1992a.
[a] Because some owners have vessels in more than one class, there are actually fewer owners than the sum of owners in each vessel length-class.

5.5   The IFQ fleet

Transferable-rights-based systems in fisheries promote consolidation of capital and resources invested in the fisheries as more efficient operations purchase the rights of less efficient operations. In economic theory, the less efficient operations, having sold their rights to participate, will retire from the fishery and devote their capital and resources to other industries in which returns are greater. Thus, one measure of the success of the IFQ programme is the level of consolidation that has occurred and the removal of excess vessels from the fishery.

Economic theory is able to "assume away" questions such as the initial allocation of rights, the malleability of capital and other resources, and the sometimes conflicting, non-economic goals and objectives of policy-makers. If these other issues are considered, then it becomes clear that the success of the IFQ programmes should be judged by additional measures, such as participant satisfaction.

Table 8
Sablefish catch by vessel length-class 1985 to 1990

Vessel length-class	Volume (000's lb)					
	1985	1986	1987	1988	1989	1990
≤ 60 feet	9 360	18 304	28 662	30 504	28 715	33 321
> 60 feet	11 520	16 969	23 220	23 246	21 503	14 492
Unknown length	0	87	172	94	91	114
Freezer-vessels	7611	7172	7 642	9 774	9 929	8 230
All vessel-owners	**28 491**	**42 532**	**59 695**	**63 618**	**60 239**	**56 157**
	Percent of total volume					
< 60 feet	32.9	43.0	48.0	47.9	47.7	59.3
> 60 feet	40.4	39.9	38.9	36.5	35.7	25.8
Unknown length	0.0	0.2	0.3	0.1	0.2	0.2
Freezer-vessels	26.7	16.9	12.8	15.4	16.5	14.7
All	**100.0**	**100.0**	**100.0**	**100.0**	**100.0**	**100.0**

Source: NPFMC 1992a.

If the IFQ programmes for the sablefish and halibut fisheries are examined from the perspective of consolidation of capital (measured by the number of owners and vessels participating in the fisheries), it might be concluded that the right-based systems did not result in sufficient consolidation to justify the implementation costs and social disruption of the programme. The initial allocation of QS in the halibut fishery issued rights to 4816 vessel-owners - 25% more than participated in 1990. In the sablefish fishery, the number of initial recipients of QS exceeded the number of participants in the 1990 fishery by more than 50%. By 1999, the number of halibut QS owners had fallen to 3649 - 6% less than the number of 1990 participants. In the sablefish fishery, the number QS owners declined to 897 by 1999 - still 30% higher than the number of participants in 1990.

While these figures demonstrate ownership interests in the fishery, they are not a proxy for the determining the number of vessels active in the IFQ fisheries. In 1999, 1613 vessels had landings of halibut compared to 4206 in 1990. In the sablefish fishery, 433 vessels had landings in 1999 compared to 822 vessels in 1990. Clearly, consolidation has occurred. The level of consolidation, however, is far short of the potential consolidation that would occur if vessels operated full-time in the fishery.

One method of approximating potential consolidation is to estimate the minimum number of vessels needed to make the number of fishing trips made currently by the existing fleet. For example, in 1999, the 1613 active halibut vessels made a total 7921 trips during the 245-day season - an average of slightly less than 5 trips per vessel. It is not unreasonable to assume that a vessel could make 25 trips during the year if it were operating full-time, and completing one trip every ten days - or 5 times as many trips as the average vessel currently makes. Similarly for the IFQ sablefish fleet, in 1999, 433 vessels made 1994 trips - again slightly less than 5 trips per vessel on average. From this perspective, it can be argued that the IFQ fleets are up to 5 times larger than necessary and that the fisheries are still severely over-capitalized.

In summary, the IFQ system initially resulted in an increase in the number of participants. This increase was followed by a period of significant consolidation. Yet, even with the post-implementation consolidations, it can be said that the IFQ fleets are still severely over-capitalized. Despite these apparent contradictions, the IFQ system has generated significant benefits to participants and is widely regarded as a success.

5.6 Participation patterns under the sablefish and halibut IFQ programmes

This section provides a detailed examination of participation patterns under IFQs for sablefish and halibut during the first five years of the programme's existence. The section presents and discusses a set of nine tables that show the initial allocation of QS, the consolidation of QS, the number of participating vessels, monthly harvest rates, and the number of trips in both fisheries.

Table 9

Initial recipients of halibut quota shares by vessel class

Vessel length-class	Number of recipients	Percent of total recipients	Percent of total halibut QS in class
≤ 35 ft	2 242	46	8
36-60ft	2 285	47	53
> 60ft	320	7	37
Freezer-vessels	42	2	3
Total	**4 816** [a]	**100**	**100**

Source: NMFS 2001.
[a] Some persons have received QS in multiple vessel classes. The numbers shown here are unique owners.

Table 10

Initial recipients of sablefish quota share by vessel class

Vessel length-class	Number of recipients	Percent of total recipients	Percent of total sablefish QS in class
≤ 60ft	805	76	37
> 60ft	204	19	42
Freezer-vessels	52	5	21
Total	**1052** [a]	100	100

Source: NMFS 2001.
[a] Some persons have received QS in multiple vessel classes. The numbers shown here are unique owners.

The tables clearly demonstrate differences in allocations received by participants with small vessels or with large vessels, as well as differences in participation levels in the two fisheries. In the halibut fishery 46% of all initial recipients owned vessels less than 35 feet. These small-vessel owners received only 8% of the total QS issued. Owners of halibut vessels between 35 and 60 feet were 47% of the initial allocation recipients and received 52% of the QS. Summing shows that 93% of the halibut QS-holders owned vessels less than 60 feet in length. In the sablefish fishery, 76% of the initial recipients of QS-owned vessels less than 60 feet and were allocated 37% of the QS available. These tables may give the appearance that the initial allocation was less than fair for small-vessel owners, but the allocation was based on participation and longevity in the fishery, and many small-vessel owners had very limited records of participation. It is also important to reiterate that the percentage of QS allocated to each vessel-class is essentially fixed at the initial allocation levels (shown in the tables) because of restrictions on transfers between vessel classes built into the programme[24]. Thus, regardless of the consolidation of the fleet through QS transfers, no less than 8% of the halibut IFQs will be issued to owners of vessels less than 35 feet in length.

The market for QS and IFQs was quickly established after the programme was implemented. Many fishers have found through friends and personal connections, parties with whom to transact deals. Still, more than 50% of the transactions have been between parties with no business or personal relationship to one another. To facilitate the trade of QS and IFQs, NMFS maintains a website with a full listing of all QS-holders and persons determined to be eligible to purchase QS. While these listings have aided in market formation, transactions have been facilitated by a number of methods. Several well-established brokers participate in the market, facilitating almost 50% of all transactions. Also, trade journals frequently have advertisements for available shares (CEFC 1999a; CFEC 1999b).

Table 11 and Table 12 show the number of approved transactions of QS and IFQs in the halibut and sablefish fisheries, respectively. In the halibut fishery, approximately 4800 entities received an initial allocation. The 1279 QS transactions in the first year were more than 25% of the number of initial shareholders. Similarly, the sablefish fishery issued initial allocations to approximately 1000 entities. The more than 400 transfers in the first year represent more than 40% of that number. The high number of transactions suggests that even from the onset the market as been efficient with low transaction costs.

---

[24] The regulations permit IFQs designated for use on large vessels to be used on smaller vessels, but IFQs designated for use on small vessels may not be used on larger vessels.

Table 11

Number of approved halibut-quota shares and IFQ transfers 1995 to 1999

Transfer type	Number of transfers				
	1995	1996	1997	1998	1999
Regular QS/IFQs	1 217	1 397	1 004	539	611
IFQs Only ("lease") [a]	31	61	53	43	39
Sweep-up of small blocks [b]	31	63	441	148	150
Total	**1 279**	**1 521**	**1 498**	**730**	**800**

Source: NMFS 2000.

[a] IFQ transfers were permitted on catcher-vessels only in 1995 1996, and 1997.

[b] Small blocks are blocks of fewer than 3000 shares that can be "swept up" into a single block.

Table 12

Number of approved sablefish-quota shares and IFQ transfers 1995 to 1999

Transfer type	Number of transfers				
	1995	1996	1997	1998	1999
Regular QS/IFQ	352	351	388	185	237
IFQ Only ("Lease") [a]	76	51	51	57	53
Sweep-up of small blocks [b]	15	20	82	33	22
Total	**443**	**422**	**521**	**275**	**312**

Source: NMFS 2000.

[a] IFQ transfers were permitted on catcher-vessels only in 1995 1996, and 1997.

[b] Small blocks are blocks of fewer than 5000 shares that can be "swept up" into a single block.

Changes in the numbers of QS-holders in the halibut and sablefish IFQ fisheries since the IFQ programme was implemented are shown in Table 13 and Table 14. The number of halibut-QS-holders declined by 25% in the first 5 years. The number of sablefish-QS-holders declined slightly more than 10%. Consolidation of the both fisheries is also shown by changes in the fleet sizes. Table 16 and Table 17 show vessel participation in the halibut and sablefish fisheries before and after the IFQ programmes were implemented. Without question, the IFQ programmes reduced the number of vessels participating in each fishery. In the first year of the IFQ programme, the number of vessels participating in the halibut fishery dropped by more than one-third. By the fifth season, the number of vessels had dropped to less than one-half of the pre-IFQ vessel participation. In the sablefish fishery, the number of vessels dropped by more than 50% in the first year of the programme. Reduction in the sablefish fleet has occurred slowly since then, with fleet size remaining more than one-third of the pre-IFQ fleet size in the fifth season. This decline suggests that the concentration in the fleet has increased efficiency in the fishery, reducing over-capitalization from the pre-IFQ fishery. The decline in number of vessels is particularly notable, given the broadly inclusive method of the initial allocation. Despite the issuance of QS to all fishers who were active in the fisheries at any time during several years, the number of active vessels has declined every year since programme implementation (including the programme's first year).

The number of QS-holders is far greater than the number of vessels. This difference can be attributed to two factors. The first is that some holders of small numbers of QS do not harvest their IFQs at all. A survey of QS-holders in the first year of the sablefish IFQ programme found that two-thirds of QS-holders with less than 1000lb of IFQs harvested less than 10% of their IFQ allotment (Knapp and Hull 1996a). In 1998, more than 20% of halibut-QS-holders who received an initial allocation had not transferred their QS and did not fish them. In the same year, more than 15% of sablefish-QS-holders that received an initial allocation had not transferred their QS and did not fish them. These owners can be assumed to have relatively small QS holdings, since more than 90% of the TAC in the sablefish fishery and more than 85% of the TAC in the halibut fishery was harvested in 1998[25]. The prevalence of holders of small QS in this group is obvious and predictable, as those with small

---

[25] Many of the small QS-holdings for areas of the Gulf of Alaska (GOA) are the result of the compensation that was provided to vessel-owners who qualified for QS in areas of the Bering Sea and Aleutian Islands where the CDQ program was implemented. The CDQ program reduced the TACs of sablefish and halibut available to QS-owners. NPFMC compensated initial recipients of QS in the Bering Sea and Aleutian Islands with small amounts of QS in each area of the GOA, and they could either sell or use these QS. The compensation program had the effect of spreading at least some of the burden of the CDQ program to all initial QS-recipients. However, the small allocations of QS received in compensation were often too small to be fished efficiently and many compensation shares are believed to go unharvested.

holdings are likely to be unable to cover the cost of a trip to fish their quota and are unlikely to see a significant benefit to transferring their shares after covering the costs of the transfer.

Another reason for the difference in the number of QS-holders and the number of vessels in the fishery is that not all QS-holders use their own vessels for fishing their QS. Some join with other QS-holders and fish their quotas on a single vessel. This practice allows QS-holders to reduce the overhead costs of vessel maintenance and reduces crew costs by consolidating multiple QS-holders on a single vessel. The tendency of QS-holders to share a trip also reduces overcapitalization in the fleet. This result is not necessarily inconsistent with the initial allocation and the policy objectives. A goal of the programme was to let fishers determine their own level of activity in the fishery and to reward only those who have invested in the fishery. Those purchasing QS and those willing to work with other QS-holders have shown a willingness to continue in the fishery.

Although there are no official counts of crew members who participate in sablefish and halibut trips, the willingness of QS-holders to team with others is thought to reduce the number of active crew members in these fisheries. With IFQs and the resulting change from a "race for fish," the number of crew members on a typical trip is believed to have fallen from a range of 3 to 6 to a range of 2 to 4 (including the skipper)[26]. One survey of sablefish fishers in the year following programme implementation suggested that a slight decline in the number of crew occurred under the programme (Knapp and Hull 1996a).

While several of those fishers with the smallest QS holdings do not participate, many fishers remain active in the fishery and have significant holdings. In 1999, 240 sablefish fishers (more than 25% of the total QS-holders in the sablefish fishery) held QS, entitling them to more than 25 000lb of catch. More than one-seventh of the QS-holders in the halibut fishery held QS, entitling them to more than 25 000lb of catch. In the same year, more than half of the QS-holders in the halibut fishery were entitled to at least 3000lb of catch, and more than half of the QS-holders in the sablefish fishery were entitled to more than 5000lb.

Particularly in the sablefish fishery, fishers have increased their interest in the fishery by purchasing QS in several different regulatory areas. Doing so enables them to make more of their livelihood from the fishery and further increases fleet consolidation. This tendency is demonstrated by the amount by which the sum of the QS owners in the various regions exceeds the number of unique holders of QS (see Table 13 and Table 14). In the sablefish fishery, the sum of the number of QS-holders in the various regulatory areas is slightly less than double the number of unique QS-holders, suggesting that a large number of fishers holds QS for more than one regulatory area. In the halibut fishery, the number of unique QS-holders is approximately one-third larger than the sum of the number of QS-holders in the various regulatory areas. By owning shares in multiple regulatory areas, fishers have further reduced overcapitalization in the fisheries.

Table 13
Consolidation of halibut QS: number of persons holding shares by area and size of holdings,
Initial issuance through December 31 1999

Number of QS [a]	Number of QS-holders				
	Initial	**End 1996**	**End 1997**	**End 1998**	**End 1999**
3000 or less	2 522	2 244	1 936	1 832	1 672
3001-10 000	1 158	925	878	865	853
10 001-25 000	648	629	613	613	586
More than 25 000	500	523	537	536	538
Total (unique persons)	**4 816**	**4 321**	**3 964**	**3 846**	**3 649**

Source: NMFS 2000.
[a] Holding sizes were calculated based on 1997 IFQ pounds.

Since IFQ programme implementation, the TACs in both fisheries have changed significantly. These changes cannot be attributed to the change in management programmes but do affect the interests of fishers in the fisheries. The TAC in the halibut fishery has grown from 37 million pounds in 1995 (the first year of the IFQ

---

[26] One concern expressed during periods of public comment was that the number of crew members necessary to conduct the fishery would be reduced. As a form of compensation for the loss of employment opportunity, NPFMC made the provision that the only persons who could purchase QS that were not initial recipients had to be "bona fide" crew members with at least 150 days of fishing experience. With this provision, crew members who might otherwise lose their jobs, can establish themselves in the fishery, and because the owner of the QS is required to be onboard when the IFQs are fished, these crew members can guarantee themselves a position.

programme) to 58 million pounds in 1999, an increase of more than 50%[27]. This change alone has considerably increased the value of the fishery to those holding QS. The TAC increase for halibut, combined with consolidation of QS in the fishery, has resulted in a significant increase in the average value of QS held by individual fishers.

Table 14

Consolidation of sablefish QS: Number of persons holding shares by area and size of holdings, initial issuance through 31 December 1999

Number of QS [a]	Number of quota share holders				
	Initial	**End 1996**	**End 1997**	**End 1998**	**End 1999**
5000 or Less	541	497	446	417	403
5001-10 000	109	102	113	115	114
10 001-25 000	146	145	144	141	140
More than 25 000	254	252	244	246	240
Total (unique persons)	**1052**	**996**	**947**	**919**	**897**

Source: NMFS 2000.

[a] Holding sizes were calculated based on 1997 IFQ pounds.

Table 15

Number of vessels landing halibut by area 1992 to 1999

Halibut management area	Before IFQ programme				Last 5 IFQ seasons			
	Number of unique vessels with landings							
	1992	**1993**	**1994**	**1995**	**1996**	**1997**	**1998**	**1999**
2C	1 775	1 562	1 461	1 105	1 029	993	836	840
3A	1 924	1 529	1 712	1 145	1 104	1 076	899	892
3B	478	401	320	332	350	357	325	323
4A	190	165	176	140	147	142	120	121
4B	82	65	74	57	64	69	47	51
4c	62	58	64	35	41	46	30	36
4d	26	19	39	27	33	33	22	29
Total (unique vessels)	3 452	3 393	3 450	2 057	1 962	1 925	1 601	1 613

Source: NMFS 2000.

Table 16

Number of vessels landing sablefish by area 1992 to 1999

Sablefish management area	Before IFQ programme				Last 5 IFQ seasons			
	Number of unique vessels with landings							
	1992	**1993**	**1994**	**1995**	**1996**	**1997**	**1998**	**1999**
Southeast Alaska	507	391	488	378	378	326	296	283
West Yakutat	266	196	249	228	218	218	176	162
Central Gulf	588	462	562	326	294	273	241	226
Western Gulf	103	29	19	86	81	79	66	63
Aleutian Islands	27	33	33	53	50	47	26	27
Bering Sea	72	40	31	55	49	41	28	20
Total (Unique vessels)	1 123	915	1 139	517	503	504	449	433

Source: NMFS 2000.

---

[27] The growth of the halibut TAC is not directly related to the change in management to IFQs. The halibut TAC increased because of improvements in the ability of IPHC scientists to estimate total halibut biomass and improvement in their ability to model fishery effects (NPFMC 1997).

In the same period, the TAC in the sablefish fishery declined. In 1995, the TAC was 45 million pounds. In 1998, the TAC had declined to less than 30 million pounds[28]. This decline has reduced the effect of the concentration of QS in the sablefish fishery and has expanded the reliance of sablefish fishers on other species. This dependence on other species is clearly illustrated by the results of a survey of vessel-owners during the first year of the IFQ programme. The survey found that only 61% of sablefish harvested were taken on a trip targeting only sablefish (Knapp and Hull 1996a).

Table 17
Monthly harvests rates of IFQ halibut 1995-1999

	1995	**1996**	**1997**	**1998**	**1999**
TAC (lb)	37 422 000	37 422 000	51 116 000	55 708 000	58 390 000
Month ending	**Percent of IFQ total allowable catch harvested by month**				
April 14	4	12	10	8	12
May 14	7	13	12	11	14
June 14	16	14	21	13	18
July 14	9	13	10	13	11
August 14	10	11	13	14	12
September 14	16	17	13	13	13
October 14	14	10	10	12	10
November 15	12	6	8	9	7
Total	**87**	**95**	**96**	**93**	**97**

Source: NMFS 2000.

Perhaps the greatest change in the move from the open-access fishery to the IFQ fishery is the change in the timing of harvests. In the open-access fishery, the harvest in the most active areas took place within a few 24-hour openings. After institution of the IFQ programme, fishers were free to harvest their shares at any time during the April to November season. Table 17 and Table 18 show the distribution of harvests across the season for 1995 to 1999 in the halibut and sablefish fisheries, respectively. In both fisheries, fishers have distributed their harvests across the entire season. The lowest harvests occur in the first few and last few months of the season, when the weather is the most threatening. Allowing fishers to choose when to harvest their IFQs has further established the halibut and sablefish fisheries as part-time fisheries. Fishers now have greater choice of when to harvest their shares, to avoid conflicts with other seasons.

Table 18
Monthly harvest rates of IFQ sablefish 1995-1999

	1995	**1996**	**1997**	**1998**	**1999**
TAC (lb)	45 658 049	35 319 897	30 233 885	29 845 875	27 154 059
Month ending	**Percent of IFQ total allowable catch harvested by month**				
April 14	10	15	11	7	8
May 14	22	24	24	18	20
June 14	22	20	20	19	20
July 14	11	10	10	12	17
August 14	4	8	11	11	7
September 14	8	8	7	9	8
October 14	7	5	8	11	9
November 15	7	4	4	6	5
Total	**91**	**94**	**95**	**93**	**94**

Source: NMFS 2000.

---

[28] As in the halibut fishery, the decline in the sablefish TAC is not directly related to the management regime, but is considered a function of changes in ocean conditions. It is likely, however, that the reduction of lost gear and sablefish mortality resulting from lost gear (deadloss) has allowed the sablefish TAC to remain higher than it would have been under the previous management regime (Lowe 2001).

Specific data showing the relative dependence on IFQ species are not regularly generated by NMFS and were therefore unavailable for this analysis. The assertion that most participants in the IFQ fisheries rely on sablefish and halibut as only part of their annual round of fisheries is supported by Table 19, which shows the number of trips in the fisheries and the average number of trips per vessel for 1995 to 1999. Over the five-year period shown, the number of halibut trips has ranged between 7030 and 8205, with an average of 7517. The average number of trips per vessel has steadily increased in the halibut fishery, from 3.4 in 1995 to 4.9 in 1999. Under the very conservative assumption that a vessel can make 1 trip every 10 days (including time in port), the average halibut vessel is actively engaged in halibut fishing for less than 50 days per year. The same is true of the sablefish fishery, in which the average for 1995 to 1999 was 4.7 trips per vessel.

Table 19
Number of trips and trips per vessel in the IFQ fisheries 1995-1999

	1995	1996	1997	1998	1999
Number of trips with IFQ halibut landings	7030	7275	8205	7153	7921
Average number of halibut trips per vessel	3.4	3.7	4.3	4.5	4.9
Number of trips with IFQ sablefish landings	2706	2367	2,153	2082	1994
Average number of sablefish trips per vessel	5.2	4.7	4.3	4.6	4.6

Source: NMFS 2001.

In summary, the initial allocation in the IFQ programmes disbursed interests in the fisheries. The programmes, however, have since allowed for considerable fleet-consolidation. Despite the consolidation that has occurred, the IFQ fisheries still occupy the average vessel for only a small part of the fishing-year. Participation in other fisheries has been aided by permitting fishers to spread harvests over the nine-month season, allowing them to choose the most opportune time to fish their IFQs.

6.   CONSEQUENCES OF THE IFQ PROGRAMME
6.1   Extent of changes

The previous section examined the number of owners and vessels participating in IFQ fisheries. The effects of the IFQ programme are felt much more broadly and include not only fishers but also processors, consumers, and fishery managers. This section examines several of these more indirect consequences.

6.2   Availability of fresh catch

Under the abbreviated seasons of the pre-IFQ fishery, fresh halibut and sablefish were available for only a very short period every year. Under the IFQ programme, fishers are permitted to fish their IFQs at any time during a season that begins in mid-March and ends in mid-November. The distribution of the catch throughout the year shows that fishers have taken advantage of the ability to choose when to fish their IFQs during the long season. A survey of registered buyers in the first year of the programme found that the production of fresh halibut rose 18 to 38% under the programme, benefiting consumers and marketers of fresh fish (Knapp and Hull 1996b).

6.3   Choice of processors and competition

Before the IFQ programme, processors wishing to be active in halibut and sablefish markets had to be able to handle large quantities of fish at a time. Fishers needed to bring their catch to processors able to handle it. Since the rush of fish to processors was intense, fishers had little choice of processors. Since processors were in high demand due to the intensity of the processing immediately after the short season, fishers had little leverage for negotiating a price for their catch. This circumstance was a greater concern in the halibut fishery, where almost all of the vessels participating in the opening tried to offload their catch immediately after the 24-hour period.

The processors operating prior to IFQs developed their businesses to accommodate the huge rush of fish that occurred with the short intense seasons. With IFQs, new niche processors catering to the fresh-market entered the fishery, and the larger traditional processors found themselves with excess capacity. With the additional competition to buy fish, vessel-owners are more able search out buyers willing to the pay the highest price. Existing processors, on the other hand, find themselves burdened with excess capacity and a loss of bargaining power[29].

6.4   Price of fish

Changes in fish prices reflect both changes in the relationships with processors and changes in the amount of fresh fish available to the market. The IFQ system has allowed fishers to time their catch to receive the best prices. In a survey of sablefish fishers in the first year of the programme, more than 75% said that price was important in determining when to fish IFQs (Knapp and Hull 1996a). In addition, most processors reported an increase in both ex-vessel and wholesale prices of sablefish in the first year of the IFQ programme. Fishers have stated that one reason they are satisfied with the programme is that they are receiving more for their catch (Matthews 1997). This response is consistent with results of a survey of sablefish fishers in which more than 50% of fishers surveyed reported a higher price for their harvests (Knapp and Hull 1996a).

Table 20 shows average ex-vessel prices and harvest volume for both halibut and sablefish 1988 to 1998. In both fisheries, prices have risen slightly since the implementation of the IFQ programme in 1995. Changes in harvest quantities and changes in the economy that affect supply and demand for fish (such as the Asian economic crisis in 1998) may account for some of the changes in prices and volumes. However, there is consensus that prices realized by fishers for halibut and sablefish have improved under IFQ management.

Table 20
Average halibut and sablefish ex-vessel prices 1988 to 1998

	Halibut										
	1988	1989	1990	1991	1992	1993	1994	1995	1996	1997	1998
Average price per pound ($)	1.22	1.50	1.79	2.00	0.97	1.24	1.931	2.01	2.24	2.11	1.27
Pounds (millions of pounds)	60.2	55.0	52.1	48.5	50.5	46.3	42.3	32.2	35.4	49.6	52.0
	Sablefish										
Average price per pound ($)	0.98	0.96	0.90	1.07	1.21	1.07	1.53	2.05	2.09	2.37	1.56
Volume (millions of pounds)	68.6	65.8	59.6	49.1	44.5	44.6	43.3	40.1	34.7	32.0	29.5

NPFMC 2000.

6.5   Safety

Fishing conditions in the pre-IFQ fisheries were dangerous - particularly in the halibut fishery. Several factors led to these unsafe conditions, including:

i.   The seasonal opening-dates were set by managers well in advance, and occurred regardless of weather conditions
ii.   Inexperienced fishers were drawn into the fishery by the ease of entry, the ease of catching commercial quantities, the relatively low costs of participating, and the relatively high price for fish
iii.   The chance for a big pay off for one day of fishing led to increased risk taking
iv.   The number of vessels crowded the fishing grounds, leading to a greater risk of conflicts and crashes
v.   The threat of pending entry limitations meant fishers had to participate regardless of weather conditions in order to qualify for future fishing rights.

With IFQs, the conditions that increased the risks in the fisheries have abated. Fishers have the flexibility to choose when to fish, and there are fewer inexperienced participants. In the first-year survey of sablefish fishers, more than 90% reported weather as an important factor in determining when to fish quota (Knapp and Hull 1996a). Table 21 shows safety statistics for the last eight years from the U.S. Coast Guard (USCG), which monitors safety at sea in Alaska. The statistics show a substantial drop in search-and-rescue missions for the IFQ fishery since implementation of the program. The number of fishers' lives lost has also declined slightly since the programme was implemented. Vessel-sinkings rose in the first 2 years of the programme but declined to a level similar to the pre-IFQ level in 1997 to 1999.

---

[29] Traditional processors in Alaska now recognize the consequences of IFQs and are exercising their political power to prevent new IFQ systems that do not include provisions that guarantee processor rights.

Table 21
Safety statistics from the pre-IFQ and IFQ fisheries

No. of Incidents	Pre-IFQ[a]	With IFQ programme in place				
		1995	1996	1997	1998	1999
Search and rescue missions	28	15	7	9	9	10
Vessel sinkings	2	4	7	2	0	2
Lives lost	2	1	1	2	1	1

Source: NMFS 2000.
[a] Average of 1992 1993, and 1994.

6.6   Gear loss and deadloss

Fishers participating in the pre-IFQ fishery, particularly in halibut, set more gear than they could retrieve so as to ensure that they could remain active during the entire opening. If fishing was poor and there were few fish per set, then most of the gear could be retrieved during the opening. However, if fishing was good, the numbers of fish would slow the retrieval, and not all of the gear that was set could be retrieved during the short season. Because it was illegal to retrieve gear after the season ended if there were fish onboard, the gear was abandoned[30]. Additionally, if gear fouled, the fishers would cut it free rather than waste precious time to untangle the lines.

Estimates of gear loss in the fisheries, both before and after IFQs, are inexact. In 1990, it was estimated that $2 million worth of halibut gear was lost (NPFMC 1992a). While the lost gear created direct costs to the fisher in terms of replacement costs, a bigger problem was the loss of fish that had been hooked by the abandoned gear but were never retrieved (deadloss). It is assumed that lost gear has at least the same catch rate as gear that is retrieved. Therefore, the lost gear generated at least 2 million pounds of halibut deadloss in 1990 worth at least $3.2 million (NPFMC 1992a)[31] .

With IFQs, estimates of gear loss have fallen significantly and are now considered inconsequential (Williams 2000). No longer do fishers set more gear than they need. Instead, they focus on catching their IFQ while minimizing costs, which translates to baiting and setting as little gear as possible. Furthermore, if gear becomes fouled, spending time to untangle it no longer reduces the amount of income earned, but instead reduces the cost of replacement gear.

6.7   Changes in cooperation and gear conflicts

Cooperation among fishers under the IFQ programme has increased operational efficiency in the halibut and sablefish fisheries. Under IFQ management, fishers do not incur any loss by advising others of favored fishing areas, since all have an assured portion of the catch that is unaffected by the success of others (Tremaine 2000; Matthews 1997). In addition, spreading out the season has reduced crowding on the grounds and has had the effect of allowing fishers to avoid conflict with others fishers. In the past, gear conflicts were thought to cost fishers substantial time and money (Pautzke and Oliver 1997).

QS-holders also appear to be cooperating by joining forces and fishing on a single vessel the IFQs of several persons. Simply comparing the number of active IFQ-holders with the number of vessels in the fisheries reveals significantly more active IFQ-holders than vessels. This cooperation is believed to have a significant impact on fishing costs.

6.8   Crew size and crew share

The IFQ programme has affected crews in ways that suggest a reduction in the number of active crew members under the programme. This decline, however, has not necessarily reduced the returns to crew members active in the fisheries. It is in fact estimated that payments per individual crew member have increased under the IFQ programme.

Data on the number of crew members in the fisheries are not collected by management agencies, so it is not possible to generate statistically accurate estimates of the changes in the number of crew or in the payments to crew members. However, the number of crew members on a typical trip is generally believed to have declined from a range of 3 to 6 before IFQs, to a range of 2 to 4 (including the skipper). Combined with the decrease in the number of active vessels, estimates of the number of active crew members in the halibut fishery

---

[30] Abandoned gear could be retrieved after the season ended if special permission from fishery managers was obtained. If gear was retrieved after the season, all hooked fish were required to be returned to the ocean.

[31] Biologists at IPHC accounted for deadloss in their stock assessment models and reduced the TACs so that gear losses would not result in overfishing (Williams 2000).

fell from 10 500 in 1994 to 3200 in 1999[32]. While the number of crew members has declined, if the number of trips per vessel is factored in, total halibut crew member trips appear to be approximately equal for the two years, at about 15 500[33].

Payments to crew members are typically between 5 and 10% of gross returns. Using an assumed crew payment percentage of 6% for each crew member and the ex-vessel prices and volumes shown in Table 20, it is estimated that the average total payments to halibut crew members during the three years before IFQs were $11 million, compared to an estimated $9 million for the first four years under IFQs. These estimates translate to $1095 per crew member in the pre-IFQ fishery and $2512 per crew member in the IFQ fishery. For sablefish, estimated average total crew payments in the pre-IFQ fishery were equal to average total crew payments in the IFQ fishery at $17 million per year[34]. The decline in the number of active sablefish vessels and reductions in crew size after IFQ implementation is estimated to have increased average payments per individual crew member from $3,165 in the pre-IFQ fishery to $8342 in the IFQ fishery.

6.9 Administrative burdens

Both buyers and fishers have stated that the programme imposes extra administrative burdens and costs on their operations. To ensure that catch is monitored, fishers are required to communicate to NMFS by radio their approximate catch at least 6 hours before landing. The requirement is particularly troublesome for small vessels that make short day trips, forcing some to approximate their catch before leaving the port. Tracking requirements and paperwork are also thought to be overly burdensome by both fishers and processors (Matthews 1997).

6.10 Enforcement

One concern in changing management of the fishery to an IFQ system was that the enforcement might be more costly and complicated than management by season closures. The short season in the pre-IFQ fisheries simplified monitoring and enforcement. Since fishers were permitted unlimited catch during the open season, the enforcement officers needed only to monitor activity in the fishery without regard to catch-limits. Under the IFQ programme, the season is protracted, with fishers active throughout a 9-month period. Different limits are imposed on different fishers by their IFQ holdings. Consequently, enforcement agents must monitor each fisher's harvest level throughout the season to prevent 'overages'.

Monitoring is conducted both at the dock and at sea. Dockside-monitoring has been simplified by requiring that all buyers be registered with NMFS. Buyers are required to report all purchases to NMFS through an automated system that records prices and weights of deliveries. In addition, requiring 6-hour advance notice of any delivery provides monitoring agents with the opportunity to be onsite at the time of landing. While some reports of leakage (unreported catch) have been received by enforcement agents, the problem is believed to be minor (Matthews 1997). At-sea boardings under the IFQ programme are much less burdensome than under the 'fishing derby' regime of the pre-IFQ fishery. Fishers in the IFQ fisheries are far less concerned about or constrained by the time away from fishing during an at-sea boarding, given the expansive time period allowed for fishing in the IFQ fishery (Matthews 1997).

High-grading - the discard of low-value catch (typically smaller fish) in favor of high-value catch - was also a concern at the outset of the IFQ programme. Catch-rates and the price structure of the industry, however, were thought to erase any incentive for high-grading in the fisheries. The relative rate of catch of low-value fish was thought to be high enough, and the price difference between low- and high-value fish was thought to be small enough, that the expense of replacing discarded low-value fish would result in little or no increase in net revenues to the fisher (NPFMC 1992b).

In addition to the monitoring by NMFS and USCG, the system is thought to have created some self-monitoring. Some fishers believe that by providing them with an interest in the fishery, the system has led to self-policing. Fishers now consider violations by other fishers as devaluing their interest in the fishery by reducing the TAC in future years (Matthews 1997).

Enforcement activities over the last two years appear in Table 22. In the 1999 IFQ season, NMFS enforcement officers boarded 26% of all vessels landing fish under the two programmes. Enforcement officers found approximately 95% compliance on these vessels, with most violations being logbook violations and overages (NMFS 2000a).

---

[32] These estimates use the number of active halibut vessels from Table 15 and assume that average crew size fell from 3 in 1994 to 2 in 1999.

[33] The number of trips per vessel in 1994 is assumed to have been 1.5 and the number of trips in 1999 was 4.9, as estimated in Table 19.

[34] Assumes 5-person crews before IFQs and 4-person crews after IFQs, with 6% of gross revenue paid to each crew member.

Table 22
Enforcement activities in the sablefish and halibut IFQ programmes 1998 to 1999

Year	USCG at-sea boardings	Dockside inspections	Violations detected	Offloads monitored	Cases initiated
		Number of incidents by agency and type			
	USCG	**Alaska enforcement division of NMFS**			
1998	276	463	38	413	196
1999	236	158	22	339	258

Sources: NMFS 1999; NMFS 2000a.

7.   SUMMARY AND CONCLUSIONS

Prior to the implementation of the IFQ programme, the consensus among stakeholders in Alaska's halibut and sablefish fisheries was that the existing management regime was not working. Fishers often compromised their safety by fishing the season regardless of the weather, as the one-day openings were a major disincentive to waiting for the weather to improve before starting fishing. TAC-management was imprecise because of the absence of limitations on individual fishers. Vessels were over-equipped as fishers attempted to maximize their catch during the short opening. Gear losses were excessive as fishers abandoned gear that had not been retrieved by the end of the fishing periods. Deadloss was also excessive because of the persistence of abandoned gear, which continued to catch fish long after the season ended. With only a few short openings, fresh fish were often unavailable. After considering several alternatives and after several years of evaluation and negotiation, managers of the fishery adopted and implemented the IFQ programme in the halibut and sablefish fisheries.

The IFQ programme clearly improved conditions in the fisheries. Fishers have the flexibility of fishing their IFQs at the times that they choose, allowing them to operate safely and time their fishing to obtain the best price for their harvests. Less efficient operations have sold their harvest-rights to more efficient operations, and vessel-owners are now fishing cooperatively-sharing "hot-spot" information and fishing IFQs from a single vessel rather than from two or more. The longer seasons, cooperation, and efficiency-gains from rights-transfers have reduced the cost of crew and gear. Since harvests are no longer restricted by short seasons, gear losses and deadloss have been reduced to minimum levels.

Consumers also have benefited from the IFQ programme. Fresh halibut is available for most of the year, and sablefish is becoming more widely known. Before the programme, fresh halibut was available for only a short period each year. The slower pace of fishing has resulted in better fish-handling, and the overall quality of the catch has improved. Although consumer prices may be slightly higher, the improvement in product quality and the availability of fresh fish have compensated for the price rise.

The programme, however, is not without its critics. Fishers excluded from the system or those who received small initial allocations believe that the system widened socio-economic disparities in the industry. Crew members are particularly frustrated by the system because they were not awarded fishing-rights, and for many, the IFQs eliminated a short but lucrative employment opportunity. Shoreside processors who were operating in the pre-IFQ fishery are also dissatisfied with the new system. The exclusive allocation of harvest-rights to fishers weakened the processors' bargaining position and created opportunities for new small-scale competitors.

Concerns have also been raised that the sablefish and halibut IFQ-system is overly complex, prevents an "optimal" level of consolidation, and remains extremely over-capitalized and inefficient. The response to these concerns is that the IFQ use, transfer, and ownership restrictions were incorporated into the programme primarily to retain the character of the fishery. Use of IFQs is restricted to vessels of the same size as the vessel from which the entitlement to the initial allocation arose. Restrictions allow transfer only to holders of an initial allocation or bona fide crew members. In addition, limits on the number of shares that can be owned or used in by a single vessel or owner prevent fleet-consolidation and the concentration of interest in the fisheries. These restrictions assured continuity between the pre-IFQ fisheries and the IFQ fisheries. The restrictions, however, may be criticized for preventing the development of an economically optimal fishery, which some economists believe would enable a minimum number of vessels to target a single species year-round. The programme was developed to serve the needs of existing fishers as well as maintain social conditions and promote economic efficiency. Consequently, the lack of economic efficiency should be considered a policy choice rather than a defect in the IFQ programme.

Any programme that attempts to allocate interests in a fishery must limit the interests of some people and will therefore draw criticism. In addition, these limitations will be tailored to serve identified policy objectives, which may draw further disagreement. The IFQ programme may be criticized on both of these grounds. Yet, the programme allows new entrants, improves efficiency by allowing the purchase of QS, within limits designed to preserve the character of the fishery. The programme represents a compromise intended to serve and balance several policy and personal interests. By any objective measure, the programme resulted in marked improvements in Alaska's halibut and sablefish fisheries. Safety, TAC-management, over-capitalization, gear loss, deadloss, and product price and quality all improved with IFQ management. The objectives of the programme in large part were met, and therefore the system should be considered a success.

8.   REFERENCES

CFEC - Alaska Commercial Fisheries Entry Commission 1999a. Changes Under Alaska's Halibut IFQ Program 1995 to 1998. Juneau, Alaska. November 1999.

CFEC - Alaska Commercial Fisheries Entry Commission 1999b. Changes Under Alaska's Sablefish IFQ Program 1995 to 1998. Juneau, Alaska. November 1999.

Gharrett, J. 2000. Data Operations Manager, National Marine Fisheries Service, Restricted Access Management Division. (Personal communication with Northern Economics, Inc., 3 December 2000).

Hartley, M. and M. Fina 2001 Allocation of individual vessel quota in the Alaskan Pacific halibut and sablefish fisheries. 251-265. *In:* Shotton, R. (Ed.) Case studies on the allocation of transferable quota rights in fisheries. FAO Fish. Tech. Pap. No. 411, FAO, Rome. 373 pp.

IPHC - International Pacific Halibut Commission 1987. The Pacific Halibut: Biology, Fishery and Management. Seattle, Washington.

Kingeter, John 2000. Assistant Special Agent in Charge for IFQ Enforcement, Alaska; National Marine Fisheries Management Service. (Pers. comm. with Northern Economics, Inc., 3 December 2000).

Knapp, G. and D. Hull 1996a. University of Alaska Anchorage, Institute for Social and Economic Research. The First Year of the Alaska IFQ Program: A Survey of Sablefish Quota Share Holders. September 1996.

Knapp, G., and D. Hull 1996b. University of Alaska Anchorage, Institute for Social and Economic Research. The First Year of the Alaska IFQ Program: A Survey of Halibut and Sablefish Registered Buyers. September 1996.

Lowe, S. 2001. NMFS Fishery Biologist. (Personal communication with Northern Economics, Inc., January 2001).

Matthews, D. 1997. Beyond IFQ Implementation: A Study of Enforcement Issues in the Alaska Individual Fishing Quota Program. Silver Spring, Maryland. April 1997.

NMFS - National Marine Fisheries Service 1994. Restricted Access Management Division. The IFQ Program, Insights and Updates. February 1994.

NMFS - National Marine Fisheries Service 1995. Restricted Access Management Division. The IFQ Program, Underway. February 1995.

NMFS - National Marine Fisheries Service 2000a. Alaska Enforcement Division. Enforcement Report for the Period 10/1/2000 through 12/5/2000. Juneau, Alaska. December 2000.

NMFS - National Marine Fisheries Service 2000b. Restricted Access Management Division. 2000 Report to the Fleet. July 2000.

NMFS - National Marine Fisheries Service 2001. Restricted Access Management Division. Individual Fishing Quota (IFQ)/Community Development (CDQ) Halibut Program Reports on the Internet at http://www.fakr.noaa.gov/ram/ ifqreports.htm, February 2001.

NPFMC - North Pacific Fishery Management Council 1992a. Final Supplemental Environmental Impact Statement/ Environmental Impact Statement for the Individual Fishing Quota Management Alternative for Fixed Gear Sablefish and Halibut Fisheries. September 1992.

NPFMC - North Pacific Fishery Management Council 1992b. True North. October 1992.

NPFMC - North Pacific Fishery Management Council 1997. Draft Environmental Assessment/Regulatory Impact Review/ Initial Regulatory Flexibility Analysis for Proposed Regulatory Amendments to Implement Management Alternatives for the Guided Sport Fishery for Halibut off Alaska. April 1997.

NPFMC - North Pacific Fishery Management Council 2000. (Electronic data provided to Northern Economics, Inc. by Special Request, November 2000).

Pautzke, C.G. and C.W. Oliver 1997. North Pacific Fishery Management Council. Development of the Individual Fishing Quota Program for Sablefish and Halibut Longline Fisheries off Alaska.

Smith, P. 2001. NMFS, Chief, Restricted Access Management Division. (Personal Communication with Northern Economics, Inc., 27 February 2001).

Tremaine, D. 2000. Former NPFMC Staff Economist and contributing Author to NPMFC 1992, Personal Communication with Northern Economics, Inc. November 2000.

Williams, Greg 2000, IPHC Biologist, (Personal Communication with Northern Economics, Inc., October 2000).

THE EFFECT OF INTRODUCING INDIVIDUAL HARVEST QUOTAS UPON FLEET CAPACITY IN THE MARINE FISHERIES OF BRITISH COLUMBIA

G.R. Munro
University of British Columbia
#997, 1873 East Mall, Vancouver, B.C. Canada V6T 1Z1
<GordonR_Munro@telus.net>

1. INTRODUCTION

The purpose of this report is to assess the impact of individual harvest quota schemes upon fleet-capacity in British Columbia marine fisheries. British Columbia now has many Individual Harvest Quota (IQ) schemes, the first one dating back to 1976[1]. The availability of data on these IQ schemes, however, varies greatly. In light of this fact, this report focuses on what are arguably the two most important of the B.C. fisheries into which IQ schemes have been introduced: Pacific halibut (*Hippoglossus stenolepis*) and sablefish (*Anoplopoma fimbria*)[2]. With respect to these two fisheries, most emphasis will be given to Pacific halibut, once again because of relative availability of data.

The introduction of individual harvest quotas (IQs) in the B.C. Pacific halibut fishery has been extensively studied by economists. In particular, there is a recent study on the impact of IQs on fleet-capacity in the fishery, using techniques which gained the approval of the recent FAO Technical Consultation on the Measurement of Fishing Capacity, Mexico City, November/December 1999 (FAO 2000). It will be argued that this study introduces what must be seen as the best available approach to assessing the impact of IQs on fleet-capacity.

The aforementioned study, when supported by the economics of the fishing capacity problems and the economics of IQs, allows some key conclusions to be drawn, which can be seen to have application to all IQ fisheries. To give but one example, it will be shown that a high degree of concern on the part of the resource managers about undue "concentration of ownership" can lead to the IQ scheme being accompanied by regulations that will seriously undermine the IQ scheme's ability to ameliorate any excess fleet-capacity existing in the fishery.

Prior to turning to the two British Columbia fisheries, it is necessary to set forth some basic concepts and economic principles. In so doing, papers arising from the FAO meeting of the Technical Working Group on the Management of Fishing Capacity, April 1998 (Gréboval 1999), and the aforementioned FAO Technical Consultation on the Measurement of Capacity, November/December 1999, will be drawn upon extensively.

2. BASIC CONCEPTS AND ECONOMIC PRINCIPLES
2.1 Open-access fisheries: "pure" and "regulated"

It is now generally accepted that the problem of excess fleet-capacity in marine capture fisheries has its roots in the open-access, "common pool," nature of these fisheries, in which property rights to the harvests, let alone the resources, are ill-defined, or simply non-existent (FAO 1998; Gréboval and Munro 1999). Among open-access capture fisheries, one can make a further broad distinction between "pure" and "regulated" open-access fisheries (Gréboval and Munro *ibid.*)[3]. A "pure" open-access fishery is one in which there is a complete absence of property-rights and government regulations, and which is open to all would be exploiters of the resource. The "pure" open-access nature of the fishery gives the fishers a powerful incentive to discount heavily future returns from the resource, with over-exploitation of the resource (from society's perspective) being the inevitable result. High-seas fisheries, prior to the U.N. Conference on the Straddling Fish Stocks and Highly Migratory Fish Stocks 1993–1995, offered several examples, *e.g.* the Alaska pollock fishery of the Bering Sea 'Donut Hole'.

In "regulated" open-access fisheries, by contrast, there is intervention upon the part of resource managers (*i.e.* government) to prevent "excessive" exploitation of the resource. Normally this is done by regulating the global harvest season-by-season. The resource managers do not, however, succeed in exercising effective control over fleet-capacity engaged in the fishery. The restricted seasonal harvest becomes the "common pool"

---

[1] The first IQ scheme in British Columbia was introduced into the herring spawn on kelp fishery in 1976. Subsequently, IQ schemes have been introduced into the following fisheries, in addition to Pacific halibut and sablefish: geoduck clam; red sea urchins; green sea urchins; sea cucumber; groundfish trawl; abalone (fishery closed since 1990), (Mylchreest, Pers. comm.).

[2] One could argue that the groundfish trawl-fishery is also very important. The IQ scheme introduced into this fishery is very complex, involving some 55 quota species. Moreover, it has only been in effect since April, 1997 – less than two years. The IQ scheme simply does not have enough history to make any reasonable assessment feasible (Turris 2000).

[3] The terms: pure open-access, and regulated open-access, owe their origin to James Wilen (1987).

with the fishers then having an incentive to compete for shares of the harvest. Those fishers who do not compete will have their harvest-shares diminish steadily. For reasons to be explained, the all-but-inevitable consequence of competition for shares of the limited harvest is the emergence of excess fleet-capacity in the fisheries.

Both the Pacific halibut and sablefish fisheries of British Columbia did, prior to the implementation of IQ schemes, have all of the attributes of "regulated" open-access fisheries (Casey *et al.* 1995; Turris 2000). For our purposes, assessing excess fleet-capacity in "regulated" open-access fisheries is, in contrast to "pure" open-access fisheries, a *relatively* straightforward process (Gréboval and Munro *ibid.*). Concepts and measures of capacity, which are routine in that branch of Economics known as Industrial Organization, can be applied without undue difficulty.

2.2 Measures of excess capacity in standard economics and the "malleability" of capital

In the standard economic theory of the firm, such as is encountered in Industrial Organization, one thinks of "capacity" (plant plus variable inputs such as labour service) in terms of the firm's ability to produce goods (or services) per period of time. The question then becomes whether the firm's capacity is "optimal," given the firm's desired level of output. If the firm's capacity is excessively large (or excessively small), the firm's costs will exceed the minimum.

Capital, be it physical-capital or human-capital, is deemed to be perfectly "malleable" if it can be quickly shifted out of a particular activity (*e.g.* manufacturing automobiles) with negligible danger of capital loss[4]. If the capital cannot be so shifted, then it will be said to exhibit some degree of "non-malleability." If excess (or deficient) capacity is to be meaningful, then some part of the relevant capital (physical or human) must be other than "perfectly" malleable. If this were not the case, then a firm with rational managers would never have excess or deficient capacity, other than momentarily (Arrow 1968). The typical firm can, in fact, adjust its capacity only slowly. Non-malleable capital is the norm, not the exception.

In any event, in order to measure the gap (if any) between a firm's actual level of capacity and the optimal level of capacity, economists make use of the concept of capacity utilization. This is measured in terms of the firm's actual level of output, and the output level for which the given capacity would be optimal. Let these two levels of output be denoted by Y and Y^* respectively. One can then construct a "capacity" utilization coefficient" simply as follows:

$$CU = Y/Y^*$$

If $CU = 1$, the firm is experiencing neither excess nor deficient capacity. If $CU < 1$, then the firm is experiencing "excess" capacity. Obviously if $CU > 1$, the firm is experiencing deficient capacity (Gréboval and Munro 1999, Kirkley and Squires 1999, Kirkley 1999).

Excess, or deficient, capacity is seen in the standard economic theory of the firm as a "short run" phenomenon. If a firm suffers from either, and if underlying conditions remain unchanged, the firm will, through time, seek to rid itself of the excess or deficiency. By way of contrast, in the case of regulated open-access fisheries, excess fleet-capacity, when it emerges, proves to be a problem that is chronic and persistent, rather than short term and self-correcting.

The "capacity utilization coefficient" as a measure of excess capacity can, nonetheless, be readily adapted to the case of regulated open-access fisheries. Output, Y, is now measured in terms of harvest. Actual output is the seasonal limited harvest determined by the resource managers, the total allowable catch (TAC), or equivalent thereof. The equivalent to Y^* is the level of harvest which the fleet currently in the fishery could produce, given the existing resource or biomass, if the fleet were unconstrained. Refer to this equivalent to Y^* as the *Potential Harvest* (Kirkley 1999). We thus have:

$$CU = \frac{TAC}{Potential\ Harvest}$$

Once again, if $CU < 1$, then excess fleet-capacity can be deemed to exist (Kirkley 1999).

2.3 Regulated open-access fisheries and excess capacity

The reasons for the emergence of excess capacity under conditions of a Regulated Open-Access Fishery can be readily illustrated with the aid of a simple abstract example drawn from Munro and Clark (1999). The assumptions initially introduced are admittedly extreme, but this is done to simplify the example. At a later point, some of the assumptions will be relaxed. In any event, it can be argued that the basic principles which will be set forth will hold under more realistic and complex assumptions.

---

[4] The concepts of malleable and non-malleable capital are analogous to the financial concepts of liquid and illiquid assets.

Assume that the fleet capital used in a particular regulated open-access fishery is perfectly non-malleable in that the "scrap" value of the vessels is zero – which implies that there are no alternative uses for the vessels. Assume as well that the rate of depreciation of the vessels is zero. Next assume that the vessels, crews, and owners are identical. Finally assume a constant TAC in the fishery.

Since the vessels, and their owners are assumed to be identical, a vessel-owner can be assumed to enjoy an average share of the TAC, *and* an average share of the Present Value of the operating profits associated with the constant TAC through time. It can be shown that, under these circumstances, it would pay the vessel-owners to invest in fleet capital, and thereby compete for shares of the global harvest up to the point that the unit cost of vessel capital is equal to the average Present Value of operating profits from the fishery (Munro and Clark 1999).

Now, letting K denote the fleet size, and letting c denote the unit cost of vessel capital, consider the following figure taken from Munro and Clark (1999).

Figure 1

Fleet capital versus present value

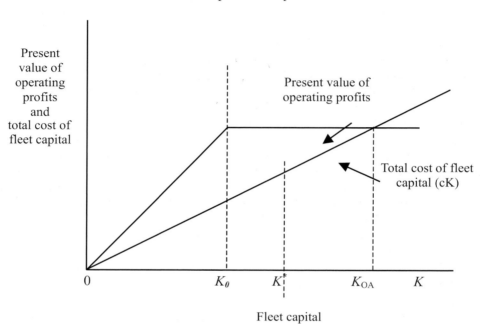

Let K_0 denote the minimum size fleet required to take the annual TAC. It is assumed that at K_0, the total cost of fleet capital, cK_0, would be less than the Present Value of operating profits. If this were not the case, the fishery would not be viable.

The present value of operating profits increases as K approaches K_0 and reaches a maximum at K_0. Given the imposition of the fixed TAC, the Present Value of operating profits remains constant for the Present Value of operating profits remains constant for all levels of K greater than K_0. Thus, if actual $K > K_0$, then it is obvious that redundant fleet capital exists. It is equally obvious that one would find that: $CU < 1$.

Suppose, for the sake of argument, that the fleet size is at the level $K = K_0$, which is not an equilibrium situation. Since $cK = cK_0$ is less than the Present Value of operating profits, the unit cost of capital is less than the average Present Value of operating profits. It will pay vessel-owners to continue competing for shares of the harvest by investing in additional capacity, even though the additional capacity is redundant. The investment behaviour, which is absurd from the point of view of society, is entirely rational from the point of view of the individual investor. Equilibrium will be achieved at a fleet level, $K = K_{OA}$, where OA denotes Open-access (Munro and Clark *ibid.*). At K_{OA}, one has $c = PV/K_{OA}$, where PV is the Present Value of Operating Profits.

2.4   The impact of individual harvest quotas (IQS) upon fleet-capacity

Now consider the impact of the introduction of an IQ scheme, assuming first that equilibrium had been achieved at $K = K_{OA}$. Assume further that the IQs are attached to the vessels, that the IQs are granted in perpetuity and that the per-vessel IQ is equal to TAC/K_{OA}. Since investment in vessels is a bygone, the interest of the vessel-owner is focussed solely upon the present value of operating profits associated with the IQ. If the IQ-holder can sell, or lease out, the IQ at a price in excess of the IQ-holder's perceived present value of the associated operating profits, then the IQ-holder will sell (lease) the IQ and remove his/her vessel from the fishery. If the IQ-holder cannot so do, then no attempt to de-activate the vessel will be made. Therefore, the very first condition that must be met, if the IQ scheme is to produce any reduction in fleet-capacity is that the IQs be transferable. If there is no opportunity to sell/lease the quotas, no removal of fleet-capacity will occur.

Transferability is a necessary, but not sufficient, condition for the IQ scheme to have an impact upon fleet-capacity. If the vessels and vessel-owners are identical, then each vessel-owner will make an identical assessment of the present value of operating profits associated with an IQ. There will obviously be no scope for trading and one should look forward to no withdrawals from the fishery. Thus a second necessary condition, which in practice is virtually always met, is that there be differences among vessels and vessel-owners in terms of harvesting efficiency.

To this point we have assumed that the fishery is in "equilibrium" when the IQ scheme is introduced. Relax this assumption and suppose that the fishery is in disequilibrium at a fleet level such as K^* in Figure 1. It is unrealistic to suppose that, if the fishery is in disequilibrium, equilibrium would be restored instantly. Suppose that the IQ scheme is introduced with $K=K^*$, and that the per vessel IQs are equal to: TAC/K^*.

Now, even if IQs are not transferable, they should produce some positive benefit. With the introduction of the IQ scheme, vessel-owners will no longer be able to capture a greater share of the TAC by investing in further capacity. Hence, the IQ scheme should at least place a cap on capacity.

Next let us relax some further assumptions, and in so doing, ask what happens to vessels that are *de-activated*. First, suppose that quota-holders are permitted to transfer quotas, but only on a season-by-season lease basis. Vessel-owners who de-activate their vessels cannot actually eliminate the vessels, since the quotas are attached to the vessels. Those who de-activate their vessels must compare the return on leasing the quota with the return from using the quota themselves plus lay-up costs.

Secondly, suppose as well that the vessel capital lacks perfect non-malleability with respect to the fishery, in that there are alternative fisheries in which the vessel can be used, *i.e.* the vessel is licensed for several fisheries. If a vessel-owner sells, or leases out, its quota, the vessel-owner then may be able to devote more time to the alternative fisheries. If this is the case, the additional operating profits to be realized from devoting more time to the alternative fisheries must enter into the vessel-owner's calculations when the vessel-owner is weighing the advantages of selling (or leasing) quota. This, in turn, raises an important policy issue.

If the alternative fisheries are non-IQ regulated open-access fisheries, then the fact that capacity exits from the fishery to which IQs have been introduced may lead to an intensification of the capacity-problem in the alternative fishery or fisheries. This is recognized in FAO reports on the fishing capacity-problem as the "spillover" effect (see, for example: FAO 1998).

In summary, the introduction of an IQ scheme can be expected, at the very least, to place a cap on the further expansion of capacity in a fishery. If the IQ scheme is to lead to the actual reduction in capacity in the fishery, then IQs must, at a minimum, be transferable. The greater the degree of transferability, *i.e.* if a vessel-owner can sell, rather than just lease, quota, the greater would be the impact upon fleet-capacity in the fishery.

Regardless of whether vessel-owners are allowed to lease, or sell, quota, one has to ask what happens to the vessels which cease to be active in the fishery. One cannot assume that they will be scrapped, or laid up. The de-activated vessels may "spillover" into other fisheries. If the other fisheries are characterized by regulated open-access, the "spillover" may intensify the excess fleet-capacity problems in these fisheries.

With these basic concepts and economic principles in hand, we turn to the British Columbia Pacific halibut fishery.

3. THE BRITISH COLUMBIA PACIFIC HALIBUT FISHERY, IVQS AND FISHING CAPACITY
3.1 Nature of the fishery and its history up to 1979

Pacific halibut, deemed to be one of the most valuable commercial species in the North Pacific (DFO 1999a), is a long lived species, which may attain up to 500lb in weight and live to over 100 years (Bell 1981). Typically, the harvested halibut range in size from 10 to 60lb. The fish generally recruit to the fishery at eight years of age (Casey *et al.* 1995).

Pacific halibut are a transboundary resource, being found predominately in British Columbia and Alaskan waters, and to a lesser extent in those off Washington. The commercial fishery in the Northeast Pacific commenced in the late 1880s with the completion of the first transcontinental railroad to Washington State. In light of the high value of halibut harvests and the slow-growing nature of the resource, it was, and is, highly vulnerable to over-exploitation. Clear signs of over-exploitation of the resource began to emerge in the years leading up to the First World War. The halibut industry, which was a Canadian/American integrated one at the time, urged the two governments to come together to conserve the resource. The two governments responded. In 1923 Canada and the United States signed a Convention and established what was to become the International Pacific Halibut Commission (IPHC). Figure 2 shows the extent of the IPHC management areas.

Figure 2
International Pacific Halibut Commission management areas

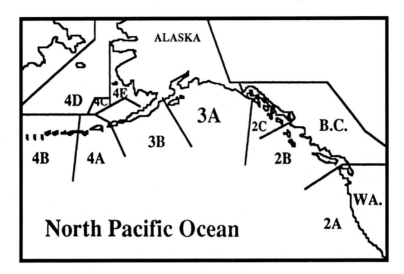

There was general agreement (among economists at least) that, up to 1979-1980, the IPHC had been reasonably successful in conserving the resource. There was also agreement, however, that, up to 1979-1980, the economic management of the fishery had suffered by virtue of the fact that there was no control over fleet-capacity. By 1980 there was clear evidence of excess capacity on both sides or the border (Crutchfield 1981). The fishery was a true regulated open-access one.

In 1977, both Canada and the United States introduced Extended Fisheries Jurisdiction out to 200 nautical miles in anticipation of the forthcoming U.N. Convention on the Law of the Sea. In response to the joint declaration of Extended Fisheries Jurisdiction, the two countries signed a Protocol to the Convention underlying the IPHC. One consequence of the Protocol was that American harvesting of halibut in the waters encompassed by Canada's Pacific Exclusive Economic Zone was eliminated[5]. Further, Canada was enabled to impose regulations on its own fleet so long as these regulations did not interfere with the conservation regulations of the IPHC. In 1979 Canada reacted to its newly obtained powers by introducing a limited-entry scheme to its fishery.

3.2   Vessel- and gear-types

The directed halibut fishery in British Columbia relies on hook-and-line (long-line) gear. Most halibut are captured with this gear but a small percentage are taken with troll gear (Casey *et al.* 1995). The vessels average between 40 and 60 feet in length and are multi-licensed. Prior to the advent of the IQ scheme, a common pattern for the halibut fisherman was to gill-net or seine, for roe-herring in the early spring, switch to long line gear for the May halibut open-season, switch to troll gear for the July-August salmon season, and switch back to long-line gear for the September halibut open-season (Casey *et al. ibid.*).

In the pre-IQ era, there were virtually no vessels that were wholly dependent upon the halibut fishery. In the last pre-IQ year, 1990, approximately one-third of the gross earnings of the vessels actively engaged in the halibut fishery was accounted for by that fishery. The lion's share of the fleet's earnings (45%) was accounted for by the Pacific salmon fisheries (DFO 1999a).

The long-line gear employed by the commercial vessels in the halibut fishery falls into two broad categories: conventional gear and snap-on gear. A detailed description of the gear-types follows (taken from DFO 1999a).

Conventional gear

Traditionally, a unit (skate) of conventional setline gear or fixed gear consists of groundline, gangions and hooks. Loops of light twine (beckets) are attached at regular intervals to the groundline. Short branch lines

---

[5] With respect to the Pacific halibut fishery, the Canadian EEZ was deemed to coincide with IPHC Fishing Area 2B – see Figure 2 (Casey *et al.* 1995).

(gangions) 3 to 4 feet long (0.9 to 1.2 meters) are attached to the beckets and a hook is attached to the end of each gangion. The most common rigs have been 3, 9, 13, 18, 21, 24, and 26 feet, (0.9, 2.7, 4.0, 5.5, 6.4, 7.3, and 7.9 meters) as those intervals facilitate baiting the hooks and coiling the lines. The skates with the baited hooks are set over a chute at the stern of the vessel.

The gear is retrieved on a power-driven wheel (gurdy). One person stands at the roller and one person coils the line after it passes the gurdy. The gear is then inspected for necessary repairs, baited, and recoiled in preparation for the next set.

Snap-on gear

Snap-on gear differs from traditional setline gear in that branchlines (gangions) are attached to the groundline with metal snaps rather than being tied to the groundline with twine. Further, the groundline used for snap-on gear is one continuous line that is simply stored on a drum after the gangions are removed, instead of being coiled. The method of attaching the hooks to the gangions is the same for snap-on gear and traditional gear. Gangions and baited hooks are stored on racks and the freshman snaps the gangions to the groundline as it unwinds from the drum during setting. Hook intervals can be changed with each set. When the gear is retrieved, the gangions are unsnapped as the groundline is rewound in the drum.

3.3 History of the fishery 1979 – 1990

At the outset of the limited-entry scheme, the Government of Canada issued 435 vessel licences, accompanied by strict limitations on their transferability. During the first year of operation, not all vessel licences were used. Some 102 licensed vessels did not participate.

Over the course of the following decade, the limited-entry scheme proved to be ineffective. To all intents and purposes, the B. C. Pacific halibut fishery continued to be a regulated open-access one (Table 1).

Table 1
British Columbia halibut fishery: number of active vessels, season length, and total harvest
1980 – 1998

Year	Number of active vessels	Season length (days)	Harvest (lbs)
1980	333	65	5 650 400
1981	337	58	5 654 900
1982	301	61	5 524 800
1983	305	24	5 416 800
1984	334	22	8 276 200
1985	363	22	9 587 900
1986	417	15	10 240 500
1987	424	16	12 251 100
1988	433	14	12 859 600
1989	435	11	10 738 700
1990	435	6	8 569 400
1991	433	214	7 189 300
1992	431	240	7 530 200
1993	351	245	10 560 100
1994	313	245	9 901 000
1995	294	245	9 499 700
1996	281	245	9 499 700
1997	279	245	12 581 400
1998	288	245	12 876 700

Source: Grafton *et al.*, 2000, and Mylchreest (Pers. comm.).

At the commencement of the limited-entry programme the annual season length was 65 days, well below the historical maximum of 240 plus days per annum. Although the harvest doubled over the ensuing decade the annual season-length steadily diminished until by 1990 it was only six days in length. The rapidly declining season-length was a clear indication of increasing fleet-capacity. The number of active vessels rose reasonably quickly to approach the upper limit of 435 set by the Canadian government. This upper limit proved to be a wholly ineffective barrier to the growth of capacity. Fishers strove, with success, to improve their gear as they competed for shares of the harvest. The resource managers had no effective control over gear improvements (Macgillivray 1996).

By the end of the decade, it was obvious to both the industry and to the Department of Fisheries and Oceans that the situation was wholly unsatisfactory. The short fishing-season produced a 'derby-style' fishery, which had, in turn, a series of malign consequences. First, fishing conditions were hazardous. Second, the resource managers were unable to keep the annual harvest within the TAC limits. Third, there was additional economic loss due to poor handling of the fish, and due to the necessity of freezing most of the catch and holding it over much of the year (Casey *et al.* 1995).

In November 1990 the Minister of Fisheries and Oceans, after a lengthy consultation with the industry, announced that the Department had decided to introduce an IQ scheme in the form of Individual Vessel Quotas (IVQs) on an experimental basis, for the 1991 and 1992 seasons (Casey *et al.* 1995). The experiment was deemed to be a success, and is still in place at the time of writing.

3.4   The Pacific halibut IVQ scheme and the consequences for fishing-capacity

Under the IVQ scheme each licensed vessel receives an allocated quota, specified as a percentage share of the TAC[6]. The fishing season was returned to its historical March-to-October term.

In the discussion of the economics of IQs and fishing-capacity, considerable emphasis was given to the importance of the transferability of quota. Hence, the transferability provisions of the IVQ scheme require close attention. These provisions have gone through three stages:

Stage I: 1991–1992. Permanent transfers of quota were allowed, but only if the transferred quota was accompanied by the vessel licence, and the transfer was made to a hitherto unlicensed vessel, which was not more than 10 feet greater in length than the vessel from which the quota and licence were stripped. Otherwise, there was an outright ban on quota-trading (Grafton *et al.* 2000). A key reason given for the ban on quota-trading was the fear of concentration of ownership of quota (Grafton *et al. ibid.*).

Stage II: 1993–1998. Over this period, restricted seasonal leasing of quota was permitted. Each vessel's quota was divided into two equal segments at the beginning of the season. A vessel-owner could lease out both shares if he so decided. A lessee vessel, however, could hold a maximum of four shares during any one season. If a vessel were to acquire the four largest shares in the fleet, its share of the TAC could not exceed 1.57% (Grafton *et al.* 2000).

While the leasing of quota shares was, *de jure*, strictly temporary, the leasing could, *de facto*, be longer. There was nothing to prevent a vessel-owner from entering into a *sub-rosa* agreement with a fellow vessel-owner, or owners, to lease the quota year-in and year-out.

Stage III: 1999 to the present. Commencing in 1999 vessel-owners were permitted to make unlimited permanent and temporary reallocations of halibut IVQs. This provision was, however, accompanied by a restriction on the amount of quota that an individual vessel could hold. An individual vessel cannot, during any one season, hold quota in excess of 1% of the TAC. A vessel to be in good standing in the fishery must hold permanent quota equal to not less than 0.01149% of the TAC. However, minimum permanent quota can be leased temporarily (Appendix 2 in DFO 1999a).

In Section 2.2 above, the concept of Capacity Utilization Coefficients (*CU*) as a measure of excess capacity was discussed. It will be recalled that, when applied to a fleet in a regulated open-access fishery, where

$$CU = \frac{TAC}{Potential\ Harvest}$$

if *CU* < 1, this can be taken as evidence of excess capacity.

A thorough investigation of excess fleet-capacity in the British Columbia halibut fleet, and the impact of IVQs on such excess capacity, would involve an attempt to measure fleet *CU*s through time. Such an investigation was undertaken by Squires *et al.* 1999). In their investigation, the authors attempted to measure capacity utilization coefficients on a per vessel, per day basis. The technique employed in estimating the coefficients consists of Data Envelopment Analysis, a technique which gained the approval of the FAO Technical Consultation on the Measurement of Fishing Capacity (FAO 2000). The summary results, for the years 1988, 1991 and 1994 are shown below in Table 2, presented as mean values.

Obviously, in the pre-IVQ year of 1988, the evidence of excess capacity was clear. Disconcertingly, the mean capacity utilization coefficient declined (sharply) between 1988 and the first year that the IVQ scheme was in place in 1991. The explanation given by the authors is that, over the period, the TAC was reduced by over 40%, while the biomass declined by only 3%. Not surprisingly, this combination had very negative consequences for the fleet *CU* (Squires *et al. ibid.*).

---

[6] Quotas were issued to vessels holding "L" licences. The quota shares were based 70% upon vessel performance between 1986 and 1987, and 30% upon vessel-length (Casey *et al.* 1995).

Table 2
Capacity utilization coefficients – British Columbia Pacific Halibut Fleet

Year	Mean capacity utilization coefficient (*CU*)
1988	0.47
1991	0.23
1994	0.55

Source: Squires *et al.* (1999) Table 3.

What there is no way of telling is whether the fishery was in "equilibrium" in the immediate pre-IVQ years. If it was not, the IVQ scheme may have served to cap the growth in capacity. In other words, if the IVQ had not been introduced, the mean fleet *CU* for 1991 might have looked even worse (Squires *et al. ibid.*).[7]

Be that as it may, from 1991 to 1994 (the last year for which the authors have data) the capacity utilization coefficient (*CU*) improved substantially. Now, with Table 2 in mind, return to Table 1 and consider the number of active vessels over time. During the first two years of the IVQ scheme, the number of active vessels was essentially unchanged from the last pre-IVQ year of 1990. From 1993 on, however, the number of active vessels has steadily declined while the *CU* per vessel has apparently improved (at least up to 1994). This is precisely what the theory would predict. The years 1991 and 1992 coincided with Stage I, in which there was an outright ban on quota-trading.

The post-1992 decline in the number of active vessels in the fishery, and the improvement in the per vessel *CU*, coincided with Stage II, when partial quota-trading was permitted. It was argued that quota-trading would not take place if fishers are identical – but then they never are. Indeed, quota-trading steadily increased, as is indicated in Table 3.

Table 3
Temporary transfers of individual quotas in the B.C. halibut fishery

Year	Number of transfers	Number of vessels involved	Percentage of total quota
1991	0	0	0
1992	0	0	0
1993	178	94	19
1994	306	154	34
1995	360	184	39
1996	413	216	44

Source: Grafton *et al.* 2000, Table 3.

While it is not possible to produce definitive proof, the evidence strongly suggests that the introduction of IVQs did lead to significant reduction in fishing-fleet capacity in the British Columbia Pacific halibut fishery, but only *after* the quota achieved some degree of tradeability[8]. This in turn suggests an obvious trade-off. If restrictions are placed on quota-trading for fear of emerging undue concentration of ownership, then the power of IQ schemes to reduce fleet-capacity will be diminished.

An awkward, and probably unanswerable, question is left. The question is: what has become of the de-activated vessels? It may be that the time they would otherwise have spent fishing for halibut is spent at the pier. On the other hand, it may be that the vessel-time has "spilled over" into non-IQ fisheries, with all that that implies.

3.5 Concentration of ownership

The evidence on shifts in the concentration of ownership in the fleet is opaque. Consider the following information provided by the DFO, which shows the number of vessel-owners in 1990 and 1999. A "person," owning a vessel, is a legal person, who can thus be either an individual or an incorporated company. A vessel can have more than one owner. Consequently the total number of vessel-owners exceeds the number of licensed vessels in each year.

---

[7] In some of the American IPHC Fishing Areas, seasons were reduced to one or two days (Casey *et al.* 1995).

[8] One would anticipate that, as part of the fleet-rationalization process, there would be a greater degree of vessel specialization among fisheries, with the consequence that the degree of fleet dependence upon the halibut fishery would increase. There is some evidence of this specialization occurring. Whereas halibut harvests accounted for roughly one-third of the gross earnings of the active halibut fleet in 1990, they accounted for 70% of the earnings of the active halibut fleet by 1998 (Department of Fisheries and Oceans, Canada).

Table 4
Vessels Owners: B.C. Pacific Halibut Fishery
1990 and 1999

Number of licences held	Vessel-owners	
	1990	1999
1	481	467
2	15	35
3	0	2
4	1	3
5	1	0
6	0	0
7	0	1
8	0	1
19	1	0

Source: Canada, Department of Fisheries and Oceans

Table 4 suggests no perceptible shift in concentration of vessel-ownership. Owners of vessels can, of course, transfer quota. Consequently, there could be a concentration of *de facto* quota-ownership, which would not be reflected in Table 4. Yet, one must recall that, under current regulations, a single vessel cannot hold quota in excess of 1% of the TAC. Thus, evidence of significant increased concentration of ownership is close to non-existent. Grafton *et al.* (2000), in their recent study of the B.C. Pacific halibut fishery, are insistent that:

"... despite the transfers [of quota] quota is neither heavily concentrated by area, individuals, or companies, and most active vessels remain owner operated[9]."

The impact of IVQs upon capacity is not just confined to the harvesting sector. In the 'derby-fishery' era, most of the catch (roughly 60%) had to be frozen. Only a few large processors had the necessary freezing capacity. With the advent of IVQs and the lengthening of the fish season, a far greater percentage of the catch (over 90%) went into the fresh-fish market (Casey *et al.* 1995). The barriers to entry to processors engaged in the fresh-fish market are much lower than are those engaged in the frozen-fish market. Casey *et al.* (1995) report a resultant *decrease* in the degree of concentration in the processing sector. Moreover, it comes as no surprise to learn that among the bitterest opponents of the proposed IVQ scheme in 1990 were the large processors (Casey *et al., ibid.*)[10].

4. THE BRITISH COLUMBIA SABLEFISH INDUSTRY

4.1 The nature of the fishery and its history prior to 1990

The British Columbia sablefish fishery has been much less intensively studied than the British Columbia Pacific halibut fishery. There have been, for example, no attempts to estimate capacity utilization coefficients (*CU*) for the fishery through time. It is for this reason that the halibut fishery was examined first, even though the introduction of IVQs to the sablefish fishery preceded the introduction of IVQs to the halibut fishery. Nonetheless, definite inferences can be drawn from the available data. The conclusions arrived at will be found to buttress those arising from the examination of the Pacific halibut fishery.

Sablefish (black cod), like halibut, is a high-valued groundfish species. The unit landed-value of the two species is roughly comparable[11]. In contrast to the Pacific halibut fishery, however, the British Columbia sablefish fishery is a relatively young one.

Prior to the 1960s sablefish was caught off British Columbia only as a by-catch in long-line and trawl fisheries. In the mid 1960s, Japanese distant-water fleets developed a directed fishery for sablefish off of British Columbia. With the advent of Extended Fisheries Jurisdiction in Canada in 1977, Japanese harvesting was eliminated (DFO 1999b). By the late 1970s, a viable Canadian fishery had been developed with the primary market being Japan.

Two forms of gear are employed: traps and long-lines. Vessels have varied in size from 30 to 120 feet (Grafton 1992). Currently the average trap-vessel is 75 feet in length, while the average long-line-vessel is 60 feet in length (Turris 2000). Thus the vessels are somewhat larger than those that are typically employed in the

---

[9] Grafton *et al.* forthcoming, p. 10.

[10] The Casey *et al.* article was published in 1995. A reversal of the de-concentration in the processing sector may have occurred since 1995. We have been unable to obtain any hard evidence that would support, or refute, the reversal hypothesis.

[11] Over the period 1996-1998, the unit landed value of halibut was approximately $Can3.10/lb, while that of sablefish was approximately $Can3.50/lb, Department of Fisheries and Oceans, Canada.

halibut fishery. Like the halibut fishery, on the other hand, the fleet was in the pre-IQ era a decidedly multi-licensed fleet. In the last pre-IQ year (1989) 48% of the active sablefish fleet's gross earnings were accounted for by species other than sablefish (Canada, Department of Fisheries and Oceans).

The history of the sablefish fishery, from the late 1970s to the introduction of an IQ scheme in 1990, is similar to that of the B.C. Pacific halibut fishery. With the escalating fishing-effort becoming increasingly evident, the Department of Fisheries and Oceans introduced a limited-entry scheme in 1981. The Department issued 48 licences (to be renewed annually). The initial season length in 1981 was 245 days.

Between 1981 and 1989, the TAC increased by 42%. Nonetheless, the season shrank from 245 days to a per-vessel season-length of 14 days by 1989 (DFO 1999b). As Table 5 indicates, the number of active vessels in the fleet was initially well below the ceiling of 48. As in the case of the halibut fishery, the number of active vessels steadily grew through the decade and approached the ceiling. In addition, capacity increased through the adoption of more and better gear. The similarity to the halibut fishery continues in that, by the late 1980s, there were complaints that the fishery was dangerous, that there was post-harvest economic loss through inferior handling of the fish and through the costs associated with short intense gluts confronting processors, followed by periods of famine. In addition, it became increasingly difficult for the resource managers to enforce the TAC (Turris 2000).

4.2   The IVQ scheme of 1990 and its impact upon capacity

Once again, as for Pacific halibut, the unsatisfactory situation led to months of discussion between the Department of Fisheries and Oceans and the industry in 1989. In late 1989, it was announced that an IQ scheme, in the form of IVQs, would be introduced for the years 1990 and 1991 on an experimental basis. The experiment was deemed to be a success and continues to the present day. As a further advance, co-management was introduced in 1993. The fishery is co-managed by the Department of Fisheries and Oceans and by the Pacific Black Cod Fishermen's Association (DFO 1999b).

Under the IVQ scheme, each licensed vessel receives a certain percentage of the TAC[12]. Quota assigned to one vessel can be re-assigned, *i.e.* leased, in unlimited quantity to other vessels in the fleet, subject to the approval of the Department of Fisheries and Oceans. The re-allocation is, however, strictly seasonal. Once again, however, it would be impossible to prevent vessel-owners from entering into *sub rosa*, *de facto*, long-term leasing agreements.

While the similarity of the histories of the sablefish and Pacific halibut fisheries is striking, there is one significant difference. In the case of the sablefish IVQ scheme, *unlimited*, leasing of quota (albeit on a strictly seasonal basis) was permitted from the inception of the scheme. It may be that the relatively small number of vessels in comparison with the halibut fishery and the resultant lack of prominence made it unnecessary for the authorities to show the same caution as they did in the Pacific halibut fishery.

A proper analysis of the impact of IVQs upon fishing-capacity in the sablefish fishery would involve estimates of fleet *CU* (capacity utilization) through time, as we now have for the halibut fishery. Such estimates, as we have indicated, do not exist. One can, nonetheless, draw inferences from Table 5. The pattern is similar to the halibut fishery in that the period of limited-entry was marked by a rapidly declining season-length. The introduction of IVQs, as in the halibut fishery, saw the season-length rise rapidly to the maximum. The number of active vessels in the fishery declined by over 50% between 1989 and 1998. It is difficult to believe that the fleet *CU* coefficient did not rise significantly over this period.

What is different from the halibut fishery is that the decline in active vessels commenced immediately upon the introduction of IVQs. The number of active vessels in the fleet declined by over one-third during the first year of the scheme[13]. The reason is straightforward: sablefish IVQs were transferable, albeit imperfectly so, from the inception of the IVQ scheme.

---

[12] Holders of "K" licences.

[13] There is no clear evidence, according to Grafton (1992), of a marked shift in fleet composition.

Table 5
British Columbia sablefish fishery: number of active vessels, season-length and total harvest
1982–1998

Year	Number of active vessels	Season-length (days)	Harvest (lbs)
1982	n/a	202	7 745 700
1983	23	148	8 906 400
1984	20	181	8 004 000
1985	27	95	8 888 500
1986	41	63	9 093 200
1987	43	45	9 506 800
1988	45	**	10 269 900
1989	47	***	10 734 600
1990	30	255	9 424 600
1991	26	365	9 990 900
1992	25	365	10 047 300
1993	21	365	12 021 200
1994	22	365	9 993 400
1995	24	365	8 176 500
1996	21	365	6 984 400
1997	24	365	8 581 900
1998	24	365	9 180 800

** In 1988 there were seven 20-day fishing periods during the year. Each vessel was allowed to fish in one.
*** In 1989 there were eight 14-day fishing periods during the year. Each vessel was allowed to fish in one.
Source: Canada, Department of Fisheries and Oceans.

While there are less data available for the sablefish fishery than exist for the halibut fishery, the sablefish results support those arising from the halibut fishery[14]. The introduction of IQs can indeed lead to a reduction in fishing-capacity given that IQs are transferable. The unanswered, but critical, question is what happens to the fishing-capacity that is removed from the fishery.

4.3   The question of ownership and concentration

The evidence on concentration of ownership is somewhat less clear than it is in the case of halibut. Table 6 is the sablefish equivalent to Table 4. Once again, a "person" owning a vessel may be an individual or an incorporated company. A vessel may have more than one owner. Hence, the number of licences appear to exceed the number of licensed vessels.

Table 6
Vessel-owners: B.C. sablefish fishery
1989 and 1999

Number of licences held	Vessel-owners	
	1989	1999
1	60	50
2	4	2
3	2	0
4	0	3
5	0	1
6	0	1

Source: Canada, Department of Fisheries and Oceans

Like the Pacific halibut fishery, there is no perceptible shift in concentration of vessel-ownership. Unlike the halibut fishery, however, there does not appear to be a strict upper limit on the amount of quota held by any one vessel[15].

---

[14] In keeping with the Pacific halibut fishery experience, there is evidence of increasing fleet specialization. In 1998, just under 90% of the active sablefish fleet's gross earnings were accounted for by sablefish harvests (Department of Fisheries and Oceans Canada).

[15] Thus, it is conceivable that there has been an increasing concentration of *de facto* quota ownership, not reflected in Table 6.

5. CONCLUSIONS

The purpose of this report has been to examine the impact of individual harvest quotas (IQs) upon fishing-capacity in British Columbia marine fisheries. Two fisheries were selected for investigation: Pacific halibut and sablefish. In both cases, individual harvest quota schemes have taken the form of individual vessel quotas (IVQs).

The B.C. Pacific halibut fishery has been extensively studied by economists. An attempt has recently been made to measure changes in fleet-capacity, using sophisticated statistical techniques in the form of Data Envelopment Analysis. The conclusion that can be drawn from this fishery are that the introduction of IQ schemes can indeed lead to a reduction in fleet-capacity – given that the IQs are tradeable. If the IQs are not tradeable, then about the best one can hope for is that the IQ scheme will serve to cap the expansion of fleet-capacity. The conclusions arising from the experience of the sablefish fishery buttress those arising from the halibut fishery.

IVQs in the Pacific halibut fishery were initially made non-tradeable, in part out of fear of undue concentration of ownership. The cost of giving high attention to increased concentration of vessel-ownership (and hence quotas) will be that of reducing the positive impact of the IQ scheme upon fleet-capacity.

In the end, the government relented and allowed increased tradeability in Pacific halibut IQs. Although it is not possible to offer conclusive proof, the evidence does suggest that the remaining restrictions and regulations, which the government has in force, have been sufficient to prevent a significant increase in the degree of concentration of vessel- (and quota-) ownership in the halibut fishery. While the evidence from the sablefish fishery is weaker, the evidence also suggests the absence of a marked degree of increase in concentration of vessel- and quota-ownership.

6. ACKNOWLEDGEMENTS

The author would like to acknowledge the invaluable assistance which he has received from Mr. Russell Mylchreest, Senior Economist, Policy and Economics Analysis Branch, Department of Fisheries and Oceans, Pacific Region, and from Ms. Stephanie McWhinnie, Department of Economics, University of British Columbia.

7. LITERATURE CITED

Arrow, K.J. 1968. "Optimal Capital Policy with Irreversible Investment," *In:* Wolfe J.N. (ed.), Value and Capital: Papers in Honour of Sir John Hicks, Edinburgh, University. of Edinburgh Press, pp. 1-20.

Bell, F.H. 1981. The Pacific Halibut: The Resource and the Fishery, Anchorage, Northwest Publishing Company.

Casey, K.E., C.M. Dewees, B.R. Turris, and J.E. Wilen 1995. "The Effects of Individual Vessel Quotas in the British Columbia Halibut Fishery," Marine Resource Economics, vol. 10, no. 3, pp. 211-230.

Crutchfield, J.A. 1981. The Pacific Halibut Fishery, Economic Council of Canada, The Public Regulation of Commercial Fisheries in Canada, Technical Report No. 17, Ottawa.

DFO - Department of Fisheries and Oceans 1999a. DFO Pacific Region, Integrated Fisheries Management Plan: Halibut 1999, Vancouver.

DFO - Department of Fisheries and Oceans 1999b. DFO Pacific Region, Management Plan January 1 1999 – July 31, 2000: Sablefish, Vancouver.

FAO 1998. Report of the Technical Working Group on the Management of Fishing Capacity, La Jolla, California, United States, 15-18 April 1998, FAO Fisheries Report No. 586, Rome.

FAO 2000. Report of the Technical Consultation on the Measurement of Fishing Capacity. Mexico City, Mexico, 29 November – 3 December 1999. FAO Fisheries Report No. 615. FAO, Rome. pp51.

Grafton, R.Q. 1992. "Rent Capture in Rights Based Fisheries," PhD dissertation, University of British Columbia, unpublished.

Grafton, R.Q., D. Squires and K.J. Fox (2000) "Private Property and Economic Efficiency: A Study of a Common-Pool Resource," Journal of Law and Economics. *43:679-713.*

Gréboval, D. 1999. Managing fishing capacity: selected papers on underlying concepts and issues. FAO Fish. Tech. Pap. No. 386. Rome. 206pp.

Gréboval, D. and G. Munro 1999. Overcapitalization and Excess Capacity in World Fisheries: Underlying Economics and Methods of Control. *In:* Gréboval 1999 (ed.). Managing Fishing Capacity: Selected papers on Underlying Concepts and Issues, FAO Fisheries Technical Paper No. 386: pp1-48.

Kirkley, J. and D. Squires 1999a. "Measuring Capacity and Capacity Utilization in Fisheries," *In:* Gréboval, D. (ed.). Managing Fishing Capacity: Selected Papers on Underlying Concepts and Issues, FAO Fisheries Technical Paper No. 386, Rome, pp.75-200.

Kirkley, J. and D. Squires 1999b. "Capacity and Capacity Utilization in Fishing Industries," FAO Technical Consultation on the Management of Fishing Capacity, Mexico City, Mexico, 29 November–3 December 1999, Background Document No. 20.

Macgillivray, P. 1996. "Canadian Experience with Individual Fishing Quotas," *In:* Gordon, D.V. and G.R. Munro (eds.). Fisheries and Uncertainty: A Precautionary Approach to Resource Management, Calgary, University of Calgary Press, pp.155-160.

Munro, G.R. and C.W. Clark, 1999. "Fishing Capacity and Resource Management Objectives," FAO Technical Consultation on the Measurement of Fishing Capacity, Mexico City, Mexico, 29 November–3 December 1999, Background Document No. 12.

Squires, D., Y. Jeon, R.Q. Grafton and J. Kirkley 1999. "Tradeable Property Rights and Overcapacity: The Case of a Fishery," FAO Technical Consultation on the Management of Fishing Capacity, Mexico City, Mexico, 29 November – 3 December 1999, Background Document No. 3.

Turris, B.R. 2000. "A Comparison of British Columbia's ITQ Fisheries for Groundfish Trawl and Sablefish: Similar Results from Programmes with Differing Objectives, Designs and Processes,". 254-261. *In:* Shotton, R. Use of property rights in fisheries management. Proceedings of the FishRights99 Conference, Fremantle, Western Australia, 11-19 November 1999. Mini-course lectures and Core Conference presentations. FAO Fish. Tech. Pap. No 404/1. FAO, Rome, pp254-261

Wilen, J.E. 1987. "Towards a Theory of the Regulated Fishery," Marine Resource Economics, vol. 1, pp.369-388.

CHANGES IN FLEET CAPACITY AND OWNERSHIP OF HARVESTING RIGHTS IN THE FISHERY FOR PATAGONIAN TOOTHFISH IN CHILE

E.P. González
Interamerican Centre for Sustainable Ecosystems Development (ICSED)
Casilla 27016, Santiago Chile,
<exequiel@icsed.org>

M.A. García and R.C. Norambuena
Fisheries Department, Under Secretariat of Fisheries, Chile
Casilla 100-V, Valparaíso, Chile
<mgarcia@subpesca.cl> and <rnorambu@subpesca.cl>

I. INTRODUCTION

The Chilean fishery for Patagonian toothfish[1] is conducted by both small-scale and industrial fishing fleets. The two fleets operate in different, but adjacent, areas off the Chilean coast (Figure 1).

The commercial fishery for Patagonian toothfish in Chile was initiated by the small-scale (artisanal) fishing sector in waters off central Chile during the 1970s and rapidly expanded to the south of the country, due to the existence of fishing grounds with higher yields (Lemaitre et al. 1991). At present, the small-scale fishery takes place from the northern border of the country in the I[st] Region (18°15'S) to the XIth Region (47°00'S). Its southern border is defined by the fishing area reserved for the industrial cod fishery (Under-Secretariat of Fisheries 1999).

Landings of Patagonian toothfish by the small-scale sector have continuously increased since the beginning of the fishery, reaching approximately 6000t in 1986. Since then, the annual landings of the small-scale sector have fluctuated between 3300 and 5600t with a slightly decreasing tendency towards the end of the 1990s (Figure 3). Data from the Chilean National Fisheries Service (SERNAPESCA) indicates that at present the official number of small-scale fishing boats operating in the fishery is approximately 120, generating employment for about 900 fishermen and 400 positions ashore

The industrial sector fishing for Patagonian toothfish began in 1992 and was a fishery managed under the "Regime of Fisheries of Incipient Development" [Pesquerías en Régimen de Desarrollo Incipiente] as defined by the Chilean *Fisheries Act* of 1991.

Figure 2 shows the area reserved for the industrial-sector fishery (a combination termed the *Unidad de Pesca*). The northern limit of this area is the parallel 47°S and the southern limit is the parallel 57°S. The eastern boundary of the fishing area is the "straight baseline"[2] and the western boundary is the line drawn parallel 70 nautical miles west of the baseline.

Figure 1
Chilean Patagonian toothfish fishery

18°20'LS

37°LS

47°LS

57°LS

┌ ┐ Fishing area for the small-
└ ┘ scale fishing fleet
┌ ┐ Fishing area for the large-
└ ┘ scale fishing fleet

---

[1] *Dissostichus eleginoides* is a long-lived fish species (up to 55 years of age) with a low fecundity rate and slow growth, reaching its first sexual maturity at 5-8 years of age and full maturity between 9-12 years.

[2] The "straight baseline" is the hypothetical straight line, drawn between the most seaward points of the coastal border of the country, which is used as geographical reference for fisheries management purposes.

The industrial fishing fleet (13 factory-vessels and 2 freezer-vessels) use deep-water long-lines and each vessel operates with an average of 10000 hooks per set. The fleet operates for 7 months (January to May and September to December) in Chilean waters and the rest of the year (April to August) in international waters. This fleet also targets Chilean hake (*Merluccius australis*) from the beginning of January to mid March of every year.

The industrial fishery generates employment for a total of 520 fishermen and 150 people in land-based support services just for the fleet – this does not include personnel occupied in processing.

The Under-Secretariat of Fisheries (1999) reports that the industrial fishing fleet not only operates in the area reserved for it (Figure 2) but also on the high-seas in the Chilean EEZ (Coast of the Argentinean Patagonia, Malvinas/Falkland Islands, Southern Georgia, and Kerguelén).

Industrial-sector landings of Patagonian toothfish in Chile started in 1991 and they consisted of harvest obtained inside the *Unidad de Pesca* and outside the Chilean EEZ (refered to hereafter as 'international waters'). After a peak volume of approximately 25 000t was landed in 1992[3], the landings of the industrial-sector have steadily decreased, reaching a new minimum of approximately 5900t in 1996 (Figure 3). Over 1997 to 1999 a moderate recovery was observed in industrial-sector landings which reached approximately 9000t in 1999.

Figure 2

Region of operations of the industrial fishing fleet in Chile waters

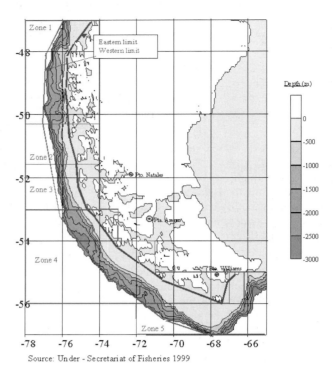

Source: Under - Secretariat of Fisheries 1999

Figure 3

Total landings of Patagonian toothfish in 1980 – 1998

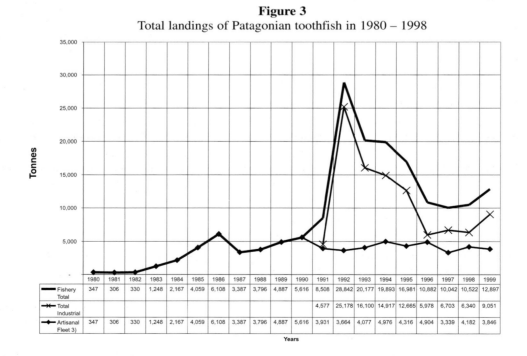

	1980	1981	1982	1983	1984	1985	1986	1987	1988	1989	1990	1991	1992	1993	1994	1995	1996	1997	1998	1999
Fishery Total	347	306	330	1,248	2,167	4,059	6,108	3,387	3,796	4,887	5,616	8,508	28,842	20,177	19,893	16,981	10,882	10,042	10,522	12,897
Total Industrial												4,577	25,178	16,100	14,917	12,665	5,978	6,703	6,340	9,051
Artisanal Fleet 3)	347	306	330	1,248	2,167	4,059	6,108	3,387	3,796	4,887	5,616	3,931	3,664	4,077	4,976	4,316	4,904	3,339	4,182	3,846

Years

---

[3] This peak in landings was generated during the research-cruise fishing conducted in 1992, which got "out of hand".

The total of industrial-sector landings correspond to captures taken in international waters as well as in the *Unidad de Pesca* area (Chilean EEZ between 47°S and 57°S). Landings taken in international waters have there represented, on average, approximately 40% of the total of the industrial-sector landings. Figure 4 shows a peak landing in 1992 that was mainly due to catches from international waters (approximately 16 800t). After this, was a dramatic drop in landings from international waters (a 85% decrease) between the years 1992 and 1994. Since 1995, landings from the international waters have shown a moderate increase, reaching approximately 5200t in 1999.

Figure 4 also shows that landings from the *Unidad de Pesca* follow a similar but lower pattern to the captures from international waters. The peak landings in the *Unidad de Pesca* took place in 1994 (approximately 12 000t) with an 84% drop in 1996, reaching only 1992t of Patagonian toothfish landed. Thereafter, the landings experienced a moderate increase, ranging around 4400t per year.

Approximately 99 % of toothfish landings are directed to processing and export markets. Statistics of Chilean seafood exports shows that annual exports of Patagonian toothfish averaged about 15 000t during the period 1992 to 1999, 87% of it as frozen products. Patagonian toothfish export- volumes show and overall decreasing behavior during this period, both as total export volume and as frozen exports (Figure 5). In fact, from 1992 to 1999 total export volumes of toothfish decreas- ed by 14%, from approximately 13780t in 1992 to 11 860t in 1999.

Nonetheless, during the same period, the total annual value of exports showed an impressive 80% increase, going from approximately $57 million in 1992 to $104.5 million in 1999 (Figure 6). This increase was due to a stronger price which grew at an annual rate of 10% during the period considered: going from $4200/t in 1992 to $8800/t in 1999 (Figure 7).

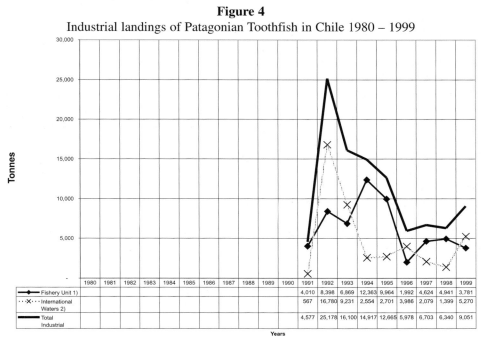

Figure 4

Industrial landings of Patagonian Toothfish in Chile 1980 – 1999

	1980	1981	1982	1983	1984	1985	1986	1987	1988	1989	1990	1991	1992	1993	1994	1995	1996	1997	1998	1999
Fishery Unit 1)												4,010	8,398	6,869	12,363	9,964	1,992	4,624	4,941	3,781
International Waters 2)												567	16,780	9,231	2,554	2,701	3,986	2,079	1,399	5,270
Total Industrial												4,577	25,178	16,100	14,917	12,665	5,978	6,703	6,340	9,051

Years

Finally, the most important foreign market for the Chilean fishery for Patagonian toothfish is Japan, which takes 58% of the total volume of Chilean Patagonian toothfish exports. Frozen products represented approximately 82% of the export value (FOB.) to Japan in 1998. The USA is the second most important foreign market for Chilean toothfish exports. Twenty-seven percent of the total export volume are fresh refrigerated products, representing 99.3% of the export value (FOB) in this market (Central Bank of Chile 1999).

2.   THE NATURE OF THE HARVESTING-RIGHT
2.1   General aspects

From legal, institutional and management perspectives, the Chilean Patagonian toothfish fishery can be divided into two types of fishing activities operating under different management schemes: the small-scale fishery (SSF) and the large-scale industrial fishery (LSF).

The SSF is conducted north of 47°S. Its two most important management regulations are: (a) limits on boat size (18m overall length, MINECON DS 43 and 439, 1986) and (b) gear restrictions (only deep-water long-lines may be used, with a limit of 12 000 hooks per cast, MINECON DS 439, 1986). No user-rights or property-rights are considered to exist in this fishery and it has been conducted under an open-access regime since its beginning. However, it may be better defined as quasi-open-access, because, although almost anyone may enter the fishery, to do so, the small-scale fishermen and their boats must be registered in the "National Registry of Small-scale

Fishermen" (or Non-industrial Registry)[4]. In 1999, the National Registry reported that a total of 12 356 fishermen targeted Patagonian toothfish in Chile. The small-scale fishing fleet included 230 non-processor boats, 1266 dinghies powered by outboard-motor, and 1088 boats powered by inboard-motors.

Figure 5
Chile Export Volume (tonnes) of Patagonian Toothfish, 1992-1999

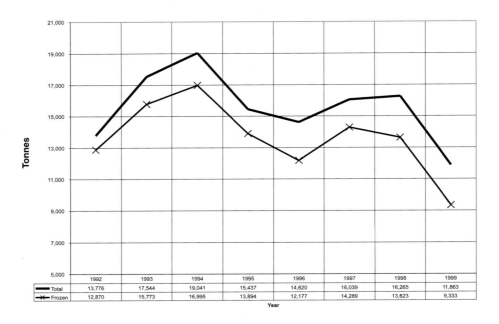

	1992	1993	1994	1995	1996	1997	1998	1999
Total	13,776	17,544	19,041	15,437	14,620	16,039	16,265	11,863
Frozen	12,870	15,773	16,995	13,894	12,177	14,289	13,623	9,333

Year

Figure 6
Chilean Export Value ($ FOB) for Patagonian toothfish, 1992-1999

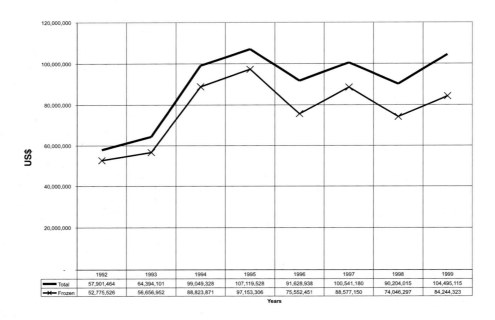

	1992	1993	1994	1995	1996	1997	1998	1999
Total	57,901,464	64,394,101	99,049,328	107,119,528	91,628,938	100,541,180	90,204,015	104,495,115
Frozen	52,775,526	56,656,952	88,823,871	97,153,306	75,552,451	88,577,150	74,046,297	84,244,323

Years

---

[4] The National Registry of Small-scale Fishermen is a record kept by the National Fisheries Service, containing a list of people authorized to conduct small-scale fishing activities in Chilean waters.

Small-scale fishing pressure on Chilean Patagonian toothfish has steadily increased over time, induced by increasing national exports to international markets, triggered by both increasing international demand, and a government policy directed at strengthening the national economy through exports. This trend continued until 1986 when a limit on boat-size and number of hooks per cast was enacted (Decree 439, 1986). These limits were triggered by decreasing yields observed by small-scale fishermen, due to the effect of their increased fishing-power and effort on stock abundance as well as to the new, and increasing, redirection of industrial fishing effort in the fleet from hake to toothfish (Pers. comm., F. Ponce, Chilean Under-Secretariat of Fisheries).

Figure 7

Apparent Price ($/t FOB) for Chilean Exports of Patagonian toothfish 1992-1999

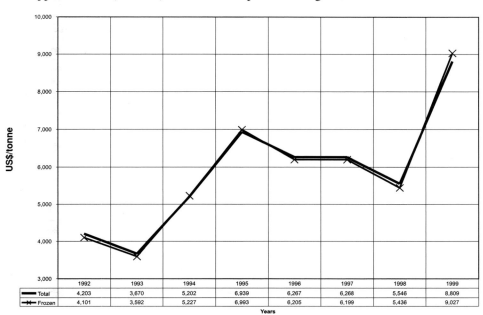

	1992	1993	1994	1995	1996	1997	1998	1999
Total	4,203	3,670	5,202	6,939	6,267	6,268	5,546	8,809
Frozen	4,101	3,592	5,227	6,993	6,205	6,199	5,436	9,027

Simultaneously with this, the *Fisheries Act* of 1991 was passed, which included a number of new regulatory measures applicable to Chilean fisheries. Consequently, the Under-Secretariat of Fisheries (USF) decided to officially initiate a LSF under one of these newly-created access-regimes: the Regime of Fisheries in Incipient Development (RFID). In this way, the Chilean USF established incentives leading to the achievement of a sustainable industrial Patagonian toothfish fishery (MINECON DS 328, 1992).

The large-scale industrial fishery (LSF) takes place south of 47°S and north of 57°S. Article 40 in the *Fisheries Act* [Ley General de Pesca y Acuicultura de 1991] states that in a fishery under a RFID regime, an annual total allowable catch (TAC) must be calculated and distributed among a number of eligible fishing companies and/or fishing operators.

In the LSF, the fishing- or harvesting-rights are allocated to fishing companies and according to the Chilean legislation they are known as "Extraordinary Fishing Permits" or EFP. These fishing-rights are divisible, transferable

Patagonian toothfish *(Dissostichus eleginoides)*
Photo: Karl Hermann Kock, Institute of Sea Fisheries, Germany

once a year, and can be leased or lent freely (Article 31, 2nd Paragraph, National *Fisheries Act* of 1991). Even though this Act does not explicitly indicate so, according to the Chilean legislation these fishing-rights may be banked.

Other regulations applying to both the SSF and the LSF are:

i. The establishment of a seasonal (biological) closure from 1 June to 31 August each year, between 53°S and 57°S (DS 273-ex).

ii. Only 'trotlines' or long-lines may be used to harvest toothfish (USF Res. 1249, 1992).

iii. toothfish harvested as by-catch may comprise no more than 2% by weight of the total harvest of southern hake or king clip, between 41° 28'6"S and 57° 00'S (MINECON, DS 679, 1993).

2.2 Responsibilities for recording

Once a fishery has been declared under a RFID, the Under-Secretariat of Fisheries must calculate an annual TAC, which is allocated among fishing companies or operators by means of a public auction (Articles 39 and 40 of the *Fisheries Act* of 1991). Successful bidders are issued an Extraordinary Fishing Permit (EFP) declaring their right to harvest annually, for a 10-year period, a maximum amount of fish equivalent to the product of the corresponding TAC and the fraction of the TAC they are awarded (*i.e.* the percentage of the TAC sold to the respective bidders). These EFPs (fishing-rights) enter into effect in the calendar year followng the auction.

SERNAPESCA is the institution in charge of keeping records of EFPs issued under the RFID-system. SERNAPESCA has created a registry of EFPs for each fishery managed under the RFID-system. This registry records all EFPs, including those awarded by public auction as well as those permits issued later (due to the division, transference or transmission of fishing-rights)[5]. The registry

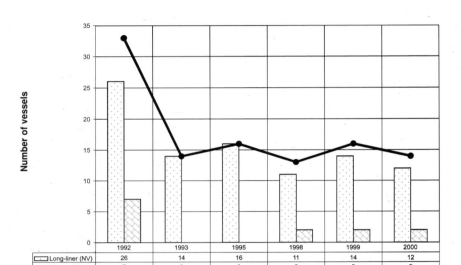

Figure 8

Fleet size (number of vessels) 1992-2000

	1992	1993	1995	1998	1999	2000
Long-liner (NV)	26	14	16	11	14	12
Trawler (NV)	7	0	0	2	2	2
Fleet (NV)	33	14	16	13	16	14

Year of Operation

records the full name and identification of the owner of the EFP and the fishing-vessel or vessels that will be used to exploit these fishing-rights. The owner of the EFPs may replace the registered vessels or add others, as long as its fishing permit is currently valid, and the regulations with respect to vessel-replacement do not apply[6]. Nonetheless, replaced or additional vessels must be registered before initiating operations in this fishery.

3. MEASUREMENT OF FLEET-CAPACITY
3.1 Total values and fleet dynamics

The fleet-capacity and subsequent changes as a consequence of the implementation of the rights-based management system are given only for the industrial-sector, which is the one managed under the RFID system. Fleet-capacity has been measured in terms of: Number of vessels, Gross Registered Tonnage (GRT) and Engine-Power (Horsepower – HP), but only for only six points in time since the beginning of the RFID-system.

Figure 8 shows the evolution of fleet-capacity due to the implementation of the RFID-system. In 1992, just before the implementation of the fishing-rights, the fleet consisted of 33 industrial vessels (26 long-liners and 7 trawlers). The long-liner fleet represented approximately 79%; the trawlers the remaining 21%. During 1993, the first year of operation of the RFID-system, the fleet experienced a drop of 58% in number, declining to 14 vessels, all long-liners. In subsequent years (1995, 1998, 1999 and 2000) the total number of vessels operating in the *Unidad de Pesca* fluctuated around 15, with an additional 1 or 2 vessels in, or out, each year. As shown in

---

[5] For leased or free lent EFPs it is enough to have an entry on the margin of the registry of the permit.

[6] Regulations for vessel substitution (DS. 64/92 and DS. 500/94) are used in those fisheries declared as fully-exploited. Limitations refer to the need that a new vessel must not exceed the total values of a combination of GRT, HP and size (LOA and D) of the vessel replaced.

Figure 8, long-liners have represented approximately 90% of the total fleet-capacity since 1993 and at present there remain only two trawlers operating in the *Unidad de Pesca*.

Figure 9 shows the evolution of the fleet in terms of its total tonnage (GRT) from 1992 to 2000. In 1992 the total was approximately 16 700 GRT: long-liners represented 90% of this with approximately 15 000 GRT and the trawlers only 10%, with approximately 1600 GRT.

In 1993 the fleet experienced a fall of 49% in total tonnage down to approximately 8500 GRT operating that year. In subsequent years (1995, 1998, 1999 and 2000) the total tonnage operating in the *Unidad de Pesca* fluctuated between 7000 and 8800 GRT with an average annual variation of 7% only. Figure 9 illustrates that long-liners have represented approximately 95% of the total annual GRT in the fleet since 1993, and shows that the importance of trawlers drastically decreased after the implementation of fishing-rights (in 1993). Thus, even though two trawlers have returned to fishing since 1998, they represent only about 7% of the total fleet capacity (GRT).

Figure 9

Fleet Gross Register Tonnage (GRT), 1992 - 2000

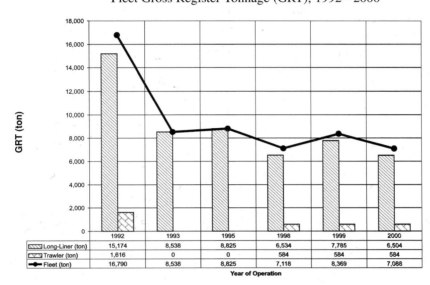

	1992	1993	1995	1998	1999	2000
Long-Liner (ton)	15,174	8,538	8,825	6,534	7,785	6,504
Trawler (ton)	1,616	0	0	584	584	584
Fleet (ton)	16,790	8,538	8,825	7,118	8,369	7,088

Year of Operation

3.2 Fleet structure

Table 1 presents the structure of the long-liner fleet in terms of GRT for six years over 1992-2000. The available data show that the most important category is the class 451–550 GRT, which includes 42 to 55% of the long-liner vessels over 1992-2000, with a peak of 63% in 1995.

Table 1

GRT structure for the long-liner fleet, 1992-2000

	Number of long-liners					
Range GRT	**1992**	**1993**	**1995**	**1998**	**1999**	**2000**
0 - 400	1	0	1	0	0	0
401 - 450	2	0	0	0	0	0
451 - 550	11	7	10	6	7	6
551 - 650	1	0	0	0	0	0
651 - 750	7	4	3	3	4	3
751 - 850	3	3	2	2	2	2
851 - 900	1	0	0	0	0	0
Total	**26**	**14**	**16**	**11**	**13**	**11**

Source: Elaborated from USF data

The second and third classes in importance are the 651-750 GRT category, which represents an annual average 27% of the fleet-capacity, and the 751-850 GRT category comprising an annual average of 16% of the total fleet-capacity.

From data presented in Table 1 it is possible to infer that the relative importance of the classes 451-550 GRT and 651-750 GRT have been moderately increasing over time, with an average annual change of 7% and 4% respectively. The relative importance of the 751-850 class has increased over time at an average annual rate of 18%.

Table 2

GRT structure for the trawler fleet, 1992-2000

Range GRT	Number of trawlers					
	1992	1993	1995	1998	1999	2000
0 - 100	3			0	0	0
101 - 200	0			0	0	0
201 - 300	3			2	2	2
301 - 400	0			0	0	0
401 - 500	0			0	0	0
501 - 600	1			0	0	0
851 - 900	0			0	0	0
Total	7			2	2	2

Source: Elaborated from USF data.

Table 2 shows the structure of the trawler fleet in terms of GRT for four years since 1992. These data show that the most important class (that of 201–300 GRT), which represented 43% of all the trawlers in 1992, and 100% since 1998. The second most important category, prior to the implementation of the RFID-system, were those under-100 GRT, being 43% of all trawlers in 1992. There have been no changes in the GRT structure of trawlers since 1998.

Table 3 shows the distribution of engine-power (measured in horse-power) of the long-liner fleet for six years since 1992. The most common class is that of 1251-1500 HP, which included 44% to 27% of the long-liners from 1992 to 2000. The classes with the next largest number of vessels are those of 1751-2000 HP (representing an average of 28% of the fleet), and those of 1001-1250 HP, contributing an average of 23% of the fleet.

The data presented in Table 3 show that for long-liners the relative importance of the categories 1251-1500 HP and 1001-1250 HP are decreasing over time, with average rates-of-change of −8% and −1% per annum respectively. The class 1751-2000 HP is increasing in relative importance over time (average rate of 27% per annum).

The most important engine-power class for the trawler fleet (not tabulated) is that of 351-500 HP, accounting for 29% of the vessels in 1992, and 100% since 1998.

The age structure of the long-liner fleet since 1992 is shown in Table 4. The data

Table 3

HP-structure for the long-liner fleet, 1992-2000

Range HP	Number of long-liners					
	1992	1993	1995	1998	1999	2000
750 - 1000	3	1	2	0	1	1
1001 - 1250	5	3	4	3	3	2
1251 - 1500	11	5	5	3	4	3
1501 - 1750	0	0	0	0	0	0
1751 - 2000	4	4	3	4	4	4
851 - 900	2	1	1	1	1	1
Total	**25**	**14**	**15**	**11**	**13**	**11**

Source: Elaborated from USF data.

show that between 55% and 45% of long-liners operating in the *Unidad de Pesca* were constructed between 1971 and 1980. Vessels constructed between 1950 and 1970 are the second most important category representing, on average, 38% of the fleet. The third most important category of vessels is those constructed between 1981 and 1990, which represent on average 11% of the fleet. Most of the important trawlers operating in the *Unidad de Pesca* were constructed between 1981 and 1990 (not tabulated).

Table 4

Age structure for the long-liner fleet, 1992-2000

Year built	Number of long-liners					
	1992	1993	1995	1998	1999	2000
1950 - 1970	9	5	6	4	5	5
1971 - 1980	14	7	8	6	7	5
1981 - 1990	3	2	2	1	1	1
1991 - 2000	0	0	0	0	0	0
Total	**26**	**14**	**16**	**11**	**13**	**11**

Source: Elaborated from USF data.

4. CONCENTRATION OF OWNERSHIP
4.1 Status prior to programme

Prior to 1992 there was no ownership of fishing-rights. In 1992, just before the implementation of the RFID system (in 1993), there were eleven fishing companies (rights-holders from 1993 on) who owned a total of 33 vessels. Five companies (45% of the total) owned 63% of the fleet, *i.e.* 21 vessels in the fishery (Figure 10). These same companies owned 75% of the total tonnage (GRT) in the fishery, *i.e.* 12 560 GRT (Figure 10).

Figure 10
Structure of fleet-ownership (numbers, and GRT) prior to the RFID-system
in the Patagonian toothfish fishery in Chile, 1992

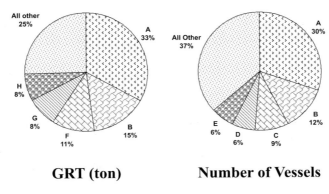

GRT (ton)   **Number of Vessels**

Right Holders: A, B, C,..

4.2 Status of the programme following implementation

The implementation of the system of fishing-rights has greatly increased the concentration of fleet-ownership in the *Unidad de Pesca*, with three fishing companies or rights-holders owning (on average over the period) 84% of the total fleet GRT, with a peak of 93% in 1998 (Figure 11). Moreover, just one fishing company owned (on average) 52% of the total GRT in the *Unidad de Pesca*, reaching a peak of 61% in the year 2000 (Figure 11).

Figure 11

Structure of fleet-ownership (GRT) during the RFID-system
in the Patagonian toothfish fishery in Chile, 1993 - 2000

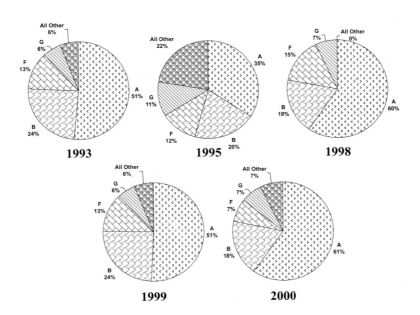

Figure 12 shows a similar pattern in of the structure of fleet-ownership (in terms of numbers of vessels operating in the *Unidad de Pesca*): The same three fishing companies owned (on average over 1993 to 2000) 80% of the total fleet, with a peak of 92% in 1998. The single most important fishing company ("A" in Figure 12) owned (on average during 1993-2000) 50% of the vessels, with a peak of 62% in 1998.

Figure 12

Structure of fleet-ownership (Number of vessels) during the RFID-system
in the Patagonian toothfish fishery in Chile, 1993 - 2000

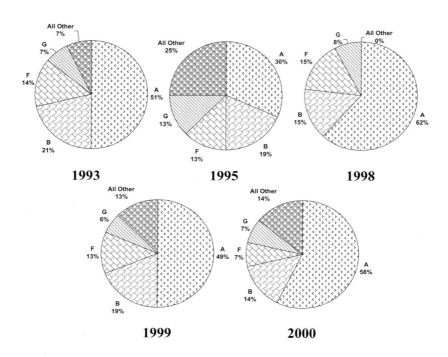

Figure 13 shows an analysis of the fishing companies' share of the TAC. The quantitative variable used corresponds to the number of fishing-rights accessed or held by each fishing company every year (including rights bought by auction, direct sale, rental and/or free lending) expressed as a percentage of the annual TAC. As Figure 13 shows, three fishing companies (A, B and C) have competed for the major part of access to the annual TAC since 1993.

4.3 Transfer of ownership
4.3.1 Types of transfers

Under the RFID-system, fishing-rights or EFPs may be transferred by auctions, direct sales, rentals, or free loans. The three methods used in the Patagonian toothfish fishery in Chile so far are: auctions (ordinary and special), direct sales and rentals.

4.3.2 Regulations, procedures and restrictions

There are no explicit regulations governing the transfer of EFPs under the RFID system in the Patagonian toothfish fishery. Nonetheless, the USF has established the following procedure in the event of future transfers of fishing-rights either by direct sales, leases or free loans.

Direct sale of fishing

If fishing-rights or EFPs are directly sold, the USF will issue a certificate that formally establishes who is the new owner and the person responsible for the financial obligation to the government (annual management fees). The certificate issued states the name and identification of the new owner of the EFPs. The procedure followed includes the following:

i.   The buyer of the EFPs must send a letter to the USF informing that he is the owner of one or more lots of fishing-rights and from whom they were bought. This enables the transfer to be formalized.. A notarized receipt of purchase must be attached to this letter along with a copy of the certificate showing the award of the EFPs to the original owner.

ii. The USF issues a new certificate in the name of the buyer showing that he is the new owner of the specified fishing-rights (EFPs).

iii.   The new owner is responsible for paying the annual payments due on the EFPs that have been bought.

Figure 13
Structure of access to fishing-rights during the RFID-system
in the Patagonian toothfish fishery, 1993 - 2000

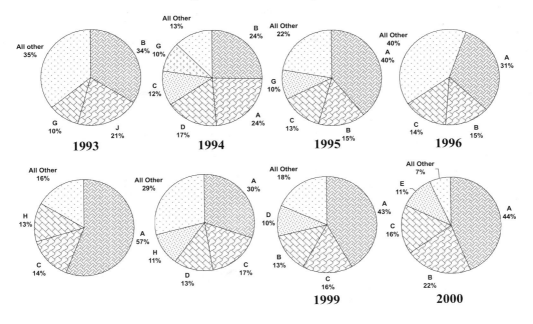

Lease or free loans of fishing rights

Since under a lease or a free loan of fishing-rights there is no transfer of ownership, new certificates are not issued by the USF. In this case, the original owner of the fishing-rights is responsible for the payment of the annual management fee to the government. SERNAPESCA is the institution that records the temporary change in the fishing-right (EFP), since it must monitor all changes so as to maintain an accurate record of amounts harvested and to conduct efficient monitoring-and-control of the fishery.

The required paperwork in this case includes the following:

i.   The owner of the EFPs must advise SERNAPESCA in writing (Department of Fisheries Statistics and Information) that he has leased or lent one or more lots of fishing-rights to the recipient, and wishes to formalize this situation. A notarized lease or free-loan contract must be attached to his letter along with a copy of the certificate of the previous registration of the EFPs to the original owner.
ii.   SERNAPESCA verifies the situation, and if everything is correct, issues a letter stating its agreement to the transfer.
iii.   The original owner remains responsible for paying the annual fees/levies on the EFPs, whether leased or lent.

Restrictions

At present, there are no major restrictions related to the transfer of fishing-rights under the RFID-system. The existing restrictions relate to the need to be duly registered in the National Industrial Fishing Registry and in the Special Auction Registry, in order to be eligible to bid in any future auction of fishing-rights. An additional restriction is that no single bidder may be awarded more than 50% of the TAC being auctioned every year. There is no explicit limit to the cumulative number of EFPs (percentage of TAC) that a single company may hold over time.

4.3.3 Prices received

Under the existing regulations and procedures for allocating and transferring fishing-rights in the *Unidad de Pesca*, the only prices recorded are auction prices (ordinary and special); these are accurate. Prices or transaction values actually used in direct sales need not be recorded as a requirement of the USF. The USF is only interested in knowing who is responsible for the annual management-fee due from the original auction of the EFPs. Thus, the price data available correspond only to auction-prices.

Average auction-prices recorded have normally ranged from approximately $600/t to $2000/t, but with two exceptional records of $15 333/t and $11 372 /t, observed in 1994 and 2000 respectively (Figure 14).

5. CONSEQUENCES OF IMPLEMENTING FISHING-RIGHTS
5.1 Changes in fleet-capacity arising from introduction of transferable property-rights

In the initial period the comparison of indexes of fleet-capacity from 1992 and 1993 (Figures 8 and 9) shows that the effects of introducing the RFID-system were: a drop of 58% in the number of vessels (33 to 14), and of 49% in the total fleet tonnage (16 700 to 8500 GRT). Further, from 1993 through until 1997 deep-water long-lines were the only fishing gear used. In the medium-term, analysis of the total fleet-capacity in the *Unidad de Pesca* between 1993 and 2000 showed a 17% reduction, with a decrease from approximately 8500 in 1993 to approximately 7100 GRT in 2000. This decline has not been uniform as the fleet

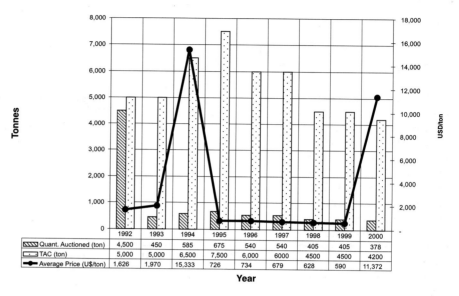

Figure 14
Auction-price ($/t), amount auctioned (tonnes), and TAC (tonnes) for the Patagonian Toothfish Fishery in Chile, 1991 – 2000

Year	1992	1993	1994	1995	1996	1997	1998	1999	2000
Quant. Auctioned (ton)	4,500	450	585	675	540	540	405	405	378
TAC (ton)	5,000	5,000	6,500	7,500	6,000	6000	4500	4500	4200
Average Price (U$/ton)	1,626	1,970	15,333	726	734	679	628	590	11,372

tonnage increased by a small amount in 1995 and 1999. An analysis of the total number of vessels during the same period shows no relevant changes in the size of the fleet of 13-16 vessels, with minor fluctuations only between years (Figure 8).

In the medium term, the analysis of the long-liner fleet (based on weighted-average per-vessel figures) shows a small increase in the average vessel-capacity of the long-liner fleet[7]. In terms of GRT, there was a 5% increase in weighted-average GRT per-vessel, increasing from 590 in 1992, to 622 in 1993. In terms of engine-power, there was a 7% increment in the weighted-average HP per vessel, going from 1305 HP in 1992 to 1393 HP in 1993.

In addition, a medium-term analysis for the long-line fleet shows a small decreasing trend (–2%) in the average individual vessel GRT, falling from 622 in 1993 to 610 in 2000. The weighted-average HP per-vessel, to the contrary, shows an increase in engine-power of the long-liners, with an equally small increase in vessel-power (2%), going from 1393 HP in 1993 to 1421 HP in 2000[8].

In spite of these small fluctuations observed after the implementation of the system of fishing-rights, it can be said that the total capacity of the fleet has been reduced, both in terms of GRT and number of vessels, and that per-vessel-capacity (weighted-average GRT and HP) has increased. The age-structure of the long-liner fleet (Table 4) does not show any significant changes between 1992 and 2000.

Even though the available data does not allow direct inferences about the change in fleet-efficiency, both industry and USF representatives agree that the fishing fleet in the *Unidad de Pesca* has been able to increase its economic efficiency through better planning of fishing-effort and when to harvest the quota.

5.2 Changes in fishing-rights ownership

An analysis of the aggregation of fishing access-rights as a percentage of the TAC (Figure 13), shows that the implementation of the RFID-system has been accompanied by increasing concentration of fishing-rights among a few of the companies participating in the fishery. Figure 15 presents the number of fishing companies with access to approximately 80% of the TAC in the *Unidad de Pesca* since 1993 (*i.e.* fishing-rights owned, or rented). The trend depicted in Figure 15 shows that approximately 80% of the TAC is being owned by a

---

[7] The weighted-average GRT and HP per vessel is estimated from the frequency-distribution of the vessels' GRT and HP ratings in the fleet for both years.

[8] The weighted-average values of GRT and HP per vessel are estimated from the frequency-distribution data presented in Tables 1 and 3 for the long-liner fleet during the five points in time considered since 1993.

decreasing number of fishing companies: available data show that nine companies had access to 78% of the TAC in 1993 but that by the year 2000 82% of the TAC was owned by only three companies.

This tendency to the concentration of fishing-rights in few hands is also evident from an analysis of the per-company access to fishing-rights in the *Unidad de Pesca* since 1993. Figure 16 shows the percentage of the annual TAC controlled by the three most important fishing companies in the *Unidad de Pesca*. It is apparent that the fraction of the TAC controlled by these three companies has increased from 52% in 1993 to 82% in 2000. The plotted trend line shows an increase in concentration of control at a decreasing rate from 1993 to 1998 and an increase in the rate since 1998.

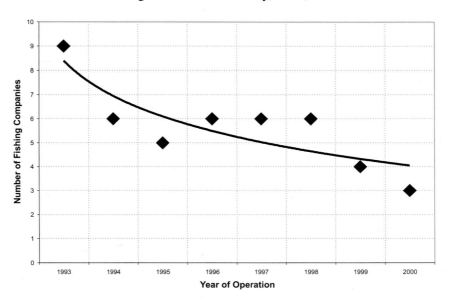

Figure 15

Number of fishing companies holding approximately 80% of the TAC in the Patagonian Toothfish fishery, Chile, 1993-2000

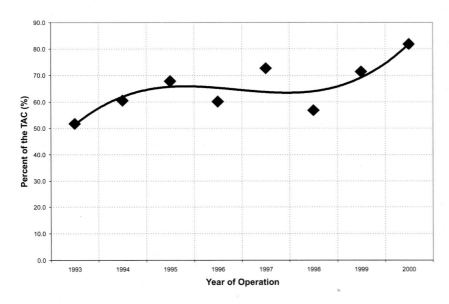

Figure 16

Percent of the TAC accessed by the first three fishing companies in the Patagonian Toothfish Fishery, 1993 - 2000

Further analysis of the data in Figure 13 shows that the three companies are acquiring their control of fishing-rights through auctions, sales or rental of EFPs. Figure 17 shows the trend in the percent of the total TAC contolled by companies A, B and C each year: Company A has increased in importance since 1993 to control 56% of the TAC in 1997, although by 2000 it still controlled only 44% of the TAC. Company B, on the other hand, decreased its EFP holdings during the first 5 years of implementation of the RFID-system, reaching a minimum of 2% of the TAC in 1997. Nonetheless, since then it has recovered its level of access, reaching 21% of the TAC in 2000. Company C has increased its participation slowly, reaching 16% of the TAC in 2000.

5.3   Fleet-capacity, fishing-effort, harvest-levels and system-effectiveness

As Section 5.1 shows, the implementation of fishing-rights in the *Unidad de Pesca* resulted in a large reduction in fleet-size and fleet-capacity. In addition, the expected effects of implementing a rights-based

Figure 17
Participation in the Annual TAC (%) of the two most important companies
in the Patagonian toothfish fishery, 1993 – 2000

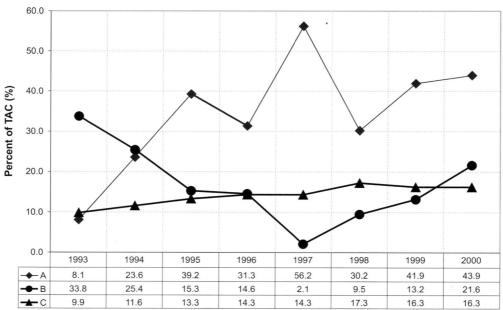

	1993	1994	1995	1996	1997	1998	1999	2000
◆ A	8.1	23.6	39.2	31.3	56.2	30.2	41.9	43.9
● B	33.8	25.4	15.3	14.6	2.1	9.5	13.2	21.6
▲ C	9.9	11.6	13.3	14.3	14.3	17.3	16.3	16.3

Year of Operation

management system are the adjustment of fishing-effort and harvest to sustainable levels, and eventually their adjustment to levels that will maximize the net economic benefits generated over time. If the starting point is a fully-exploited or over-exploited fishery, the expected adjustments will be a reduction in fishing-effort with an increase in harvest-levels and net economic benefits. If the starting point is an under-exploited fishery, expected adjustments will be an increase in fishing-effort, harvest-levels and net economic benefits. In the case of the Chilean Patagonian toothfish fishery, the starting point was an under-exploited fishery.

The available data on total fishing-effort and harvest-level for the *Unidad de Pesca* (Figure 18) show an overall trend of increasing fishing-effort with decreasing harvest-levels and catch-per-unit effort. If the data analysis is expressed on a per-vessel basis (Figure 19) over the 9 years the values show an increasing trend in fishing-effort but a decrease in harvest-levels.

Figure 18
Histograms of fishing-effort, TAC and catch for the Patagonian toothfish fishery - Chile, 1991 – 1999

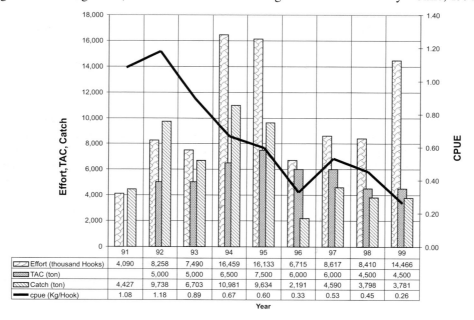

	91	92	93	94	95	96	97	98	99
▨ Effort (thousand Hooks)	4,090	8,258	7,490	16,459	16,133	6,715	8,617	8,410	14,466
▨ TAC (ton)		5,000	5,000	6,500	7,500	6,000	6,000	4,500	4,500
▨ Catch (ton)	4,427	9,738	6,703	10,981	9,634	2,191	4,590	3,798	3,781
— cpue (Kg/Hook)	1.08	1.18	0.89	0.67	0.60	0.33	0.53	0.45	0.26

Year

Figure 19
Fishing-effort per vessel, harvest per vessel, and fleet-size in the Patagonian toothfish fishery
in Chile, 1991 – 1999

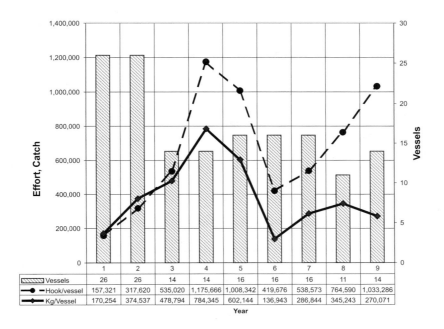

	1	2	3	4	5	6	7	8	9
Vessels	26	26	14	14	16	16	16	11	14
Hook/vessel	157,321	317,620	535,020	1,175,666	1,008,342	419,676	538,573	764,590	1,033,286
Kg/Vessel	170,254	374,537	478,794	784,345	602,144	136,943	286,844	345,243	270,071

Year

Figure 20
Size of the Patagonian toothfish stock in the *Unidad de Pesca*, 1991 - 1999

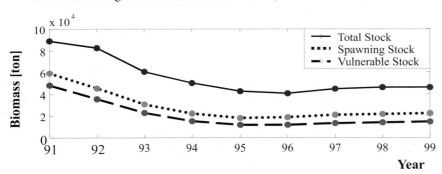

Source: Under Secretariat of Fisheries

The data for the fishing-effort and harvest (figure 19) indicates that the *Unidad de Pesca* may be heading towards over-exploitation. This may be caused by: (a) a lack of compliance by rights-holders with individual quotas, and (b) the fishing-rights having been established in a fishery based on transboundary or shared fish-stock resources.

The data available regarding abundance of the fish-stock resource over time (Figure 20) show a decreasing trend over 1992 to 1996, that stabilized at lower levels from 1997 to 1999.

5.4 Lack of compliance

Figure 18 shows that during 1992–1995 harvest-levels were greater than the annual TAC permitted for the *Unidad de Pesca*. Fisheries Officers reported that this had happened as a consequence of the lack of adequate means of enforcement of the individual quota, as well as a lack of understanding of and commitment to the RFID-management system by rights-holders. Even though since 1996 the harvest-levels have been smaller than the TACs, through comparison with export data, Fisheries Officers have been able to identify significant under-reporting of the catch statistics.

Thus, even though personnel from the USF have in the past been able to detect and correct some compliance problems, if the RFID-system is to be effective for the management of the Patagonian toothfish fishery in Chile, improvements in the operational capabilities of the Chilean National Fisheries Service are essential (González *et al.* 2001). Furthermore, communication and the means for dialogue, between Fisheries Officers and fishing rights-holders should also be improved if sufficient awareness is to be achieved regarding the benefits of the system and the need for compliance.

5.5   Transboundary stocks

Stock assessments by the Commission for the Conservation of Antarctic Marine Living Resources (CCAMLR) have shown that the Patagonian toothfish stocks in the *Unidad de Pesca* have strong migratory links with the resource distributed in the southern Atlantic and the CCAMLR area. Thus, this fishery is based upon a transboundary fish-stock, and the expected benefits of the fishing-rights system could be outweighed by the fishing being undertaken outside of the *Unidad de Pesca* area, and thus the harvest of some toothfish beyond the control of the Chilean management-system

Therefore, given the transboundary nature of the resource, for the RFID-management system to be effective, three conditions must be met in the medium- and long-term: (a) a transparent flow of information regarding the fish stocks, their population dynamics and abundance, as well as the amount of fishing-effort inside and outside the *Unidad de Pesca*, including both national and international waters; (b) the creation of co-ordinated and collaborative research-efforts regarding the fish stock, its population dynamics and biology both inside and outside the *Unidad de Pesca*, including parties from all involved countries and; (c) the implementation of co-ordinated and collaborative joint-management of the Patagonian toothfish fishery, over its full extent, inside and outside Chile, and including all parties from all involved countries.

5.6   Auction prices, fish-values, concentration of fishing-rights and equity issues

As Figure 14 shows, the average auction-prices have mostly ranged from approximately $590/t to $2000/t, but two unusually high prices were realized in the years 1994 and 2000. A comparison of the auction- and declared export-prices (FOB) of Chilean Patagonian toothfish (Figures 14, and 7) indicates that these two out-of-the-range auction prices occured in years when export prices also showed large increases[9]. These two auction-prices were higher than export-prices (three times higher in 1992, and 77% higher in 2000).

A typical Chilean industrial-scale longliner in the Patagonian toothfish fishery
A black browed albatross is flying astern of the ship, the birds on the water are petrels.
Photo: C. A. Moreno, Universidad Austral de Chile

---

[9] Apparent export price (FOB) experienced an increase of 42% in 1994, going from $3670/tonne in 1993 to $5202/tonne in 1994. The increase in year 2000 was of 59% (jumping from the $5546/tonne in1999, to $8809/tonne in 2000).

These extremely high auction-prices, seemingly inconsistent with economic rationale, may be the result of the strategic behavior of current fishing rights-holders trying to avoid the entrance of new players in the presence of increasing market prices[10]. The observed jump in the auction-prices for 1994 (Figure 14) may have been supported by high expectations of large future profits, based on the facts that: (a) market prices were rising significantly, (b) production costs at that time had not significantly increased (since the volume auctioned and awarded at this price was only a small fraction of the total production of each of the current holders of fishing-rights), and (c) there was a growing supply of product from 1991 up to 1994. The observed jump in the year 2000 may be explained by expectations of further decreases in supply and the need to keep control of the current market-share because of: (a) recovering and increasing market-prices, (b) current production costs not significantly increasing (since the volume auctioned and awarded at this price was only a small fraction of the total production by each current holder of fishing-rights), and (c) fish landings and production have had decreased since 1994.

As mentioned in Section 4.3.2, the USF and the National Fisheries Service do not record any sale- or rental-values for the harvesting-rights. Thus, since only the original auction-price is of interest to the USF, these auction-prices are used to calculate the obligatory annual payments due against the allocated EFPs. The National Fisheries Service is only interested in recording landing statistics and their origin, in order to track of harvest levels with respect to the annual TAC. Thus, no attention is paid to market-dynamics, nor to the value of the "uncaught" resources, since only the original auction-prices, along with other physical fisheries data, are used to monitor the fishery.

In recognizing the need to pay for the right of exclusive access to a specified fraction of a natural resource, one recognizes that these resources have both a current market-value and a future value (*i.e.* the value of the Patagonian toothfish stock in the water). That payment of this value is made to the state or the government, implies recognition that this natural resource belongs to all nationals, not just to those who have the means to harvest them.

Therefore, the current practice of not recording either direct sale-prices or rental-values has social equity implications. If actual prices or values are higher than the original auction-prices, society is forgoing a portion of the resource-value. If actual prices or values are lower than the original auction-prices, the rights-holders are paying too much to society for their right to use the resource base.

Finally, the process of an increasing concentration of fishing-rights in the *Unidad de Pesca* (Section 5.2, Figures 15 and 16) is regarded around the world as a negative effect of this type of management-system, and monopolies are often portrayed as the maximum expression of social inequity. This is true in the following two cases.

First, when the monopoly-operator is able to set his production at levels whereby his marginal-revenues are equal to the marginal-costs, he obtains a market-price that is higher than he would achieve under perfect competition. Nonetheless, this is not the case here since the *Unidad de Pesca* exports its entire production (representing only a small portion of total supply of Patagonain toothfish) to international markets. Thus, any variation in its supply (exports) will not affect prices in the market, as the fishery is a "price taker", selling its production at the price available on the international markets.

Second, under conditions of perfect competition, an operator produces at a level in which average revenue (*i.e.* price) is equal to the marginal-cost and average-cost at the same time, therefore he only earns producer surplus beyond the normal rent (this normal rent is already included in costs-curves). In monopolies (the case of the sole-owner in the fisheries sector) an operator produces at levels for which his marginal-revenue is equal to his marginal-cost and therefore he is captures all the extra rent (the resource-rent in the case of fisheries).

Further, when the value of the fish in the water (*i.e.* the potential resource-rent) is accrued by those operating in the fishery, equity-issues may be raised if these rights-holders are just a few, or even a single owner. But, if a mechanism to recover and distribute this value among members of society is set in place, as in Chile, equity-issues do not arise regarding the concentration of ownership, provided that the control of concentration is appropriate (González *et al.* 2001). Thus, appropriate recording of the values of sales and rentals, and the proper setting of floor-prices for auctions, become important issues in order to correctly determine the values of fishing-rights over time. If properly determined, these values should equal the resource-rent to be collected and distributed by the government.

---

[10] Figure 6 shows an increasing trend in export values since 1991, which reached a new high in 1999.

6.   LITERATURE CITED

González, E., E.R. Norambuena and M. García 2001. Initial Allocation of Harvesting Rights in the Chilean fishery for Patagonian Toothfish (*Dissostichus eleginoides*). 304-321. *In:* Shotton, R. (Ed.) Case studies on the Allocation of Transferable Quotas Rights in Fisheries. FAO Tech. Fish. Pap. No. 411, FAO, Rome. 373 pp.

Government of Chile 1991. General Law of Fisheries and Aquaculture. SD 430 of 1991. Santiago: Official Gazette of the Republic, Chile.

Government of Chile 1996. Regulations for the Auction of Extraordinary Fishing Permits. SD. Number 97, 1996. Santiago: Official Gazette of the Republic, Chile.

Lemaitre, C., P.S. Rubilar, P. Gebauer and C.A. Moreno 1991. Regional Catch Analysis of Long-line Fisheries of *Dissostichus eleginoides* in Chile. Document WG-FSA-91/10 CCAMLR. Hobart, Australia.

National Fisheries Service 1999. Registry of Artisanal Fishermen of Chile. Santiago: National Fisheries Service (SERNAPESCA), Ministry of Economy.

Under-Secretariat of Fisheries 1999. Total Allowable Catch Quota for the Deep Water Cod Fishery, Year 2000. Valparaíso: Fishery Resources Department, Under Secretariat of Fisheries, Ministry of Economy. Technical Report N° 67.